A$_E^P$x CALCULUS Q3
Version 4.0

Gregory Hartman, Ph.D.

Department of Applied Mathematics

Virginia Military Institute

Contributing Authors

Troy Siemers, Ph.D.

Department of Applied Mathematics

Virginia Military Institute

Brian Heinold, Ph.D.

Department of Mathematics and Computer Science

Mount Saint Mary's University

Dimplekumar Chalishajar, Ph.D.

Department of Applied Mathematics

Virginia Military Institute

Editor

Jennifer Bowen, Ph.D.

Department of Mathematics and Computer Science

The College of Wooster

Contents

PREFACE
A Note on Using this Text

Thank you for reading this short preface. Allow us to share a few key points about the text so that you may better understand what you will find beyond this page.

This text is Part 3 of a four–text series on Calculus, split to better accommodate the needs of schools on the quarter system. The first part covers limits and derivatives, found in Chapters 1 through 4. The second text covers integration and its applications, along with an introduction to differential equations, found in Chapters 5 through 7 (where the differential equation material is contained in an appendix). This third text covers sequences and series, along with parametric equations, polar coordinates and vector–valued functions, found in Chapters 8 through 11. The fourth text in the series covers functions of more than one variable, including partial derivatives, iterated integration and vector analysis, found in Chapters 12 through 14. All four volumes are available separately for free at www.apexcalculus.com. These four texts (without the differential equations appendix) are intended to work together and make one cohesive text, *APEX Calculus*, which can also be downloaded from the website.

Printing the entire text as one volume makes for a large, heavy, cumbersome book. One can certainly only print the pages they currently need, but some prefer to have a nice, bound copy of the text. Therefore this text has been split into these four manageable parts, each of which can be purchased for about $15 at Amazon.com. The book has also been divided into a three–part series, more appropriate for schools on the semester system. Visit www.apexcalculus.com for more information.

A result of this splitting is that sometimes a concept is said to be explored in a "later section," though that section does not actually appear in this particular text. Also, the index makes reference to topics and page numbers that do not appear in this text. This is done intentionally to show the reader what topics are available for study. Downloading the .pdf of *APEX Calculus* will ensure that you have all the content.

For Students: How to Read this Text

Mathematics textbooks have a reputation for being hard to read. High–level mathematical writing often seeks to say much with few words, and this style often seeps into texts of lower–level topics. This book was written with the goal of being easier to read than many other calculus textbooks, without becoming too verbose.

Each chapter and section starts with an introduction of the coming material, hopefully setting the stage for "why you should care," and ends with a look ahead to see how the just–learned material helps address future problems.

Please read the text; it is written to explain the concepts of Calculus. There are numerous examples to demonstrate the meaning of definitions, the truth of theorems, and the application of mathematical techniques. When you encounter a sentence you don't understand, read it again. If it still doesn't make sense, read on anyway, as sometimes confusing sentences are explained by later sentences.

You don't have to read every equation. The examples generally show "all" the steps needed to solve a problem. Sometimes reading through each step is helpful; sometimes it is confusing. When the steps are illustrating a new technique, one probably should follow each step closely to learn the new technique. When the steps are showing the mathematics needed to find a number to be used later, one can usually skip ahead and see how that number is being used, instead of getting bogged down in reading how the number was found.

Most proofs have been omitted. In mathematics, *proving* something is always true is extremely important, and entails much more than testing to see if it works twice. However, students often are confused by the details of a proof, or become concerned that they should have been able to construct this proof on their own. To alleviate this potential problem, we do not include the proofs to most theorems in the text. The interested reader is highly encouraged to find proofs online or from their instructor. In most cases, one is very capable of understanding what a theorem *means* and *how to apply it* without knowing fully *why* it is true.

Interactive, 3D Graphics

New to Version 3.0 was the addition of interactive, 3D graphics in the .pdf version. Nearly all graphs of objects in space can be rotated, shifted, and zoomed in/out so the reader can better understand the object illustrated.

As of this writing, the only pdf viewers that support these 3D graphics are Adobe Reader & Acrobat (and only the versions for PC/Mac/Unix/Linux computers, not tablets or smartphones). To activate the interactive mode, click on the image. Once activated, one can click/drag to rotate the object and use the scroll wheel on a mouse to zoom in/out. (A great way to investigate an image is to first zoom in on the page of the pdf viewer so the graphic itself takes up much of the screen, then zoom inside the graphic itself.) A CTRL-click/drag pans the object left/right or up/down. By right-clicking on the graph one can access a menu of other options, such as changing the lighting scheme or perspective. One can also revert the graph back to its default view. If you wish to deactivate the interactivity, one can right-click and choose the "Disable Content" option.

Thanks

There are many people who deserve recognition for the important role they have played in the development of this text. First, I thank Michelle for her support and encouragement, even as this "project from work" occupied my time and attention at home. Many thanks to Troy Siemers, whose most important contributions extend far beyond the sections he wrote or the 227 figures he coded in Asymptote for 3D interaction. He provided incredible support, advice and encouragement for which I am very grateful. My thanks to Brian Heinold and Dimplekumar Chalishajar for their contributions and to Jennifer Bowen for reading through so much material and providing great feedback early on. Thanks to Troy, Lee Dewald, Dan Joseph, Meagan Herald, Bill Lowe, John David, Vonda Walsh, Geoff Cox, Jessica Libertini and other faculty of VMI who have given me numerous suggestions and corrections based on their experience with teaching from the text. (Special thanks to Troy, Lee & Dan for their patience in teaching Calc III while I was still writing the Calc III material.) Thanks to Randy Cone for encouraging his tutors of VMI's Open Math Lab to read through the text and check the solutions, and thanks to the tutors for spending their time doing so. A very special thanks to Kristi Brown and Paul Janiczek who took this opportunity far above & beyond what I expected, meticulously checking every solution and carefully reading every example. Their comments have been extraordinarily helpful. I am also thankful for the support provided by Wane Schneiter, who as my Dean provided me with extra time to work on this project. Thanks to Ross

Magi and Jonathan Duncan for their help in making this quarter–system version. Finally, a huge heap of thanks is to be bestowed on the numerous people I do not know who took the time to email me corrections and suggestions. I am blessed to have so many people give of their time to make this book better.

AP̲E̲X – Affordable Print and Electronic teXts

AP̲E̲X is a consortium of authors who collaborate to produce high–quality, low–cost textbooks. The current textbook–writing paradigm is facing a potential revolution as desktop publishing and electronic formats increase in popularity. However, writing a good textbook is no easy task, as the time requirements alone are substantial. It takes countless hours of work to produce text, write examples and exercises, edit and publish. Through collaboration, however, the cost to any individual can be lessened, allowing us to create texts that we freely distribute electronically and sell in printed form for an incredibly low cost. Having said that, nothing is entirely free; someone always bears some cost. This text "cost" the authors of this book their time, and that was not enough. *APEX Calculus* would not exist had not the Virginia Military Institute, through a generous Jackson–Hope grant, given the lead author significant time away from teaching so he could focus on this text.

Each text is available as a free .pdf, protected by a Creative Commons Attribution - Noncommercial 4.0 copyright. That means you can give the .pdf to anyone you like, print it in any form you like, and even edit the original content and redistribute it. If you do the latter, you must clearly reference this work and you cannot sell your edited work for money.

We encourage others to adapt this work to fit their own needs. One might add sections that are "missing" or remove sections that your students won't need. The source files can be found at github.com/APEXCalculus.

You can learn more at www.vmi.edu/APEX.

Version 4.0

Key changes from Version 3.0 to 4.0:

- Numerous typographical and "small" mathematical corrections (again, thanks to all my close readers!).

- "Large" mathematical corrections and adjustments. There were a number of places in Version 3.0 where a definition/theorem was not correct as stated. See www.apexcalculus.com for more information.

- More useful numbering of Examples, Theorems, etc. "Definition 11.4.2" refers to the second definition of Chapter 11, Section 4.

- The addition of Section 13.7: Triple Integration with Cylindrical and Spherical Coordinates

- The addition of Chapter 14: Vector Analysis.

8: SEQUENCES AND SERIES

This chapter introduces **sequences** and **series**, important mathematical constructions that are useful when solving a large variety of mathematical problems. The content of this chapter is considerably different from the content of the chapters before it. While the material we learn here definitely falls under the scope of "calculus," we will make very little use of derivatives or integrals. Limits are extremely important, though, especially limits that involve infinity.

One of the problems addressed by this chapter is this: suppose we know information about a function and its derivatives at a point, such as $f(1) = 3$, $f'(1) = 1, f''(1) = -2, f'''(1) = 7$, and so on. What can I say about $f(x)$ itself? Is there any reasonable approximation of the value of $f(2)$? The topic of Taylor Series addresses this problem, and allows us to make excellent approximations of functions when limited knowledge of the function is available.

8.1 Sequences

We commonly refer to a set of events that occur one after the other as a *sequence* of events. In mathematics, we use the word *sequence* to refer to an ordered set of numbers, i.e., a set of numbers that "occur one after the other."

For instance, the numbers 2, 4, 6, 8, ..., form a sequence. The order is important; the first number is 2, the second is 4, etc. It seems natural to seek a formula that describes a given sequence, and often this can be done. For instance, the sequence above could be described by the function $a(n) = 2n$, for the values of $n = 1, 2, \ldots$ To find the 10th term in the sequence, we would compute $a(10)$. This leads us to the following, formal definition of a sequence.

Definition 8.1.1   Sequence

A **sequence** is a function $a(n)$ whose domain is $\mathbb{N}$. The **range** of a sequence is the set of all distinct values of $a(n)$.

The **terms** of a sequence are the values $a(1)$, $a(2)$, ..., which are usually denoted with subscripts as a_1, a_2, ...

A sequence $a(n)$ is often denoted as $\{a_n\}$.

Notation: We use $\mathbb{N}$ to describe the set of natural numbers, that is, the integers 1, 2, 3, ...

Factorial: The expression 4! refers to the number $4 \cdot 3 \cdot 2 \cdot 1 = 24$.

In general, $n! = n \cdot (n-1) \cdot (n-2) \cdots 2 \cdot 1$, where n is a natural number.

We define $0! = 1$. While this does not immediately make sense, it makes many mathematical formulas work properly.

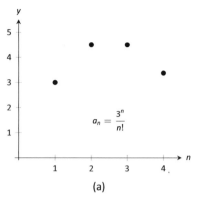

(a)

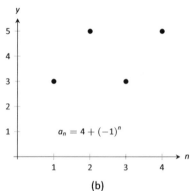

(b)

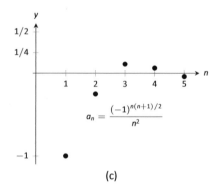

(c)

Figure 8.1.1: Plotting sequences in Example 8.1.1.

Example 8.1.1   **Listing terms of a sequence**

List the first four terms of the following sequences.

1. $\{a_n\} = \left\{ \dfrac{3^n}{n!} \right\}$ 2. $\{a_n\} = \{4 + (-1)^n\}$ 3. $\{a_n\} = \left\{ \dfrac{(-1)^{n(n+1)/2}}{n^2} \right\}$

SOLUTION

1. $a_1 = \dfrac{3^1}{1!} = 3$; $a_2 = \dfrac{3^2}{2!} = \dfrac{9}{2}$; $a_3 = \dfrac{3^3}{3!} = \dfrac{9}{2}$; $a_4 = \dfrac{3^4}{4!} = \dfrac{27}{8}$

   We can plot the terms of a sequence with a scatter plot. The "x"-axis is used for the values of n, and the values of the terms are plotted on the y-axis. To visualize this sequence, see Figure 8.1.1(a).

2. $a_1 = 4 + (-1)^1 = 3$; $a_2 = 4 + (-1)^2 = 5$;

   $a_3 = 4 + (-1)^3 = 3$; $a_4 = 4 + (-1)^4 = 5$. Note that the range of this sequence is finite, consisting of only the values 3 and 5. This sequence is plotted in Figure 8.1.1(b).

3. $a_1 = \dfrac{(-1)^{1(2)/2}}{1^2} = -1$; $a_2 = \dfrac{(-1)^{2(3)/2}}{2^2} = -\dfrac{1}{4}$;

   $a_3 = \dfrac{(-1)^{3(4)/2}}{3^2} = \dfrac{1}{9}$ $a_4 = \dfrac{(-1)^{4(5)/2}}{4^2} = \dfrac{1}{16}$;

   $a_5 = \dfrac{(-1)^{5(6)/2}}{5^2} = -\dfrac{1}{25}$.

   We gave one extra term to begin to show the pattern of signs is "$-, -, +, +, -, -, \ldots$" due to the fact that the exponent of -1 is a special quadratic. This sequence is plotted in Figure 8.1.1(c).

Example 8.1.2   **Determining a formula for a sequence**

Find the nth term of the following sequences, i.e., find a function that describes each of the given sequences.

1. $2, 5, 8, 11, 14, \ldots$

2. $2, -5, 10, -17, 26, -37, \ldots$

3. $1, 1, 2, 6, 24, 120, 720, \ldots$

4. $\dfrac{5}{2}, \dfrac{5}{2}, \dfrac{15}{8}, \dfrac{5}{4}, \dfrac{25}{32}, \cdots$

Notes:

SOLUTION   We should first note that there is never exactly one function that describes a finite set of numbers as a sequence. There are many sequences that start with 2, then 5, as our first example does. We are looking for a simple formula that describes the terms given, knowing there is possibly more than one answer.

1. Note how each term is 3 more than the previous one. This implies a linear function would be appropriate: $a(n) = a_n = 3n + b$ for some appropriate value of b. As we want $a_1 = 2$, we set $b = -1$. Thus $a_n = 3n - 1$.

2. First notice how the sign changes from term to term. This is most commonly accomplished by multiplying the terms by either $(-1)^n$ or $(-1)^{n+1}$. Using $(-1)^n$ multiplies the odd terms by (-1); using $(-1)^{n+1}$ multiplies the even terms by (-1). As this sequence has negative even terms, we will multiply by $(-1)^{n+1}$.

    After this, we might feel a bit stuck as to how to proceed. At this point, we are just looking for a pattern of some sort: what do the numbers 2, 5, 10, 17, etc., have in common? There are many correct answers, but the one that we'll use here is that each is one more than a perfect square. That is, $2 = 1^2 + 1$, $5 = 2^2 + 1$, $10 = 3^2 + 1$, etc. Thus our formula is $a_n = (-1)^{n+1}(n^2 + 1)$.

3. One who is familiar with the factorial function will readily recognize these numbers. They are 0!, 1!, 2!, 3!, etc. Since our sequences start with $n = 1$, we cannot write $a_n = n!$, for this misses the 0! term. Instead, we shift by 1, and write $a_n = (n - 1)!$.

4. This one may appear difficult, especially as the first two terms are the same, but a little "sleuthing" will help. Notice how the terms in the numerator are always multiples of 5, and the terms in the denominator are always powers of 2. Does something as simple as $a_n = \frac{5n}{2^n}$ work?

    When $n = 1$, we see that we indeed get $5/2$ as desired. When $n = 2$, we get $10/4 = 5/2$. Further checking shows that this formula indeed matches the other terms of the sequence.

A common mathematical endeavor is to create a new mathematical object (for instance, a sequence) and then apply previously known mathematics to the new object. We do so here. The fundamental concept of calculus is the limit, so we will investigate what it means to find the limit of a sequence.

Notes:

Definition 8.1.2   Limit of a Sequence, Convergent, Divergent

Let $\{a_n\}$ be a sequence and let L be a real number. Given any $\varepsilon > 0$, if an m can be found such that $|a_n - L| < \varepsilon$ for all $n > m$, then we say the **limit of $\{a_n\}$, as n approaches infinity, is L**, denoted

$$\lim_{n\to\infty} a_n = L.$$

If $\lim\limits_{n\to\infty} a_n$ exists, we say the sequence **converges**; otherwise, the sequence **diverges**.

This definition states, informally, that if the limit of a sequence is L, then if you go far enough out along the sequence, all subsequent terms will be *really close* to L. Of course, the terms "far enough" and "really close" are subjective terms, but hopefully the intent is clear.

This definition is reminiscent of the $\varepsilon{-}\delta$ proofs of Chapter 1. In that chapter we developed other tools to evaluate limits apart from the formal definition; we do so here as well.

Theorem 8.1.1   Limit of a Sequence

Let $\{a_n\}$ be a sequence and let $f(x)$ be a function whose domain contains the positive real numbers where $f(n) = a_n$ for all n in $\mathbb{N}$.

If $\lim\limits_{x\to\infty} f(x) = L$, then $\lim\limits_{n\to\infty} a_n = L$.

Theorem 8.1.1 allows us, in certain cases, to apply the tools developed in Chapter 1 to limits of sequences. Note two things *not* stated by the theorem:

1. If $\lim\limits_{x\to\infty} f(x)$ does not exist, we cannot conclude that $\lim\limits_{n\to\infty} a_n$ does not exist. It may, or may not, exist. For instance, we can define a sequence $\{a_n\} = \{\cos(2\pi n)\}$. Let $f(x) = \cos(2\pi x)$. Since the cosine function oscillates over the real numbers, the limit $\lim\limits_{x\to\infty} f(x)$ does not exist.

   However, for every positive integer n, $\cos(2\pi n) = 1$, so $\lim\limits_{n\to\infty} a_n = 1$.

2. If we cannot find a function $f(x)$ whose domain contains the positive real numbers where $f(n) = a_n$ for all n in $\mathbb{N}$, we cannot conclude $\lim\limits_{n\to\infty} a_n$ does not exist. It may, or may not, exist.

Notes:

Example 8.1.3   Determining convergence/divergence of a sequence

Determine the convergence or divergence of the following sequences.

1. $\{a_n\} = \left\{ \dfrac{3n^2 - 2n + 1}{n^2 - 1000} \right\}$ 2. $\{a_n\} = \{\cos n\}$ 3. $\{a_n\} = \left\{ \dfrac{(-1)^n}{n} \right\}$

Solution

1. Using Theorem 1.6.1, we can state that $\lim\limits_{x\to\infty} \dfrac{3x^2 - 2x + 1}{x^2 - 1000} = 3$. (We could have also directly applied l'Hôpital's Rule.) Thus the sequence $\{a_n\}$ converges, and its limit is 3. A scatter plot of every 5 values of a_n is given in Figure 8.1.2 (a). The values of a_n vary widely near $n = 30$, ranging from about -73 to 125, but as n grows, the values approach 3.

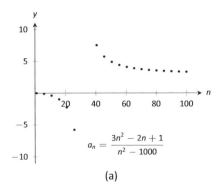

(a)

2. The limit $\lim\limits_{x\to\infty} \cos x$ does not exist as $\cos x$ oscillates (and takes on every value in $[-1, 1]$ infinitely many times). Thus we cannot apply Theorem 8.1.1.

   The fact that the cosine function oscillates strongly hints that $\cos n$, when n is restricted to $\mathbb{N}$, will also oscillate. Figure 8.1.2 (b), where the sequence is plotted, implies that this is true. Because only discrete values of cosine are plotted, it does not bear strong resemblance to the familiar cosine wave. The proof of the following statement is beyond the scope of this text, but it is true: there are infinitely many integers n that are arbitrarily (i.e., *very*) close to an even multiple of π, so that $\cos n \approx 1$. Similarly, there are infinitely many integers m that are arbitrarily close to an odd multiple of π, so that $\cos m \approx -1$. As the sequence takes on values near 1 and -1 infinitely many times, we conclude that $\lim\limits_{n\to\infty} a_n$ does not exist.

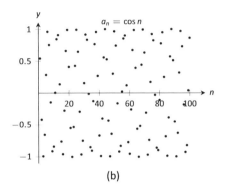

(b)

3. We cannot actually apply Theorem 8.1.1 here, as the function $f(x) = (-1)^x/x$ is not well defined. (What does $(-1)^{\sqrt{2}}$ mean? In actuality, there is an answer, but it involves *complex analysis*, beyond the scope of this text.)

   Instead, we invoke the definition of the limit of a sequence. By looking at the plot in Figure 8.1.2 (c), we would like to conclude that the sequence converges to $L = 0$. Let $\varepsilon > 0$ be given. We can find a natural number m

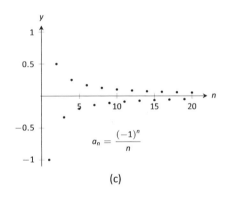

(c)

Figure 8.1.2: Scatter plots of the sequences in Example 8.1.3.

Notes:

such that $1/m < \varepsilon$. Let $n > m$, and consider $|a_n - L|$:

$$|a_n - L| = \left| \frac{(-1)^n}{n} - 0 \right|$$

$$= \frac{1}{n}$$

$$< \frac{1}{m} \quad \text{(since } n > m\text{)}$$

$$< \varepsilon.$$

We have shown that by picking m large enough, we can ensure that a_n is arbitrarily close to our limit, $L = 0$, hence by the definition of the limit of a sequence, we can say $\lim\limits_{n\to\infty} a_n = 0$.

In the previous example we used the definition of the limit of a sequence to determine the convergence of a sequence as we could not apply Theorem 8.1.1. In general, we like to avoid invoking the definition of a limit, and the following theorem gives us tool that we could use in that example instead.

Theorem 8.1.2   Absolute Value Theorem

Let $\{a_n\}$ be a sequence. If $\lim\limits_{n\to\infty} |a_n| = 0$, then $\lim\limits_{n\to\infty} a_n = 0$

Example 8.1.4   Determining the convergence/divergence of a sequence
Determine the convergence or divergence of the following sequences.

$$\text{1. } \{a_n\} = \left\{ \frac{(-1)^n}{n} \right\} \qquad \text{2. } \{a_n\} = \left\{ \frac{(-1)^n(n+1)}{n} \right\}$$

SOLUTION

1. This appeared in Example 8.1.3. We want to apply Theorem 8.1.2, so consider the limit of $\{|a_n|\}$:

$$\lim_{n\to\infty} |a_n| = \lim_{n\to\infty} \left| \frac{(-1)^n}{n} \right|$$

$$= \lim_{n\to\infty} \frac{1}{n}$$

$$= 0.$$

Since this limit is 0, we can apply Theorem 8.1.2 and state that $\lim\limits_{n\to\infty} a_n = 0$.

Notes:

2. Because of the alternating nature of this sequence (i.e., every other term is multiplied by -1), we cannot simply look at the limit $\lim_{x\to\infty} \dfrac{(-1)^x(x+1)}{x}$. We can try to apply the techniques of Theorem 8.1.2:

$$\lim_{n\to\infty} |a_n| = \lim_{n\to\infty} \left| \frac{(-1)^n(n+1)}{n} \right|$$
$$= \lim_{n\to\infty} \frac{n+1}{n}$$
$$= 1.$$

We have concluded that when we ignore the alternating sign, the sequence approaches 1. This means we cannot apply Theorem 8.1.2; it states the the limit must be 0 in order to conclude anything.

Since we know that the signs of the terms alternate *and* we know that the limit of $|a_n|$ is 1, we know that as n approaches infinity, the terms will alternate between values close to 1 and -1, meaning the sequence diverges. A plot of this sequence is given in Figure 8.1.3.

We continue our study of the limits of sequences by considering some of the properties of these limits.

Theorem 8.1.3   Properties of the Limits of Sequences

Let $\{a_n\}$ and $\{b_n\}$ be sequences such that $\lim_{n\to\infty} a_n = L$, $\lim_{n\to\infty} b_n = K$, and let c be a real number.

1. $\lim_{n\to\infty} (a_n \pm b_n) = L \pm K$ 3. $\lim_{n\to\infty} (a_n/b_n) = L/K, K \neq 0$

2. $\lim_{n\to\infty} (a_n \cdot b_n) = L \cdot K$ 4. $\lim_{n\to\infty} c \cdot a_n = c \cdot L$

Example 8.1.5   Applying properties of limits of sequences

Let the following sequences, and their limits, be given:

- $\{a_n\} = \left\{ \dfrac{n+1}{n^2} \right\}$, and $\lim_{n\to\infty} a_n = 0$;

- $\{b_n\} = \left\{ \left(1 + \dfrac{1}{n}\right)^n \right\}$, and $\lim_{n\to\infty} b_n = e$; and

- $\{c_n\} = \{n \cdot \sin(5/n)\}$, and $\lim_{n\to\infty} c_n = 5$.

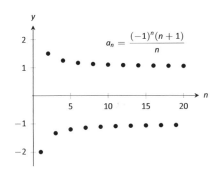

Figure 8.1.3: A plot of a sequence in Example 8.1.4, part 2.

Notes:

Evaluate the following limits.

$$1.\ \lim_{n\to\infty} (a_n + b_n) \qquad 2.\ \lim_{n\to\infty} (b_n \cdot c_n) \qquad 3.\ \lim_{n\to\infty} (1000 \cdot a_n)$$

Solution   We will use Theorem 8.1.3 to answer each of these.

1. Since $\lim\limits_{n\to\infty} a_n = 0$ and $\lim\limits_{n\to\infty} b_n = e$, we conclude that $\lim\limits_{n\to\infty} (a_n + b_n) = 0 + e = e$. So even though we are adding something to each term of the sequence b_n, we are adding something so small that the final limit is the same as before.

2. Since $\lim\limits_{n\to\infty} b_n = e$ and $\lim\limits_{n\to\infty} c_n = 5$, we conclude that $\lim\limits_{n\to\infty} (b_n \cdot c_n) = e \cdot 5 = 5e$.

3. Since $\lim\limits_{n\to\infty} a_n = 0$, we have $\lim\limits_{n\to\infty} 1000 a_n = 1000 \cdot 0 = 0$. It does not matter that we multiply each term by 1000; the sequence still approaches 0. (It just takes longer to get close to 0.)

There is more to learn about sequences than just their limits. We will also study their range and the relationships terms have with the terms that follow. We start with some definitions describing properties of the range.

Definition 8.1.3   Bounded and Unbounded Sequences

A sequence $\{a_n\}$ is said to be **bounded** if there exist real numbers m and M such that $m < a_n < M$ for all n in $\mathbb{N}$.

A sequence $\{a_n\}$ is said to be **unbounded** if it is not bounded.

A sequence $\{a_n\}$ is said to be **bounded above** if there exists an M such that $a_n < M$ for all n in $\mathbb{N}$; it is **bounded below** if there exists an m such that $m < a_n$ for all n in $\mathbb{N}$.

It follows from this definition that an unbounded sequence may be bounded above or bounded below; a sequence that is both bounded above and below is simply a bounded sequence.

Example 8.1.6   Determining boundedness of sequences
Determine the boundedness of the following sequences.

$$1.\ \{a_n\} = \left\{\frac{1}{n}\right\} \qquad 2.\ \{a_n\} = \{2^n\}$$

Notes:

SOLUTION

1. The terms of this sequence are always positive but are decreasing, so we have $0 < a_n < 2$ for all n. Thus this sequence is bounded. Figure 8.1.4(a) illustrates this.

2. The terms of this sequence obviously grow without bound. However, it is also true that these terms are all positive, meaning $0 < a_n$. Thus we can say the sequence is unbounded, but also bounded below. Figure 8.1.4(b) illustrates this.

The previous example produces some interesting concepts. First, we can recognize that the sequence $\{1/n\}$ converges to 0. This says, informally, that "most" of the terms of the sequence are "really close" to 0. This implies that the sequence is bounded, using the following logic. First, "most" terms are near 0, so we could find some sort of bound on these terms (using Definition 8.1.2, the bound is ε). That leaves a "few" terms that are not near 0 (i.e., a *finite* number of terms). A finite list of numbers is always bounded.

This logic implies that if a sequence converges, it must be bounded. This is indeed true, as stated by the following theorem.

Theorem 8.1.4	**Convergent Sequences are Bounded**

Let $\{a_n\}$ be a convergent sequence. Then $\{a_n\}$ is bounded.

In Example 8.1.5 we saw the sequence $\{b_n\} = \left\{(1 + 1/n)^n\right\}$, where it was stated that $\lim\limits_{n\to\infty} b_n = e$. (Note that this is simply restating part of Theorem 1.3.5.) Even though it may be difficult to intuitively grasp the behavior of this sequence, we know immediately that it is bounded.

Another interesting concept to come out of Example 8.1.6 again involves the sequence $\{1/n\}$. We stated, without proof, that the terms of the sequence were decreasing. That is, that $a_{n+1} < a_n$ for all n. (This is easy to show. Clearly $n < n + 1$. Taking reciprocals flips the inequality: $1/n > 1/(n+1)$. This is the same as $a_n > a_{n+1}$.) Sequences that either steadily increase or decrease are important, so we give this property a name.

Note: Keep in mind what Theorem 8.1.4 does *not* say. It does not say that bounded sequences must converge, nor does it say that if a sequence does not converge, it is not bounded.

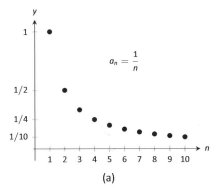

(a)

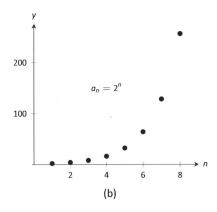

(b)

Figure 8.1.4: A plot of $\{a_n\} = \{1/n\}$ and $\{a_n\} = \{2^n\}$ from Example 8.1.6.

Notes:

Note: It is sometimes useful to call a monotonically increasing sequence *strictly increasing* if $a_n < a_{n+1}$ for all n; i.e, we remove the possibility that subsequent terms are equal.

A similar statement holds for *strictly decreasing*.

Definition 8.1.4   Monotonic Sequences

1. A sequence $\{a_n\}$ is **monotonically increasing** if $a_n \leq a_{n+1}$ for all n, i.e.,
$$a_1 \leq a_2 \leq a_3 \leq \cdots a_n \leq a_{n+1} \cdots$$

2. A sequence $\{a_n\}$ is **monotonically decreasing** if $a_n \geq a_{n+1}$ for all n, i.e.,
$$a_1 \geq a_2 \geq a_3 \geq \cdots a_n \geq a_{n+1} \cdots$$

3. A sequence is **monotonic** if it is monotonically increasing or monotonically decreasing.

Example 8.1.7   Determining monotonicity

Determine the monotonicity of the following sequences.

1. $\{a_n\} = \left\{ \dfrac{n+1}{n} \right\}$

2. $\{a_n\} = \left\{ \dfrac{n^2+1}{n+1} \right\}$

3. $\{a_n\} = \left\{ \dfrac{n^2-9}{n^2-10n+26} \right\}$

4. $\{a_n\} = \left\{ \dfrac{n^2}{n!} \right\}$

SOLUTION   In each of the following, we will examine $a_{n+1} - a_n$. If $a_{n+1} - a_n \geq 0$, we conclude that $a_n \leq a_{n+1}$ and hence the sequence is increasing. If $a_{n+1} - a_n \leq 0$, we conclude that $a_n \geq a_{n+1}$ and the sequence is decreasing. Of course, a sequence need not be monotonic and perhaps neither of the above will apply.

We also give a scatter plot of each sequence. These are useful as they suggest a pattern of monotonicity, but analytic work should be done to confirm a graphical trend.

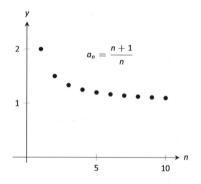

1.
$$a_{n+1} - a_n = \frac{n+2}{n+1} - \frac{n+1}{n}$$
$$= \frac{(n+2)(n) - (n+1)^2}{(n+1)n}$$
$$= \frac{-1}{n(n+1)}$$
$$< 0 \quad \text{for all } n.$$

Since $a_{n+1} - a_n < 0$ for all n, we conclude that the sequence is decreasing.

Figure 8.1.5: A plot of $\{a_n\} = \{(n+1)/n\}$ in Example 8.1.7.

Notes:

2.
$$a_{n+1} - a_n = \frac{(n+1)^2 + 1}{n+2} - \frac{n^2 + 1}{n+1}$$
$$= \frac{((n+1)^2 + 1)(n+1) - (n^2 + 1)(n+2)}{(n+1)(n+2)}$$
$$= \frac{n^2 + 4n + 1}{(n+1)(n+2)}$$
$$> 0 \quad \text{for all } n.$$

Since $a_{n+1} - a_n > 0$ for all n, we conclude the sequence is increasing; see Figure 8.1.6(a).

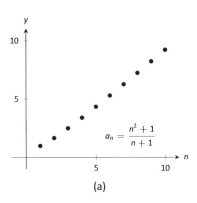

(a)

3. We can clearly see in Figure 8.1.6(b), where the sequence is plotted, that it is not monotonic. However, it does seem that after the first 4 terms it is decreasing. To understand why, perform the same analysis as done before:

$$a_{n+1} - a_n = \frac{(n+1)^2 - 9}{(n+1)^2 - 10(n+1) + 26} - \frac{n^2 - 9}{n^2 - 10n + 26}$$
$$= \frac{n^2 + 2n - 8}{n^2 - 8n + 17} - \frac{n^2 - 9}{n^2 - 10n + 26}$$
$$= \frac{(n^2 + 2n - 8)(n^2 - 10n + 26) - (n^2 - 9)(n^2 - 8n + 17)}{(n^2 - 8n + 17)(n^2 - 10n + 26)}$$
$$= \frac{-10n^2 + 60n - 55}{(n^2 - 8n + 17)(n^2 - 10n + 26)}.$$

We want to know when this is greater than, or less than, 0. The denominator is always positive, therefore we are only concerned with the numerator. For small values of n, the numerator is positive. As n grows large, the numerator is dominated by $-10n^2$, meaning the entire fraction will be negative; i.e., for large enough n, $a_{n+1} - a_n < 0$. Using the quadratic formula we can determine that the numerator is negative for $n \geq 5$.

In short, the sequence is simply not monotonic, though it is useful to note that for $n \geq 5$, the sequence is monotonically decreasing.

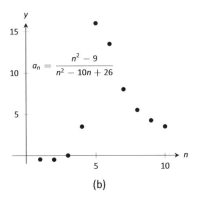

(b)

4. Again, the plot in Figure 8.1.6(c) shows that the sequence is not monotonic, but it suggests that it is monotonically decreasing after the first term. We perform the usual analysis to confirm this.

$$a_{n+1} - a_n = \frac{(n+1)^2}{(n+1)!} - \frac{n^2}{n!}$$
$$= \frac{(n+1)^2 - n^2(n+1)}{(n+1)!}$$
$$= \frac{-n^3 + 2n + 1}{(n+1)!}$$

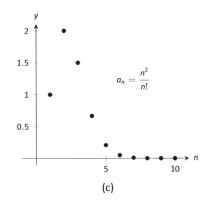

(c)

Figure 8.1.6: Plots of sequences in Example 8.1.7.

Notes:

When $n = 1$, the above expression is > 0; for $n \geq 2$, the above expression is < 0. Thus this sequence is not monotonic, but it is monotonically decreasing after the first term.

Knowing that a sequence is monotonic can be useful. Consider, for example, a sequence that is monotonically decreasing and is bounded below. We know the sequence is always getting smaller, but that there is a bound to how small it can become. This is enough to prove that the sequence will converge, as stated in the following theorem.

Theorem 8.1.5   Bounded Monotonic Sequences are Convergent

1. Let $\{a_n\}$ be a monotonically increasing sequence that is bounded above. Then $\{a_n\}$ converges.

2. Let $\{a_n\}$ be a monotonically decreasing sequence that is bounded below. Then $\{a_n\}$ converges.

Consider once again the sequence $\{a_n\} = \{1/n\}$. It is easy to show it is monotonically decreasing and that it is always positive (i.e., bounded below by 0). Therefore we can conclude by Theorem 8.1.5 that the sequence converges. We already knew this by other means, but in the following section this theorem will become very useful.

We can replace Theorem 8.1.5 with the statement "Let $\{a_n\}$ be a bounded, monotonic sequence. Then $\{a_n\}$ converges; i.e., $\lim\limits_{n \to \infty} a_n$ exists." We leave it to the reader in the exercises to show the theorem and the above statement are equivalent.

Sequences are a great source of mathematical inquiry. The On-Line Encyclopedia of Integer Sequences (http://oeis.org) contains thousands of sequences and their formulae. (As of this writing, there are 297,573 sequences in the database.) Perusing this database quickly demonstrates that a single sequence can represent several different "real life" phenomena.

Interesting as this is, our interest actually lies elsewhere. We are more interested in the *sum* of a sequence. That is, given a sequence $\{a_n\}$, we are very interested in $a_1 + a_2 + a_3 + \cdots$. Of course, one might immediately counter with "Doesn't this just add up to 'infinity'?" Many times, yes, but there are many important cases where the answer is no. This is the topic of *series*, which we begin to investigate in the next section.

Notes:

Exercises 8.1

Terms and Concepts

1. Use your own words to define a *sequence*.

2. The domain of a sequence is the _____ numbers.

3. Use your own words to describe the *range* of a sequence.

4. Describe what it means for a sequence to be *bounded*.

Problems

In Exercises 5 – 8, give the first five terms of the given sequence.

5. $\{a_n\} = \left\{ \dfrac{4^n}{(n+1)!} \right\}$

6. $\{b_n\} = \left\{ \left(-\dfrac{3}{2}\right)^n \right\}$

7. $\{c_n\} = \left\{ -\dfrac{n^{n+1}}{n+2} \right\}$

8. $\{d_n\} = \left\{ \dfrac{1}{\sqrt{5}} \left(\left(\dfrac{1+\sqrt{5}}{2}\right)^n - \left(\dfrac{1-\sqrt{5}}{2}\right)^n \right) \right\}$

In Exercises 9 – 12, determine the nth term of the given sequence.

9. $4, 7, 10, 13, 16, \ldots$

10. $3, -\dfrac{3}{2}, \dfrac{3}{4}, -\dfrac{3}{8}, \ldots$

11. $10, 20, 40, 80, 160, \ldots$

12. $1, 1, \dfrac{1}{2}, \dfrac{1}{6}, \dfrac{1}{24}, \dfrac{1}{120}, \ldots$

In Exercises 13 – 16, use the following information to determine the limit of the given sequences.

- $\{a_n\} = \left\{ \dfrac{2^n - 20}{2^n} \right\}$; $\lim\limits_{n\to\infty} a_n = 1$
- $\{b_n\} = \left\{ \left(1 + \dfrac{2}{n}\right)^n \right\}$; $\lim\limits_{n\to\infty} b_n = e^2$
- $\{c_n\} = \{\sin(3/n)\}$; $\lim\limits_{n\to\infty} c_n = 0$

13. $\{a_n\} = \left\{ \dfrac{2^n - 20}{7 \cdot 2^n} \right\}$

14. $\{a_n\} = \{3b_n - a_n\}$

15. $\{a_n\} = \left\{ \sin(3/n)\left(1 + \dfrac{2}{n}\right)^n \right\}$

16. $\{a_n\} = \left\{ \left(1 + \dfrac{2}{n}\right)^{2n} \right\}$

In Exercises 17 – 28, determine whether the sequence converges or diverges. If convergent, give the limit of the sequence.

17. $\{a_n\} = \left\{ (-1)^n \dfrac{n}{n+1} \right\}$

18. $\{a_n\} = \left\{ \dfrac{4n^2 - n + 5}{3n^2 + 1} \right\}$

19. $\{a_n\} = \left\{ \dfrac{4^n}{5^n} \right\}$

20. $\{a_n\} = \left\{ \dfrac{n-1}{n} - \dfrac{n}{n-1} \right\}, n \geq 2$

21. $\{a_n\} = \{\ln(n)\}$

22. $\{a_n\} = \left\{ \dfrac{3n}{\sqrt{n^2 + 1}} \right\}$

23. $\{a_n\} = \left\{ \left(1 + \dfrac{1}{n}\right)^n \right\}$

24. $\{a_n\} = \left\{ 5 - \dfrac{1}{n} \right\}$

25. $\{a_n\} = \left\{ \dfrac{(-1)^{n+1}}{n} \right\}$

26. $\{a_n\} = \left\{ \dfrac{1.1^n}{n} \right\}$

27. $\{a_n\} = \left\{ \dfrac{2n}{n+1} \right\}$

28. $\{a_n\} = \left\{ (-1)^n \dfrac{n^2}{2^n - 1} \right\}$

In Exercises 29 – 34, determine whether the sequence is bounded, bounded above, bounded below, or none of the above.

29. $\{a_n\} = \{\sin n\}$

30. $\{a_n\} = \{\tan n\}$

31. $\{a_n\} = \left\{ (-1)^n \dfrac{3n-1}{n} \right\}$

32. $\{a_n\} = \left\{ \dfrac{3n^2 - 1}{n} \right\}$

33. $\{a_n\} = \{n \cos n\}$

34. $\{a_n\} = \{2^n - n!\}$

0.

In Exercises 35 – 38, determine whether the sequence is monotonically increasing or decreasing. If it is not, determine if there is an m such that it is monotonic for all $n \geq m$.

35. $\{a_n\} = \left\{ \dfrac{n}{n+2} \right\}$

36. $\{a_n\} = \left\{ \dfrac{n^2 - 6n + 9}{n} \right\}$

37. $\{a_n\} = \left\{ (-1)^n \dfrac{1}{n^3} \right\}$

38. $\{a_n\} = \left\{ \dfrac{n^2}{2^n} \right\}$

Exercises 39 – 42 explore further the theory of sequences.

39. Prove Theorem 8.1.2; that is, use the definition of the limit of a sequence to show that if $\lim\limits_{n \to \infty} |a_n| = 0$, then $\lim\limits_{n \to \infty} a_n = 0$.

40. Let $\{a_n\}$ and $\{b_n\}$ be sequences such that $\lim\limits_{n \to \infty} a_n = L$ and $\lim\limits_{n \to \infty} b_n = K$.

   (a) Show that if $a_n < b_n$ for all n, then $L \leq K$.

   (b) Give an example where $L = K$.

41. Prove the Squeeze Theorem for sequences: Let $\{a_n\}$ and $\{b_n\}$ be such that $\lim\limits_{n \to \infty} a_n = L$ and $\lim\limits_{n \to \infty} b_n = L$, and let $\{c_n\}$ be such that $a_n \leq c_n \leq b_n$ for all n. Then $\lim\limits_{n \to \infty} c_n = L$

42. Prove the statement "Let $\{a_n\}$ be a bounded, monotonic sequence. Then $\{a_n\}$ converges; i.e., $\lim\limits_{n \to \infty} a_n$ exists." is equivalent to Theorem 8.1.5. That is,

   (a) Show that if Theorem 8.1.5 is true, then above statement is true, and

   (b) Show that if the above statement is true, then Theorem 8.1.5 is true.

8.2   Infinite Series

Given the sequence $\{a_n\} = \{1/2^n\} = 1/2,\ 1/4,\ 1/8,\ \ldots$, consider the following sums:

$$
\begin{array}{rcccc}
a_1 & = & 1/2 & = & 1/2 \\
a_1 + a_2 & = & 1/2 + 1/4 & = & 3/4 \\
a_1 + a_2 + a_3 & = & 1/2 + 1/4 + 1/8 & = & 7/8 \\
a_1 + a_2 + a_3 + a_4 & = & 1/2 + 1/4 + 1/8 + 1/16 & = & 15/16
\end{array}
$$

In general, we can show that

$$
a_1 + a_2 + a_3 + \cdots + a_n = \frac{2^n - 1}{2^n} = 1 - \frac{1}{2^n}.
$$

Let S_n be the sum of the first n terms of the sequence $\{1/2^n\}$. From the above, we see that $S_1 = 1/2$, $S_2 = 3/4$, etc. Our formula at the end shows that $S_n = 1 - 1/2^n$.

Now consider the following limit: $\lim\limits_{n\to\infty} S_n = \lim\limits_{n\to\infty}\left(1 - 1/2^n\right) = 1$. This limit can be interpreted as saying something amazing: *the sum of all the terms of the sequence $\{1/2^n\}$ is 1.*

This example illustrates some interesting concepts that we explore in this section. We begin this exploration with some definitions.

Definition 8.2.1   **Infinite Series, n^{th} Partial Sums, Convergence, Divergence**

Let $\{a_n\}$ be a sequence.

1. The sum $\displaystyle\sum_{n=1}^{\infty} a_n$ is an **infinite series** (or, simply **series**).

2. Let $\displaystyle S_n = \sum_{i=1}^{n} a_i$; the sequence $\{S_n\}$ is the sequence of nth **partial sums** of $\{a_n\}$.

3. If the sequence $\{S_n\}$ converges to L, we say the series $\displaystyle\sum_{n=1}^{\infty} a_n$ **converges** to L, and we write $\displaystyle\sum_{n=1}^{\infty} a_n = L$.

4. If the sequence $\{S_n\}$ diverges, the series $\displaystyle\sum_{n=1}^{\infty} a_n$ **diverges**.

Notes:

Using our new terminology, we can state that the series $\displaystyle\sum_{n=1}^{\infty} 1/2^n$ converges, and $\displaystyle\sum_{n=1}^{\infty} 1/2^n = 1$.

We will explore a variety of series in this section. We start with two series that diverge, showing how we might discern divergence.

Example 8.2.1   **Showing series diverge**

1. Let $\{a_n\} = \{n^2\}$. Show $\displaystyle\sum_{n=1}^{\infty} a_n$ diverges.

2. Let $\{b_n\} = \{(-1)^{n+1}\}$. Show $\displaystyle\sum_{n=1}^{\infty} b_n$ diverges.

Solution

1. Consider S_n, the nth partial sum.

$$S_n = a_1 + a_2 + a_3 + \cdots + a_n$$
$$= 1^2 + 2^2 + 3^2 \cdots + n^2.$$

By Theorem 5.3.1, this is

$$= \frac{n(n+1)(2n+1)}{6}.$$

Since $\displaystyle\lim_{n\to\infty} S_n = \infty$, we conclude that the series $\displaystyle\sum_{n=1}^{\infty} n^2$ diverges. It is instructive to write $\displaystyle\sum_{n=1}^{\infty} n^2 = \infty$ for this tells us *how* the series diverges: it grows without bound.

A scatter plot of the sequences $\{a_n\}$ and $\{S_n\}$ is given in Figure 8.2.1(a). The terms of $\{a_n\}$ are growing, so the terms of the partial sums $\{S_n\}$ are growing even faster, illustrating that the series diverges.

Notes:

2. The sequence $\{b_n\}$ starts with $1, -1, 1, -1, \ldots$. Consider some of the partial sums S_n of $\{b_n\}$:

$$S_1 = 1$$
$$S_2 = 0$$
$$S_3 = 1$$
$$S_4 = 0$$

This pattern repeats; we find that $S_n = \begin{cases} 1 & n \text{ is odd} \\ 0 & n \text{ is even} \end{cases}$. As $\{S_n\}$ oscillates, repeating $1, 0, 1, 0, \ldots$, we conclude that $\lim\limits_{n\to\infty} S_n$ does not exist, hence $\sum\limits_{n=1}^{\infty} (-1)^{n+1}$ diverges.

A scatter plot of the sequence $\{b_n\}$ and the partial sums $\{S_n\}$ is given in Figure 8.2.1(b). When n is odd, $b_n = S_n$ so the marks for b_n are drawn oversized to show they coincide.

While it is important to recognize when a series diverges, we are generally more interested in the series that converge. In this section we will demonstrate a few general techniques for determining convergence; later sections will delve deeper into this topic.

Geometric Series

One important type of series is a *geometric series*.

Definition 8.2.2 Geometric Series

A **geometric series** is a series of the form

$$\sum_{n=0}^{\infty} r^n = 1 + r + r^2 + r^3 + \cdots + r^n + \cdots$$

Note that the index starts at $n = 0$, not $n = 1$.

We started this section with a geometric series, although we dropped the first term of 1. One reason geometric series are important is that they have nice convergence properties.

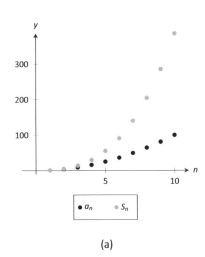

(a)

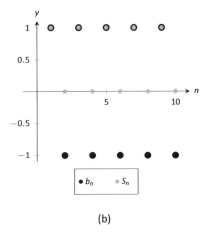

(b)

Figure 8.2.1: Scatter plots relating to Example 8.2.1.

Notes:

421

> **Theorem 8.2.1   Geometric Series Test**
>
> Consider the geometric series $\displaystyle\sum_{n=0}^{\infty} r^n$.
>
> 1. The nth partial sum is: $S_n = \dfrac{1 - r^{n+1}}{1 - r}, r \neq 1$.
>
> 2. The series converges if, and only if, $|r| < 1$. When $|r| < 1$,
>
> $$\sum_{n=0}^{\infty} r^n = \frac{1}{1-r}.$$

According to Theorem 8.2.1, the series

$$\sum_{n=0}^{\infty} \frac{1}{2^n} = \sum_{n=0}^{\infty} \left(\frac{1}{2}\right)^2 = 1 + \frac{1}{2} + \frac{1}{4} + \cdots$$

converges as $r = 1/2$, and $\displaystyle\sum_{n=0}^{\infty} \frac{1}{2^n} = \frac{1}{1 - 1/2} = 2$. This concurs with our introductory example; while there we got a sum of 1, we skipped the first term of 1.

Example 8.2.2   Exploring geometric series
Check the convergence of the following series. If the series converges, find its sum.

1. $\displaystyle\sum_{n=2}^{\infty} \left(\frac{3}{4}\right)^n$ 2. $\displaystyle\sum_{n=0}^{\infty} \left(\frac{-1}{2}\right)^n$ 3. $\displaystyle\sum_{n=0}^{\infty} 3^n$

SOLUTION

1. Since $r = 3/4 < 1$, this series converges. By Theorem 8.2.1, we have that

$$\sum_{n=0}^{\infty} \left(\frac{3}{4}\right)^n = \frac{1}{1 - 3/4} = 4.$$

However, note the subscript of the summation in the given series: we are to start with $n = 2$. Therefore we subtract off the first two terms, giving:

$$\sum_{n=2}^{\infty} \left(\frac{3}{4}\right)^n = 4 - 1 - \frac{3}{4} = \frac{9}{4}.$$

This is illustrated in Figure 8.2.2.

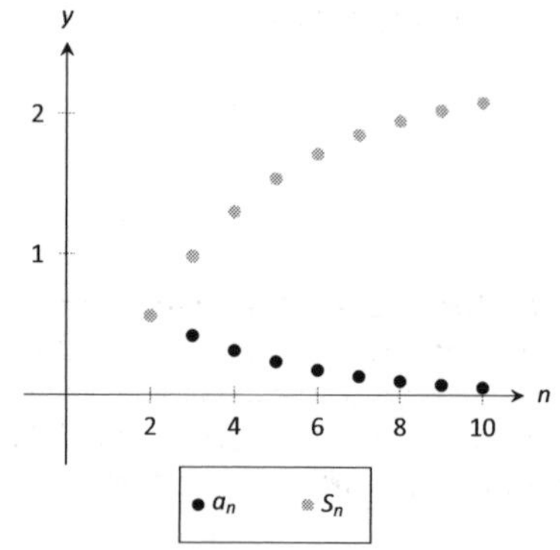

Figure 8.2.2: Scatter plots relating to the series in Example 8.2.2.

Notes:

2. Since $|r| = 1/2 < 1$, this series converges, and by Theorem 8.2.1,

$$\sum_{n=0}^{\infty}\left(\frac{-1}{2}\right)^n = \frac{1}{1-(-1/2)} = \frac{2}{3}.$$

The partial sums of this series are plotted in Figure 8.2.3(a). Note how the partial sums are not purely increasing as some of the terms of the sequence $\{(-1/2)^n\}$ are negative.

3. Since $r > 1$, the series diverges. (This makes "common sense"; we expect the sum

$$1 + 3 + 9 + 27 + 81 + 243 + \cdots$$

to diverge.) This is illustrated in Figure 8.2.3(b).

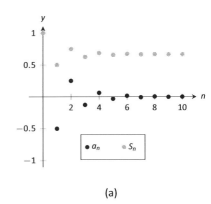

(a)

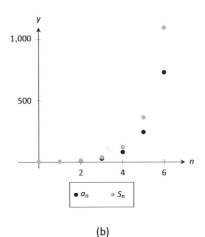

(b)

Figure 8.2.3: Scatter plots relating to the series in Example 8.2.2.

p–Series

Another important type of series is the *p-series*.

Definition 8.2.3 p–Series, General p–Series

1. A *p*–**series** is a series of the form

$$\sum_{n=1}^{\infty}\frac{1}{n^p},\qquad \text{where } p > 0.$$

2. A **general *p*–series** is a series of the form

$$\sum_{n=1}^{\infty}\frac{1}{(an+b)^p},\qquad \text{where } p > 0 \text{ and } a, b \text{ are real numbers.}$$

Like geometric series, one of the nice things about p–series is that they have easy to determine convergence properties.

Theorem 8.2.2 p–Series Test

A general *p*–series $\sum_{n=1}^{\infty}\dfrac{1}{(an+b)^p}$ will converge if, and only if, $p > 1$.

Note: Theorem 8.2.2 assumes that $an + b \neq 0$ for all n. If $an + b = 0$ for some n, then of course the series does not converge regardless of p as not all of the terms of the sequence are defined.

Notes:

Example 8.2.3   Determining convergence of series

Determine the convergence of the following series.

1. $\displaystyle\sum_{n=1}^{\infty} \frac{1}{n}$ 3. $\displaystyle\sum_{n=1}^{\infty} \frac{1}{\sqrt{n}}$ 5. $\displaystyle\sum_{n=11}^{\infty} \frac{1}{\left(\frac{1}{2}n - 5\right)^3}$

2. $\displaystyle\sum_{n=1}^{\infty} \frac{1}{n^2}$ 4. $\displaystyle\sum_{n=1}^{\infty} \frac{(-1)^n}{n}$ 6. $\displaystyle\sum_{n=1}^{\infty} \frac{1}{2^n}$

SOLUTION

1. This is a *p*–series with $p = 1$. By Theorem 8.2.2, this series diverges.

   This series is a famous series, called the *Harmonic Series*, so named because of its relationship to *harmonics* in the study of music and sound.

2. This is a *p*–series with $p = 2$. By Theorem 8.2.2, it converges. Note that the theorem does not give a formula by which we can determine *what* the series converges to; we just know it converges. A famous, unexpected result is that this series converges to $\pi^2/6$.

3. This is a *p*–series with $p = 1/2$; the theorem states that it diverges.

4. This is not a *p*–series; the definition does not allow for alternating signs. Therefore we cannot apply Theorem 8.2.2. (Another famous result states that this series, the *Alternating Harmonic Series*, converges to ln 2.)

5. This is a general *p*–series with $p = 3$, therefore it converges.

6. This is not a *p*–series, but a geometric series with $r = 1/2$. It converges.

Later sections will provide tests by which we can determine whether or not a given series converges. This, in general, is much easier than determining *what* a given series converges to. There are many cases, though, where the sum can be determined.

Example 8.2.4   Telescoping series

Evaluate the sum $\displaystyle\sum_{n=1}^{\infty} \left(\frac{1}{n} - \frac{1}{n+1} \right)$.

SOLUTION   It will help to write down some of the first few partial sums

Notes:

of this series.

$$S_1 = \frac{1}{1} - \frac{1}{2} \qquad\qquad = 1 - \frac{1}{2}$$

$$S_2 = \left(\frac{1}{1} - \frac{1}{2}\right) + \left(\frac{1}{2} - \frac{1}{3}\right) \qquad\qquad = 1 - \frac{1}{3}$$

$$S_3 = \left(\frac{1}{1} - \frac{1}{2}\right) + \left(\frac{1}{2} - \frac{1}{3}\right) + \left(\frac{1}{3} - \frac{1}{4}\right) \qquad\qquad = 1 - \frac{1}{4}$$

$$S_4 = \left(\frac{1}{1} - \frac{1}{2}\right) + \left(\frac{1}{2} - \frac{1}{3}\right) + \left(\frac{1}{3} - \frac{1}{4}\right) + \left(\frac{1}{4} - \frac{1}{5}\right) \qquad = 1 - \frac{1}{5}$$

Note how most of the terms in each partial sum are canceled out! In general, we see that $S_n = 1 - \dfrac{1}{n+1}$. The sequence $\{S_n\}$ converges, as $\lim\limits_{n\to\infty} S_n = \lim\limits_{n\to\infty}\left(1 - \dfrac{1}{n+1}\right) = 1$, and so we conclude that $\sum\limits_{n=1}^{\infty}\left(\dfrac{1}{n} - \dfrac{1}{n+1}\right) = 1$. Partial sums of the series are plotted in Figure 8.2.4.

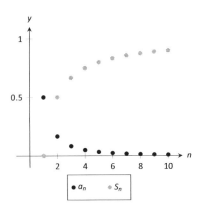

Figure 8.2.4: Scatter plots relating to the series of Example 8.2.4.

The series in Example 8.2.4 is an example of a **telescoping series**. Informally, a telescoping series is one in which most terms cancel with preceding or following terms, reducing the number of terms in each partial sum. The partial sum S_n did not contain n terms, but rather just two: 1 and $1/(n+1)$.

When possible, seek a way to write an explicit formula for the nth partial sum S_n. This makes evaluating the limit $\lim\limits_{n\to\infty} S_n$ much more approachable. We do so in the next example.

Example 8.2.5   Evaluating series
Evaluate each of the following infinite series.

1. $\displaystyle\sum_{n=1}^{\infty} \frac{2}{n^2 + 2n}$ 2. $\displaystyle\sum_{n=1}^{\infty} \ln\left(\frac{n+1}{n}\right)$

SOLUTION

1. We can decompose the fraction $2/(n^2 + 2n)$ as

$$\frac{2}{n^2 + 2n} = \frac{1}{n} - \frac{1}{n+2}.$$

(See Section 6.5, Partial Fraction Decomposition, to recall how this is done, if necessary.)

Notes:

Expressing the terms of $\{S_n\}$ is now more instructive:

$$S_1 = 1 - \frac{1}{3} \qquad\qquad = 1 - \frac{1}{3}$$

$$S_2 = \left(1 - \frac{1}{3}\right) + \left(\frac{1}{2} - \frac{1}{4}\right) \qquad\qquad = 1 + \frac{1}{2} - \frac{1}{3} - \frac{1}{4}$$

$$S_3 = \left(1 - \frac{1}{3}\right) + \left(\frac{1}{2} - \frac{1}{4}\right) + \left(\frac{1}{3} - \frac{1}{5}\right) \qquad\qquad = 1 + \frac{1}{2} - \frac{1}{4} - \frac{1}{5}$$

$$S_4 = \left(1 - \frac{1}{3}\right) + \left(\frac{1}{2} - \frac{1}{4}\right) + \left(\frac{1}{3} - \frac{1}{5}\right) + \left(\frac{1}{4} - \frac{1}{6}\right) \qquad\qquad = 1 + \frac{1}{2} - \frac{1}{5} - \frac{1}{6}$$

$$S_5 = \left(1 - \frac{1}{3}\right) + \left(\frac{1}{2} - \frac{1}{4}\right) + \left(\frac{1}{3} - \frac{1}{5}\right) + \left(\frac{1}{4} - \frac{1}{6}\right) + \left(\frac{1}{5} - \frac{1}{7}\right) \qquad\qquad = 1 + \frac{1}{2} - \frac{1}{6} - \frac{1}{7}$$

We again have a telescoping series. In each partial sum, most of the terms cancel and we obtain the formula $S_n = 1 + \dfrac{1}{2} - \dfrac{1}{n+1} - \dfrac{1}{n+2}$. Taking limits allows us to determine the convergence of the series:

$$\lim_{n\to\infty} S_n = \lim_{n\to\infty} \left(1 + \frac{1}{2} - \frac{1}{n+1} - \frac{1}{n+2}\right) = \frac{3}{2}, \quad \text{so} \quad \sum_{n=1}^{\infty} \frac{1}{n^2 + 2n} = \frac{3}{2}.$$

This is illustrated in Figure 8.2.5(a).

2. We begin by writing the first few partial sums of the series:

$$S_1 = \ln(2)$$

$$S_2 = \ln(2) + \ln\left(\frac{3}{2}\right)$$

$$S_3 = \ln(2) + \ln\left(\frac{3}{2}\right) + \ln\left(\frac{4}{3}\right)$$

$$S_4 = \ln(2) + \ln\left(\frac{3}{2}\right) + \ln\left(\frac{4}{3}\right) + \ln\left(\frac{5}{4}\right)$$

At first, this does not seem helpful, but recall the logarithmic identity: $\ln x + \ln y = \ln(xy)$. Applying this to S_4 gives:

$$S_4 = \ln(2) + \ln\left(\frac{3}{2}\right) + \ln\left(\frac{4}{3}\right) + \ln\left(\frac{5}{4}\right) = \ln\left(\frac{2}{1} \cdot \frac{3}{2} \cdot \frac{4}{3} \cdot \frac{5}{4}\right) = \ln(5).$$

We can conclude that $\{S_n\} = \{\ln(n+1)\}$. This sequence does not converge, as $\lim\limits_{n\to\infty} S_n = \infty$. Therefore $\sum\limits_{n=1}^{\infty} \ln\left(\dfrac{n+1}{n}\right) = \infty$; the series diverges. Note in Figure 8.2.5(b) how the sequence of partial sums grows

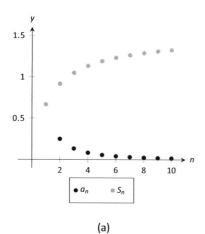

(a)

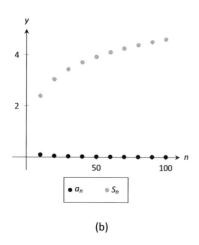

(b)

Figure 8.2.5: Scatter plots relating to the series in Example 8.2.5.

Notes:

slowly; after 100 terms, it is not yet over 5. Graphically we may be fooled into thinking the series converges, but our analysis above shows that it does not.

We are learning about a new mathematical object, the series. As done before, we apply "old" mathematics to this new topic.

Theorem 8.2.3   Properties of Infinite Series

Let $\displaystyle\sum_{n=1}^{\infty} a_n = L$, $\displaystyle\sum_{n=1}^{\infty} b_n = K$, and let c be a constant.

1. Constant Multiple Rule: $\displaystyle\sum_{n=1}^{\infty} c \cdot a_n = c \cdot \sum_{n=1}^{\infty} a_n = c \cdot L$.

2. Sum/Difference Rule: $\displaystyle\sum_{n=1}^{\infty} \left(a_n \pm b_n\right) = \sum_{n=1}^{\infty} a_n \pm \sum_{n=1}^{\infty} b_n = L \pm K$.

Before using this theorem, we provide a few "famous" series.

Key Idea 8.2.1   Important Series

1. $\displaystyle\sum_{n=0}^{\infty} \frac{1}{n!} = e.$ (Note that the index starts with $n = 0$.)

2. $\displaystyle\sum_{n=1}^{\infty} \frac{1}{n^2} = \frac{\pi^2}{6}.$

3. $\displaystyle\sum_{n=1}^{\infty} \frac{(-1)^{n+1}}{n^2} = \frac{\pi^2}{12}.$

4. $\displaystyle\sum_{n=0}^{\infty} \frac{(-1)^n}{2n+1} = \frac{\pi}{4}.$

5. $\displaystyle\sum_{n=1}^{\infty} \frac{1}{n}$ diverges.   (This is called the *Harmonic Series*.)

6. $\displaystyle\sum_{n=1}^{\infty} \frac{(-1)^{n+1}}{n} = \ln 2.$ (This is called the *Alternating Harmonic Series*.)

Notes:

Example 8.2.6   **Evaluating series**

Evaluate the given series.

1. $\displaystyle\sum_{n=1}^{\infty} \frac{(-1)^{n+1}(n^2 - n)}{n^3}$ 2. $\displaystyle\sum_{n=1}^{\infty} \frac{1000}{n!}$ 3. $\displaystyle\frac{1}{16} + \frac{1}{25} + \frac{1}{36} + \frac{1}{49} + \cdots$

SOLUTION

1. We start by using algebra to break the series apart:

$$\sum_{n=1}^{\infty} \frac{(-1)^{n+1}(n^2 - n)}{n^3} = \sum_{n=1}^{\infty}\left(\frac{(-1)^{n+1}n^2}{n^3} - \frac{(-1)^{n+1}n}{n^3}\right)$$

$$= \sum_{n=1}^{\infty} \frac{(-1)^{n+1}}{n} - \sum_{n=1}^{\infty} \frac{(-1)^{n+1}}{n^2}$$

$$= \ln(2) - \frac{\pi^2}{12} \approx -0.1293.$$

This is illustrated in Figure 8.2.6(a).

2. This looks very similar to the series that involves e in Key Idea 8.2.1. Note, however, that the series given in this example starts with $n = 1$ and not $n = 0$. The first term of the series in the Key Idea is $1/0! = 1$, so we will subtract this from our result below:

$$\sum_{n=1}^{\infty} \frac{1000}{n!} = 1000 \cdot \sum_{n=1}^{\infty} \frac{1}{n!}$$

$$= 1000 \cdot (e - 1) \approx 1718.28.$$

This is illustrated in Figure 8.2.6(b). The graph shows how this particular series converges very rapidly.

3. The denominators in each term are perfect squares; we are adding $\displaystyle\sum_{n=4}^{\infty} \frac{1}{n^2}$ (note we start with $n = 4$, not $n = 1$). This series will converge. Using the

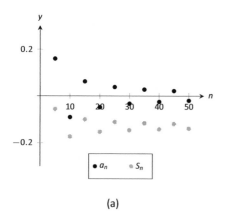

(a)

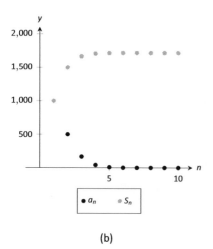

(b)

Figure 8.2.6: Scatter plots relating to the series in Example 8.2.6.

Notes:

formula from Key Idea 8.2.1, we have the following:

$$\sum_{n=1}^{\infty} \frac{1}{n^2} = \sum_{n=1}^{3} \frac{1}{n^2} + \sum_{n=4}^{\infty} \frac{1}{n^2}$$

$$\sum_{n=1}^{\infty} \frac{1}{n^2} - \sum_{n=1}^{3} \frac{1}{n^2} = \sum_{n=4}^{\infty} \frac{1}{n^2}$$

$$\frac{\pi^2}{6} - \left(\frac{1}{1} + \frac{1}{4} + \frac{1}{9} \right) = \sum_{n=4}^{\infty} \frac{1}{n^2}$$

$$\frac{\pi^2}{6} - \frac{49}{36} = \sum_{n=4}^{\infty} \frac{1}{n^2}$$

$$0.2838 \approx \sum_{n=4}^{\infty} \frac{1}{n^2}$$

It may take a while before one is comfortable with this statement, whose truth lies at the heart of the study of infinite series: *it is possible that the sum of an infinite list of nonzero numbers is finite.* We have seen this repeatedly in this section, yet it still may "take some getting used to."

As one contemplates the behavior of series, a few facts become clear.

1. In order to add an infinite list of nonzero numbers and get a finite result, "most" of those numbers must be "very near" 0.

2. If a series diverges, it means that the sum of an infinite list of numbers is not finite (it may approach $\pm\infty$ or it may oscillate), and:

    (a) The series will still diverge if the first term is removed.

    (b) The series will still diverge if the first 10 terms are removed.

    (c) The series will still diverge if the first $1,000,000$ terms are removed.

    (d) The series will still diverge if any finite number of terms from anywhere in the series are removed.

These concepts are very important and lie at the heart of the next two theorems.

Theorem 8.2.4 n^{th}**–Term Test for Divergence**

Consider the series $\displaystyle\sum_{n=1}^{\infty} a_n$. If $\displaystyle\lim_{n \to \infty} a_n \neq 0$, then $\displaystyle\sum_{n=1}^{\infty} a_n$ diverges.

Notes:

Important! This theorem *does not state* that if $\lim\limits_{n\to\infty} a_n = 0$ then $\sum\limits_{n=1}^{\infty} a_n$ converges. The standard example of this is the Harmonic Series, as given in Key Idea 8.2.1. The Harmonic Sequence, $\{1/n\}$, converges to 0; the Harmonic Series, $\sum\limits_{n=1}^{\infty} \dfrac{1}{n}$, diverges.

Looking back, we can apply this theorem to the series in Example 8.2.1. In that example, the nth terms of both sequences do not converge to 0, therefore we can quickly conclude that each series diverges.

One can rewrite Theorem 8.2.4 to state "If a series converges, then the underlying sequence converges to 0." While it is important to understand the truth of this statement, in practice it is rarely used. It is generally far easier to prove the convergence of a sequence than the convergence of a series.

Theorem 8.2.5   Infinite Nature of Series

The convergence or divergence of an infinite series remains unchanged by the addition or subtraction of any finite number of terms. That is:

1. A divergent series will remain divergent with the addition or subtraction of any finite number of terms.

2. A convergent series will remain convergent with the addition or subtraction of any finite number of terms. (Of course, the *sum* will likely change.)

Consider once more the Harmonic Series $\sum\limits_{n=1}^{\infty} \dfrac{1}{n}$ which diverges; that is, the sequence of partial sums $\{S_n\}$ grows (very, very slowly) without bound. One might think that by removing the "large" terms of the sequence that perhaps the series will converge. This is simply not the case. For instance, the sum of the first 10 million terms of the Harmonic Series is about 16.7. Removing the first 10 million terms from the Harmonic Series changes the nth partial sums, effectively subtracting 16.7 from the sum. However, a sequence that is growing without bound will still grow without bound when 16.7 is subtracted from it.

The equations below illustrate this. The first line shows the infinite sum of the Harmonic Series split into the sum of the first 10 million terms plus the sum of "everything else." The next equation shows us subtracting these first 10 million terms from both sides. The final equation employs a bit of "psuedo–math":

Notes:

subtracting 16.7 from "infinity" still leaves one with "infinity."

$$\sum_{n=1}^{\infty} \frac{1}{n} = \sum_{n=1}^{10,000,000} \frac{1}{n} + \sum_{n=10,000,001}^{\infty} \frac{1}{n}$$

$$\sum_{n=1}^{\infty} \frac{1}{n} - \sum_{n=1}^{10,000,000} \frac{1}{n} = \sum_{n=10,000,001}^{\infty} \frac{1}{n}$$

$$\infty \quad - \quad 16.7 \quad = \quad \infty.$$

This section introduced us to series and defined a few special types of series whose convergence properties are well known: we know when a p-series or a geometric series converges or diverges. Most series that we encounter are not one of these types, but we are still interested in knowing whether or not they converge. The next three sections introduce tests that help us determine whether or not a given series converges.

Notes:

Exercises 8.2

Terms and Concepts

1. Use your own words to describe how sequences and series are related.

2. Use your own words to define a *partial sum*.

3. Given a series $\sum_{n=1}^{\infty} a_n$, describe the two sequences related to the series that are important.

4. Use your own words to explain what a geometric series is.

5. T/F: If $\{a_n\}$ is convergent, then $\sum_{n=1}^{\infty} a_n$ is also convergent.

6. T/F: If $\{a_n\}$ converges to 0, then $\sum_{n=0}^{\infty} a_n$ converges.

Problems

In Exercises 7 – 14, a series $\sum_{n=1}^{\infty} a_n$ is given.

(a) Give the first 5 partial sums of the series.

(b) Give a graph of the first 5 terms of a_n and S_n on the same axes.

7. $\sum_{n=1}^{\infty} \dfrac{(-1)^n}{n}$

8. $\sum_{n=1}^{\infty} \dfrac{1}{n^2}$

9. $\sum_{n=1}^{\infty} \cos(\pi n)$

10. $\sum_{n=1}^{\infty} n$

11. $\sum_{n=1}^{\infty} \dfrac{1}{n!}$

12. $\sum_{n=1}^{\infty} \dfrac{1}{3^n}$

13. $\sum_{n=1}^{\infty} \left(-\dfrac{9}{10}\right)^n$

14. $\sum_{n=1}^{\infty} \left(\dfrac{1}{10}\right)^n$

In Exercises 15 – 20, use Theorem 8.2.4 to show the given series diverges.

15. $\sum_{n=1}^{\infty} \dfrac{3n^2}{n(n+2)}$

16. $\sum_{n=1}^{\infty} \dfrac{2^n}{n^2}$

17. $\sum_{n=1}^{\infty} \dfrac{n!}{10^n}$

18. $\sum_{n=1}^{\infty} \dfrac{5^n - n^5}{5^n + n^5}$

19. $\sum_{n=1}^{\infty} \dfrac{2^n + 1}{2^{n+1}}$

20. $\sum_{n=1}^{\infty} \left(1 + \dfrac{1}{n}\right)^n$

In Exercises 21 – 30, state whether the given series converges or diverges.

21. $\sum_{n=1}^{\infty} \dfrac{1}{n^5}$

22. $\sum_{n=0}^{\infty} \dfrac{1}{5^n}$

23. $\sum_{n=0}^{\infty} \dfrac{6^n}{5^n}$

24. $\sum_{n=1}^{\infty} n^{-4}$

25. $\sum_{n=1}^{\infty} \sqrt{n}$

26. $\sum_{n=1}^{\infty} \dfrac{10}{n!}$

27. T/F: If $\{a_n\}$ converges to 0, then $\sum_{n=0}^{\infty} a_n$ converges.

28. $\sum_{n=1}^{\infty} \dfrac{2}{(2n+8)^2}$

29. $\sum_{n=1}^{\infty} \dfrac{1}{2n}$

30. $\displaystyle\sum_{n=1}^{\infty} \frac{1}{2n-1}$

In Exercises 31 – 46, a series is given.

    (a) Find a formula for S_n, the nth partial sum of the series.

    (b) Determine whether the series converges or diverges. If it converges, state what it converges to.

31. $\displaystyle\sum_{n=0}^{\infty} \frac{1}{4^n}$

32. $\displaystyle\sum_{n=1}^{\infty} 2$

33. $1^3 + 2^3 + 3^3 + 4^3 + \cdots$

34. $\displaystyle\sum_{n=1}^{\infty} (-1)^n n$

35. $\displaystyle\sum_{n=0}^{\infty} \frac{5}{2^n}$

36. $\displaystyle\sum_{n=1}^{\infty} e^{-n}$

37. $1 - \dfrac{1}{3} + \dfrac{1}{9} - \dfrac{1}{27} + \dfrac{1}{81} + \cdots$

38. $\displaystyle\sum_{n=1}^{\infty} \frac{1}{n(n+1)}$

39. $\displaystyle\sum_{n=1}^{\infty} \frac{3}{n(n+2)}$

40. $\displaystyle\sum_{n=1}^{\infty} \frac{1}{(2n-1)(2n+1)}$

41. $\displaystyle\sum_{n=1}^{\infty} \ln\left(\frac{n}{n+1}\right)$

42. $\displaystyle\sum_{n=1}^{\infty} \frac{2n+1}{n^2(n+1)^2}$

43. $\dfrac{1}{1\cdot 4} + \dfrac{1}{2\cdot 5} + \dfrac{1}{3\cdot 6} + \dfrac{1}{4\cdot 7} + \cdots$

44. $2 + \left(\dfrac{1}{2} + \dfrac{1}{3}\right) + \left(\dfrac{1}{4} + \dfrac{1}{9}\right) + \left(\dfrac{1}{8} + \dfrac{1}{27}\right) + \cdots$

45. $\displaystyle\sum_{n=2}^{\infty} \frac{1}{n^2-1}$

46. $\displaystyle\sum_{n=0}^{\infty} \left(\sin 1\right)^n$

47. Break the Harmonic Series into the sum of the odd and even terms:
$$\sum_{n=1}^{\infty} \frac{1}{n} = \sum_{n=1}^{\infty} \frac{1}{2n-1} + \sum_{n=1}^{\infty} \frac{1}{2n}.$$

The goal is to show that each of the series on the right diverge.

    (a) Show why $\displaystyle\sum_{n=1}^{\infty} \frac{1}{2n-1} > \sum_{n=1}^{\infty} \frac{1}{2n}$.

    (Compare each nth partial sum.)

    (b) Show why $\displaystyle\sum_{n=1}^{\infty} \frac{1}{2n-1} < 1 + \sum_{n=1}^{\infty} \frac{1}{2n}$

    (c) Explain why (a) and (b) demonstrate that the series of odd terms is convergent, if, and only if, the series of even terms is also convergent. (That is, show both converge or both diverge.)

    (d) Explain why knowing the Harmonic Series is divergent determines that the even and odd series are also divergent.

48. Show the series $\displaystyle\sum_{n=1}^{\infty} \frac{n}{(2n-1)(2n+1)}$ diverges.

8.3 Integral and Comparison Tests

Knowing whether or not a series converges is very important, especially when we discuss Power Series in Section 8.6. Theorems 8.2.1 and 8.2.2 give criteria for when Geometric and p-series converge, and Theorem 8.2.4 gives a quick test to determine if a series diverges. There are many important series whose convergence cannot be determined by these theorems, though, so we introduce a set of tests that allow us to handle a broad range of series. We start with the Integral Test.

Integral Test

We stated in Section 8.1 that a sequence $\{a_n\}$ is a function $a(n)$ whose domain is $\mathbb{N}$, the set of natural numbers. If we can extend $a(n)$ to $\mathbb{R}$, the real numbers, and it is both positive and decreasing on $[1, \infty)$, then the convergence of $\sum_{n=1}^{\infty} a_n$ is the same as $\int_1^{\infty} a(x)\, dx$.

Note: Theorem 8.3.1 does not state that the integral and the summation have the same value.

Theorem 8.3.1   Integral Test

Let a sequence $\{a_n\}$ be defined by $a_n = a(n)$, where $a(n)$ is continuous, positive and decreasing on $[1, \infty)$. Then $\sum_{n=1}^{\infty} a_n$ converges, if, and only if,

$$\int_1^{\infty} a(x)\, dx \text{ converges.}$$

We can demonstrate the truth of the Integral Test with two simple graphs. In Figure 8.3.1(a), the height of each rectangle is $a(n) = a_n$ for $n = 1, 2, \ldots$, and clearly the rectangles enclose more area than the area under $y = a(x)$. Therefore we can conclude that

$$\int_1^{\infty} a(x)\, dx < \sum_{n=1}^{\infty} a_n. \tag{8.1}$$

In Figure 8.3.1(b), we draw rectangles under $y = a(x)$ with the Right-Hand rule, starting with $n = 2$. This time, the area of the rectangles is less than the area under $y = a(x)$, so $\sum_{n=2}^{\infty} a_n < \int_1^{\infty} a(x)\, dx$. Note how this summation starts with $n = 2$; adding a_1 to both sides lets us rewrite the summation starting with

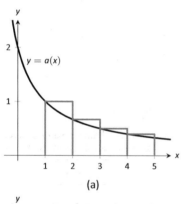

(a)

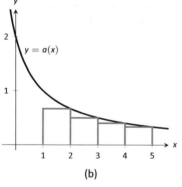

(b)

Figure 8.3.1: Illustrating the truth of the Integral Test.

Notes:

$n = 1$:

$$\sum_{n=1}^{\infty} a_n < a_1 + \int_{1}^{\infty} a(x)\, dx. \qquad (8.2)$$

Combining Equations (8.1) and (8.2), we have

$$\sum_{n=1}^{\infty} a_n < a_1 + \int_{1}^{\infty} a(x)\, dx < a_1 + \sum_{n=1}^{\infty} a_n. \qquad (8.3)$$

From Equation (8.3) we can make the following two statements:

1. If $\displaystyle\sum_{n=1}^{\infty} a_n$ diverges, so does $\displaystyle\int_{1}^{\infty} a(x)\, dx$ (because $\displaystyle\sum_{n=1}^{\infty} a_n < a_1 + \int_{1}^{\infty} a(x)\, dx$)

2. If $\displaystyle\sum_{n=1}^{\infty} a_n$ converges, so does $\displaystyle\int_{1}^{\infty} a(x)\, dx$ (because $\displaystyle\int_{1}^{\infty} a(x)\, dx < \sum_{n=1}^{\infty} a_n$.)

Therefore the series and integral either both converge or both diverge. Theorem 8.2.5 allows us to extend this theorem to series where $a(n)$ is positive and decreasing on $[b, \infty)$ for some $b > 1$.

Example 8.3.1   Using the Integral Test

Determine the convergence of $\displaystyle\sum_{n=1}^{\infty} \frac{\ln n}{n^2}$. (The terms of the sequence $\{a_n\} = \{\ln n/n^2\}$ and the nth partial sums are given in Figure 8.3.2.)

SOLUTION   Figure 8.3.2 implies that $a(n) = (\ln n)/n^2$ is positive and decreasing on $[2, \infty)$. We can determine this analytically, too. We know $a(n)$ is positive as both $\ln n$ and n^2 are positive on $[2, \infty)$. To determine that $a(n)$ is decreasing, consider $a'(n) = (1 - 2\ln n)/n^3$, which is negative for $n \geq 2$. Since $a'(n)$ is negative, $a(n)$ is decreasing.

Applying the Integral Test, we test the convergence of $\displaystyle\int_{1}^{\infty} \frac{\ln x}{x^2}\, dx$. Integrating this improper integral requires the use of Integration by Parts, with $u = \ln x$ and $dv = 1/x^2\, dx$.

$$\int_{1}^{\infty} \frac{\ln x}{x^2}\, dx = \lim_{b\to\infty} \int_{1}^{b} \frac{\ln x}{x^2}\, dx$$

$$= \lim_{b\to\infty} \left. -\frac{1}{x} \ln x \right|_{1}^{b} + \int_{1}^{b} \frac{1}{x^2}\, dx$$

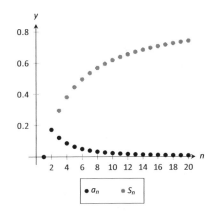

Figure 8.3.2: Plotting the sequence and series in Example 8.3.1.

Notes:

$$= \lim_{b \to \infty} -\frac{1}{x} \ln x - \frac{1}{x} \Big|_1^b$$

$$= \lim_{b \to \infty} 1 - \frac{1}{b} - \frac{\ln b}{b}. \quad \text{Apply L'Hôpital's Rule:}$$

$$= 1.$$

Since $\int_1^\infty \frac{\ln x}{x^2} \, dx$ converges, so does $\displaystyle\sum_{n=1}^\infty \frac{\ln n}{n^2}$.

Theorem 8.2.2 was given without justification, stating that the general p-series $\displaystyle\sum_{n=1}^\infty \frac{1}{(an+b)^p}$ converges if, and only if, $p > 1$. In the following example, we prove this to be true by applying the Integral Test.

Example 8.3.2 **Using the Integral Test to establish Theorem 8.2.2.**

Use the Integral Test to prove that $\displaystyle\sum_{n=1}^\infty \frac{1}{(an+b)^p}$ converges if, and only if, $p > 1$.

SOLUTION Consider the integral $\int_1^\infty \frac{1}{(ax+b)^p} \, dx$; assuming $p \neq 1$,

$$\int_1^\infty \frac{1}{(ax+b)^p} \, dx = \lim_{c \to \infty} \int_1^c \frac{1}{(ax+b)^p} \, dx$$

$$= \lim_{c \to \infty} \frac{1}{a(1-p)} (ax+b)^{1-p} \Big|_1^c$$

$$= \lim_{c \to \infty} \frac{1}{a(1-p)} \left((ac+b)^{1-p} - (a+b)^{1-p} \right).$$

This limit converges if, and only if, $p > 1$. It is easy to show that the integral also diverges in the case of $p = 1$. (This result is similar to the work preceding Key Idea 6.8.1.)

Therefore $\displaystyle\sum_{n=1}^\infty \frac{1}{(an+b)^p}$ converges if, and only if, $p > 1$.

We consider two more convergence tests in this section, both *comparison* tests. That is, we determine the convergence of one series by comparing it to another series with known convergence.

Notes:

Direct Comparison Test

Theorem 8.3.2   Direct Comparison Test

Let $\{a_n\}$ and $\{b_n\}$ be positive sequences where $a_n \leq b_n$ for all $n \geq N$, for some $N \geq 1$.

1. If $\displaystyle\sum_{n=1}^{\infty} b_n$ converges, then $\displaystyle\sum_{n=1}^{\infty} a_n$ converges.

2. If $\displaystyle\sum_{n=1}^{\infty} a_n$ diverges, then $\displaystyle\sum_{n=1}^{\infty} b_n$ diverges.

Note: A sequence $\{a_n\}$ is a **positive sequence** if $a_n > 0$ for all n.

Because of Theorem 8.2.5, any theorem that relies on a positive sequence still holds true when $a_n > 0$ for all but a finite number of values of n.

Example 8.3.3   Applying the Direct Comparison Test

Determine the convergence of $\displaystyle\sum_{n=1}^{\infty} \frac{1}{3^n + n^2}$.

SOLUTION   This series is neither a geometric or p-series, but seems related. We predict it will converge, so we look for a series with larger terms that converges. (Note too that the Integral Test seems difficult to apply here.)

Since $3^n < 3^n + n^2$, $\dfrac{1}{3^n} > \dfrac{1}{3^n + n^2}$ for all $n \geq 1$. The series $\displaystyle\sum_{n=1}^{\infty} \frac{1}{3^n}$ is a

convergent geometric series; by Theorem 8.3.2, $\displaystyle\sum_{n=1}^{\infty} \frac{1}{3^n + n^2}$ converges.

Example 8.3.4   Applying the Direct Comparison Test

Determine the convergence of $\displaystyle\sum_{n=1}^{\infty} \frac{1}{n - \ln n}$.

SOLUTION   We know the Harmonic Series $\displaystyle\sum_{n=1}^{\infty} \frac{1}{n}$ diverges, and it seems that the given series is closely related to it, hence we predict it will diverge.

Since $n \geq n - \ln n$ for all $n \geq 1$, $\dfrac{1}{n} \leq \dfrac{1}{n - \ln n}$ for all $n \geq 1$.

The Harmonic Series diverges, so we conclude that $\displaystyle\sum_{n=1}^{\infty} \frac{1}{n - \ln n}$ diverges as well.

Notes:

The concept of direct comparison is powerful and often relatively easy to apply. Practice helps one develop the necessary intuition to quickly pick a proper series with which to compare. However, it is easy to construct a series for which it is difficult to apply the Direct Comparison Test.

Consider $\displaystyle\sum_{n=1}^{\infty} \frac{1}{n + \ln n}$. It is very similar to the divergent series given in Example 8.3.4. We suspect that it also diverges, as $\dfrac{1}{n} \approx \dfrac{1}{n + \ln n}$ for large n. However, the inequality that we naturally want to use "goes the wrong way": since $n \le n + \ln n$ for all $n \ge 1$, $\dfrac{1}{n} \ge \dfrac{1}{n + \ln n}$ for all $n \ge 1$. The given series has terms *less than* the terms of a divergent series, and we cannot conclude anything from this.

Fortunately, we can apply another test to the given series to determine its convergence.

Limit Comparison Test

Theorem 8.3.3   Limit Comparison Test

Let $\{a_n\}$ and $\{b_n\}$ be positive sequences.

1. If $\displaystyle\lim_{n \to \infty} \frac{a_n}{b_n} = L$, where L is a positive real number, then $\displaystyle\sum_{n=1}^{\infty} a_n$ and $\displaystyle\sum_{n=1}^{\infty} b_n$ either both converge or both diverge.

2. If $\displaystyle\lim_{n \to \infty} \frac{a_n}{b_n} = 0$, then if $\displaystyle\sum_{n=1}^{\infty} b_n$ converges, then so does $\displaystyle\sum_{n=1}^{\infty} a_n$.

3. If $\displaystyle\lim_{n \to \infty} \frac{a_n}{b_n} = \infty$, then if $\displaystyle\sum_{n=1}^{\infty} b_n$ diverges, then so does $\displaystyle\sum_{n=1}^{\infty} a_n$.

Theorem 8.3.3 is most useful when the convergence of the series from $\{b_n\}$ is known and we are trying to determine the convergence of the series from $\{a_n\}$.

We use the Limit Comparison Test in the next example to examine the series $\displaystyle\sum_{n=1}^{\infty} \frac{1}{n + \ln n}$ which motivated this new test.

Notes:

Example 8.3.5 **Applying the Limit Comparison Test**

Determine the convergence of $\sum_{n=1}^{\infty} \dfrac{1}{n + \ln n}$ using the Limit Comparison Test.

SOLUTION We compare the terms of $\sum_{n=1}^{\infty} \dfrac{1}{n + \ln n}$ to the terms of the Harmonic Sequence $\sum_{n=1}^{\infty} \dfrac{1}{n}$:

$$\lim_{n \to \infty} \frac{1/(n + \ln n)}{1/n} = \lim_{n \to \infty} \frac{n}{n + \ln n}$$
$$= 1 \quad \text{(after applying L'Hôpital's Rule).}$$

Since the Harmonic Series diverges, we conclude that $\sum_{n=1}^{\infty} \dfrac{1}{n + \ln n}$ diverges as well.

Example 8.3.6 **Applying the Limit Comparison Test**

Determine the convergence of $\sum_{n=1}^{\infty} \dfrac{1}{3^n - n^2}$

SOLUTION This series is similar to the one in Example 8.3.3, but now we are considering "$3^n - n^2$" instead of "$3^n + n^2$." This difference makes applying the Direct Comparison Test difficult.

Instead, we use the Limit Comparison Test and compare with the series $\sum_{n=1}^{\infty} \dfrac{1}{3^n}$:

$$\lim_{n \to \infty} \frac{1/(3^n - n^2)}{1/3^n} = \lim_{n \to \infty} \frac{3^n}{3^n - n^2}$$
$$= 1 \quad \text{(after applying L'Hôpital's Rule twice).}$$

We know $\sum_{n=1}^{\infty} \dfrac{1}{3^n}$ is a convergent geometric series, hence $\sum_{n=1}^{\infty} \dfrac{1}{3^n - n^2}$ converges as well.

As mentioned before, practice helps one develop the intuition to quickly choose a series with which to compare. A general rule of thumb is to pick a series based on the dominant term in the expression of $\{a_n\}$. It is also helpful to note that factorials dominate exponentials, which dominate algebraic functions (e.g., polynomials), which dominate logarithms. In the previous example,

Notes:

the dominant term of $\dfrac{1}{3^n - n^2}$ was 3^n, so we compared the series to $\displaystyle\sum_{n=1}^{\infty} \dfrac{1}{3^n}$. It is hard to apply the Limit Comparison Test to series containing factorials, though, as we have not learned how to apply L'Hôpital's Rule to $n!$.

Example 8.3.7   Applying the Limit Comparison Test

Determine the convergence of $\displaystyle\sum_{n=1}^{\infty} \dfrac{\sqrt{n} + 3}{n^2 - n + 1}$.

SOLUTION   We naïvely attempt to apply the rule of thumb given above and note that the dominant term in the expression of the series is $1/n^2$. Knowing that $\displaystyle\sum_{n=1}^{\infty} \dfrac{1}{n^2}$ converges, we attempt to apply the Limit Comparison Test:

$$\lim_{n\to\infty} \frac{(\sqrt{n} + 3)/(n^2 - n + 1)}{1/n^2} = \lim_{n\to\infty} \frac{n^2(\sqrt{n} + 3)}{n^2 - n + 1}$$

$$= \infty \quad \text{(Apply L'Hôpital's Rule)}.$$

Theorem 8.3.3 part (3) only applies when $\displaystyle\sum_{n=1}^{\infty} b_n$ diverges; in our case, it converges. Ultimately, our test has not revealed anything about the convergence of our series.

The problem is that we chose a poor series with which to compare. Since the numerator and denominator of the terms of the series are both algebraic functions, we should have compared our series to the dominant term of the numerator divided by the dominant term of the denominator.

The dominant term of the numerator is $n^{1/2}$ and the dominant term of the denominator is n^2. Thus we should compare the terms of the given series to $n^{1/2}/n^2 = 1/n^{3/2}$:

$$\lim_{n\to\infty} \frac{(\sqrt{n} + 3)/(n^2 - n + 1)}{1/n^{3/2}} = \lim_{n\to\infty} \frac{n^{3/2}(\sqrt{n} + 3)}{n^2 - n + 1}$$

$$= 1 \quad \text{(Apply L'Hôpital's Rule)}.$$

Since the p-series $\displaystyle\sum_{n=1}^{\infty} \dfrac{1}{n^{3/2}}$ converges, we conclude that $\displaystyle\sum_{n=1}^{\infty} \dfrac{\sqrt{n} + 3}{n^2 - n + 1}$ converges as well.

We mentioned earlier that the Integral Test did not work well with series containing factorial terms. The next section introduces the Ratio Test, which does handle such series well. We also introduce the Root Test, which is good for series where each term is raised to a power.

Notes:

Exercises 8.3

Terms and Concepts

1. In order to apply the Integral Test to a sequence $\{a_n\}$, the function $a(n) = a_n$ must be _____, _____ and _____.

2. T/F: The Integral Test can be used to determine the sum of a convergent series.

3. What test(s) in this section do not work well with factorials?

4. Suppose $\sum\limits_{n=0}^{\infty} a_n$ is convergent, and there are sequences $\{b_n\}$ and $\{c_n\}$ such that $0 \le b_n \le a_n \le c_n$ for all n. What can be said about the series $\sum\limits_{n=0}^{\infty} b_n$ and $\sum\limits_{n=0}^{\infty} c_n$?

Problems

In Exercises 5 – 12, use the Integral Test to determine the convergence of the given series.

5. $\sum\limits_{n=1}^{\infty} \dfrac{1}{2^n}$

6. $\sum\limits_{n=1}^{\infty} \dfrac{1}{n^4}$

7. $\sum\limits_{n=1}^{\infty} \dfrac{n}{n^2 + 1}$

8. $\sum\limits_{n=2}^{\infty} \dfrac{1}{n \ln n}$

9. $\sum\limits_{n=1}^{\infty} \dfrac{1}{n^2 + 1}$

10. $\sum\limits_{n=2}^{\infty} \dfrac{1}{n(\ln n)^2}$

11. $\sum\limits_{n=1}^{\infty} \dfrac{n}{2^n}$

12. $\sum\limits_{n=1}^{\infty} \dfrac{\ln n}{n^3}$

In Exercises 13 – 22, use the Direct Comparison Test to determine the convergence of the given series; state what series is used for comparison.

13. $\sum\limits_{n=1}^{\infty} \dfrac{1}{n^2 + 3n - 5}$

14. $\sum\limits_{n=1}^{\infty} \dfrac{1}{4^n + n^2 - n}$

15. $\sum\limits_{n=1}^{\infty} \dfrac{\ln n}{n}$

16. $\sum\limits_{n=1}^{\infty} \dfrac{1}{n! + n}$

17. $\sum\limits_{n=2}^{\infty} \dfrac{1}{\sqrt{n^2 - 1}}$

18. $\sum\limits_{n=5}^{\infty} \dfrac{1}{\sqrt{n - 2}}$

19. $\sum\limits_{n=1}^{\infty} \dfrac{n^2 + n + 1}{n^3 - 5}$

20. $\sum\limits_{n=1}^{\infty} \dfrac{2^n}{5^n + 10}$

21. $\sum\limits_{n=2}^{\infty} \dfrac{n}{n^2 - 1}$

22. $\sum\limits_{n=2}^{\infty} \dfrac{1}{n^2 \ln n}$

In Exercises 23 – 32, use the Limit Comparison Test to determine the convergence of the given series; state what series is used for comparison.

23. $\sum\limits_{n=1}^{\infty} \dfrac{1}{n^2 - 3n + 5}$

24. $\sum\limits_{n=1}^{\infty} \dfrac{1}{4^n - n^2}$

25. $\sum\limits_{n=4}^{\infty} \dfrac{\ln n}{n - 3}$

26. $\sum\limits_{n=1}^{\infty} \dfrac{1}{\sqrt{n^2 + n}}$

27. $\sum\limits_{n=1}^{\infty} \dfrac{1}{n + \sqrt{n}}$

28. $\sum\limits_{n=1}^{\infty} \dfrac{n - 10}{n^2 + 10n + 10}$

29. $\sum\limits_{n=1}^{\infty} \sin\left(1/n\right)$

30. $\displaystyle\sum_{n=1}^{\infty} \frac{n+5}{n^3-5}$

31. $\displaystyle\sum_{n=1}^{\infty} \frac{\sqrt{n}+3}{n^2+17}$

32. $\displaystyle\sum_{n=1}^{\infty} \frac{1}{\sqrt{n}+100}$

In Exercises 33 – 40, determine the convergence of the given series. State the test used; more than one test may be appropriate.

33. $\displaystyle\sum_{n=1}^{\infty} \frac{n^2}{2^n}$

34. $\displaystyle\sum_{n=1}^{\infty} \frac{1}{(2n+5)^3}$

35. $\displaystyle\sum_{n=1}^{\infty} \frac{n!}{10^n}$

36. $\displaystyle\sum_{n=1}^{\infty} \frac{\ln n}{n!}$

37. $\displaystyle\sum_{n=1}^{\infty} \frac{1}{3^n+n}$

38. $\displaystyle\sum_{n=1}^{\infty} \frac{n-2}{10n+5}$

39. $\displaystyle\sum_{n=1}^{\infty} \frac{3^n}{n^3}$

40. $\displaystyle\sum_{n=1}^{\infty} \frac{\cos(1/n)}{\sqrt{n}}$

41. Given that $\displaystyle\sum_{n=1}^{\infty} a_n$ converges, state which of the following series converges, may converge, or does not converge.

(a) $\displaystyle\sum_{n=1}^{\infty} \frac{a_n}{n}$

(b) $\displaystyle\sum_{n=1}^{\infty} a_n a_{n+1}$

(c) $\displaystyle\sum_{n=1}^{\infty} (a_n)^2$

(d) $\displaystyle\sum_{n=1}^{\infty} n a_n$

(e) $\displaystyle\sum_{n=1}^{\infty} \frac{1}{a_n}$

8.4 Ratio and Root Tests

The nth–Term Test of Theorem 8.2.4 states that in order for a series $\sum_{n=1}^{\infty} a_n$ to converge, $\lim_{n\to\infty} a_n = 0$. That is, the terms of $\{a_n\}$ must get very small. Not only must the terms approach 0, they must approach 0 "fast enough": while $\lim_{n\to\infty} 1/n = 0$, the Harmonic Series $\sum_{n=1}^{\infty} \frac{1}{n}$ diverges as the terms of $\{1/n\}$ do not approach 0 "fast enough."

The comparison tests of the previous section determine convergence by comparing terms of a series to terms of another series whose convergence is known. This section introduces the Ratio and Root Tests, which determine convergence by analyzing the terms of a series to see if they approach 0 "fast enough."

Ratio Test

Theorem 8.4.1   Ratio Test

Let $\{a_n\}$ be a positive sequence where $\lim_{n\to\infty} \frac{a_{n+1}}{a_n} = L$.

1. If $L < 1$, then $\sum_{n=1}^{\infty} a_n$ converges.

2. If $L > 1$ or $L = \infty$, then $\sum_{n=1}^{\infty} a_n$ diverges.

3. If $L = 1$, the Ratio Test is inconclusive.

Note: Theorem 8.2.5 allows us to apply the Ratio Test to series where $\{a_n\}$ is positive for all but a finite number of terms.

The principle of the Ratio Test is this: if $\lim_{n\to\infty} \frac{a_{n+1}}{a_n} = L < 1$, then for large n, each term of $\{a_n\}$ is significantly smaller than its previous term which is enough to ensure convergence.

Example 8.4.1   Applying the Ratio Test
Use the Ratio Test to determine the convergence of the following series:

1. $\sum_{n=1}^{\infty} \frac{2^n}{n!}$ 2. $\sum_{n=1}^{\infty} \frac{3^n}{n^3}$ 3. $\sum_{n=1}^{\infty} \frac{1}{n^2 + 1}$.

Notes:

SOLUTION

1. $\displaystyle\sum_{n=1}^{\infty} \frac{2^n}{n!}$:

$$\lim_{n\to\infty} \frac{2^{n+1}/(n+1)!}{2^n/n!} = \lim_{n\to\infty} \frac{2^{n+1}n!}{2^n(n+1)!}$$

$$= \lim_{n\to\infty} \frac{2}{n+1}$$

$$= 0.$$

Since the limit is $0 < 1$, by the Ratio Test $\displaystyle\sum_{n=1}^{\infty} \frac{2^n}{n!}$ converges.

2. $\displaystyle\sum_{n=1}^{\infty} \frac{3^n}{n^3}$:

$$\lim_{n\to\infty} \frac{3^{n+1}/(n+1)^3}{3^n/n^3} = \lim_{n\to\infty} \frac{3^{n+1}n^3}{3^n(n+1)^3}$$

$$= \lim_{n\to\infty} \frac{3n^3}{(n+1)^3}$$

$$= 3.$$

Since the limit is $3 > 1$, by the Ratio Test $\displaystyle\sum_{n=1}^{\infty} \frac{3^n}{n^3}$ diverges.

3. $\displaystyle\sum_{n=1}^{\infty} \frac{1}{n^2+1}$:

$$\lim_{n\to\infty} \frac{1/\big((n+1)^2+1\big)}{1/(n^2+1)} = \lim_{n\to\infty} \frac{n^2+1}{(n+1)^2+1}$$

$$= 1.$$

Since the limit is 1, the Ratio Test is inconclusive. We can easily show this series converges using the Direct or Limit Comparison Tests, with each comparing to the series $\displaystyle\sum_{n=1}^{\infty} \frac{1}{n^2}$.

Notes:

The Ratio Test is not effective when the terms of a series *only* contain algebraic functions (e.g., polynomials). It is most effective when the terms contain some factorials or exponentials. The previous example also reinforces our developing intuition: factorials dominate exponentials, which dominate algebraic functions, which dominate logarithmic functions. In Part 1 of the example, the factorial in the denominator dominated the exponential in the numerator, causing the series to converge. In Part 2, the exponential in the numerator dominated the algebraic function in the denominator, causing the series to diverge.

While we have used factorials in previous sections, we have not explored them closely and one is likely to not yet have a strong intuitive sense for how they behave. The following example gives more practice with factorials.

Example 8.4.2   **Applying the Ratio Test**

Determine the convergence of $\displaystyle\sum_{n=1}^{\infty} \frac{n!n!}{(2n)!}$.

SOLUTION   Before we begin, be sure to note the difference between $(2n)!$ and $2n!$. When $n = 4$, the former is $8! = 8 \cdot 7 \cdot \ldots \cdot 2 \cdot 1 = 40,320$, whereas the latter is $2(4 \cdot 3 \cdot 2 \cdot 1) = 48$.

Applying the Ratio Test:

$$\lim_{n\to\infty} \frac{(n+1)!(n+1)!/\big(2(n+1)\big)!}{n!n!/(2n)!} = \lim_{n\to\infty} \frac{(n+1)!(n+1)!(2n)!}{n!n!(2n+2)!}$$

Noting that $(2n+2)! = (2n+2) \cdot (2n+1) \cdot (2n)!$, we have

$$= \lim_{n\to\infty} \frac{(n+1)(n+1)}{(2n+2)(2n+1)}$$
$$= 1/4.$$

Since the limit is $1/4 < 1$, by the Ratio Test we conclude $\displaystyle\sum_{n=1}^{\infty} \frac{n!n!}{(2n)!}$ converges.

Root Test

The final test we introduce is the Root Test, which works particularly well on series where each term is raised to a power, and does not work well with terms containing factorials.

Notes:

Note: Theorem 8.2.5 allows us to apply the Root Test to series where $\{a_n\}$ is positive for all but a finite number of terms.

Theorem 8.4.2   Root Test

Let $\{a_n\}$ be a positive sequence, and let $\lim\limits_{n\to\infty} (a_n)^{1/n} = L$.

1. If $L < 1$, then $\sum\limits_{n=1}^{\infty} a_n$ converges.

2. If $L > 1$ or $L = \infty$, then $\sum\limits_{n=1}^{\infty} a_n$ diverges.

3. If $L = 1$, the Root Test is inconclusive.

Example 8.4.3   Applying the Root Test

Determine the convergence of the following series using the Root Test:

1. $\sum\limits_{n=1}^{\infty} \left(\dfrac{3n+1}{5n-2}\right)^n$

2. $\sum\limits_{n=1}^{\infty} \dfrac{n^4}{(\ln n)^n}$

3. $\sum\limits_{n=1}^{\infty} \dfrac{2^n}{n^2}.$

SOLUTION

1. $\lim\limits_{n\to\infty} \left(\left(\dfrac{3n+1}{5n-2}\right)^n\right)^{1/n} = \lim\limits_{n\to\infty} \dfrac{3n+1}{5n-2} = \dfrac{3}{5}.$

   Since the limit is less than 1, we conclude the series converges. Note: it is difficult to apply the Ratio Test to this series.

2. $\lim\limits_{n\to\infty} \left(\dfrac{n^4}{(\ln n)^n}\right)^{1/n} = \lim\limits_{n\to\infty} \dfrac{\left(n^{1/n}\right)^4}{\ln n}.$

   As n grows, the numerator approaches 1 (apply L'Hôpital's Rule) and the denominator grows to infinity. Thus

   $$\lim\limits_{n\to\infty} \dfrac{\left(n^{1/n}\right)^4}{\ln n} = 0.$$

   Since the limit is less than 1, we conclude the series converges.

3. $\lim\limits_{n\to\infty} \left(\dfrac{2^n}{n^2}\right)^{1/n} = \lim\limits_{n\to\infty} \dfrac{2}{\left(n^{1/n}\right)^2} = 2.$

   Since this is greater than 1, we conclude the series diverges.

Each of the tests we have encountered so far has required that we analyze series from *positive* sequences. The next section relaxes this restriction by considering *alternating series*, where the underlying sequence has terms that alternate between being positive and negative.

Notes:

Exercises 8.4

Terms and Concepts

1. The Ratio Test is not effective when the terms of a sequence only contain _____ functions.

2. The Ratio Test is most effective when the terms of a sequence contains _____ and/or _____ functions.

3. What three convergence tests do not work well with terms containing factorials?

4. The Root Test works particularly well on series where each term is _____ to a _____.

Problems

In Exercises 5 – 14, determine the convergence of the given series using the Ratio Test. If the Ratio Test is inconclusive, state so and determine convergence with another test.

5. $\displaystyle\sum_{n=0}^{\infty} \frac{2n}{n!}$

6. $\displaystyle\sum_{n=0}^{\infty} \frac{5^n - 3n}{4^n}$

7. $\displaystyle\sum_{n=0}^{\infty} \frac{n! 10^n}{(2n)!}$

8. $\displaystyle\sum_{n=1}^{\infty} \frac{5^n + n^4}{7^n + n^2}$

9. $\displaystyle\sum_{n=1}^{\infty} \frac{1}{n}$

10. $\displaystyle\sum_{n=1}^{\infty} \frac{1}{3n^3 + 7}$

11. $\displaystyle\sum_{n=1}^{\infty} \frac{10 \cdot 5^n}{7^n - 3}$

12. $\displaystyle\sum_{n=1}^{\infty} n \cdot \left(\frac{3}{5}\right)^n$

13. $\displaystyle\sum_{n=1}^{\infty} \frac{2 \cdot 4 \cdot 6 \cdot 8 \cdots 2n}{3 \cdot 6 \cdot 9 \cdot 12 \cdots 3n}$

14. $\displaystyle\sum_{n=1}^{\infty} \frac{n!}{5 \cdot 10 \cdot 15 \cdots (5n)}$

In Exercises 15 – 24, determine the convergence of the given series using the Root Test. If the Root Test is inconclusive, state so and determine convergence with another test.

15. $\displaystyle\sum_{n=1}^{\infty} \left(\frac{2n + 5}{3n + 11}\right)^n$

16. $\displaystyle\sum_{n=1}^{\infty} \left(\frac{.9n^2 - n - 3}{n^2 + n + 3}\right)^n$

17. $\displaystyle\sum_{n=1}^{\infty} \frac{2^n n^2}{3^n}$

18. $\displaystyle\sum_{n=1}^{\infty} \frac{1}{n^n}$

19. $\displaystyle\sum_{n=1}^{\infty} \frac{3^n}{n^2 2^{n+1}}$

20. $\displaystyle\sum_{n=1}^{\infty} \frac{4^{n+7}}{7^n}$

21. $\displaystyle\sum_{n=1}^{\infty} \left(\frac{n^2 - n}{n^2 + n}\right)^n$

22. $\displaystyle\sum_{n=1}^{\infty} \left(\frac{1}{n} - \frac{1}{n^2}\right)^n$

23. $\displaystyle\sum_{n=1}^{\infty} \frac{1}{(\ln n)^n}$

24. $\displaystyle\sum_{n=1}^{\infty} \frac{n^2}{(\ln n)^n}$

In Exercises 25 – 34, determine the convergence of the given series. State the test used; more than one test may be appropriate.

25. $\displaystyle\sum_{n=1}^{\infty} \frac{n^2 + 4n - 2}{n^3 + 4n^2 - 3n + 7}$

26. $\displaystyle\sum_{n=1}^{\infty} \frac{n^4 4^n}{n!}$

27. $\displaystyle\sum_{n=1}^{\infty} \frac{n^2}{3^n + n}$

28. $\displaystyle\sum_{n=1}^{\infty} \frac{3^n}{n^n}$

29. $\displaystyle\sum_{n=1}^{\infty} \frac{n}{\sqrt{n^2 + 4n + 1}}$

30. $\displaystyle\sum_{n=1}^{\infty} \frac{n!n!n!}{(3n)!}$

31. $\displaystyle\sum_{n=1}^{\infty} \frac{1}{\ln n}$

32. $\displaystyle\sum_{n=1}^{\infty} \left(\frac{n+2}{n+1}\right)^n$

33. $\displaystyle\sum_{n=1}^{\infty} \frac{n^3}{\left(\ln n\right)^n}$

34. $\displaystyle\sum_{n=1}^{\infty} \left(\frac{1}{n} - \frac{1}{n+2}\right)$

8.5   Alternating Series and Absolute Convergence

All of the series convergence tests we have used require that the underlying sequence $\{a_n\}$ be a positive sequence. (We can relax this with Theorem 8.2.5 and state that there must be an $N > 0$ such that $a_n > 0$ for all $n > N$; that is, $\{a_n\}$ is positive for all but a finite number of values of n.)

In this section we explore series whose summation includes negative terms. We start with a very specific form of series, where the terms of the summation alternate between being positive and negative.

Definition 8.5.1   Alternating Series

Let $\{a_n\}$ be a positive sequence. An **alternating series** is a series of either the form

$$\sum_{n=1}^{\infty}(-1)^n a_n \qquad \text{or} \qquad \sum_{n=1}^{\infty}(-1)^{n+1} a_n.$$

Recall the terms of Harmonic Series come from the Harmonic Sequence $\{a_n\} = \{1/n\}$. An important alternating series is the **Alternating Harmonic Series**:

$$\sum_{n=1}^{\infty}(-1)^{n+1}\frac{1}{n} = 1 - \frac{1}{2} + \frac{1}{3} - \frac{1}{4} + \frac{1}{5} - \frac{1}{6} + \cdots$$

Geometric Series can also be alternating series when $r < 0$. For instance, if $r = -1/2$, the geometric series is

$$\sum_{n=0}^{\infty}\left(\frac{-1}{2}\right)^n = 1 - \frac{1}{2} + \frac{1}{4} - \frac{1}{8} + \frac{1}{16} - \frac{1}{32} + \cdots$$

Theorem 8.2.1 states that geometric series converge when $|r| < 1$ and gives the sum: $\sum_{n=0}^{\infty} r^n = \dfrac{1}{1-r}$. When $r = -1/2$ as above, we find

$$\sum_{n=0}^{\infty}\left(\frac{-1}{2}\right)^n = \frac{1}{1-(-1/2)} = \frac{1}{3/2} = \frac{2}{3}.$$

A powerful convergence theorem exists for other alternating series that meet a few conditions.

Notes:

Theorem 8.5.1   Alternating Series Test

Let $\{a_n\}$ be a positive, decreasing sequence where $\lim\limits_{n\to\infty} a_n = 0$. Then

$$\sum_{n=1}^{\infty}(-1)^n a_n \qquad \text{and} \qquad \sum_{n=1}^{\infty}(-1)^{n+1} a_n$$

converge.

The basic idea behind Theorem 8.5.1 is illustrated in Figure 8.5.1. A positive, decreasing sequence $\{a_n\}$ is shown along with the partial sums

$$S_n = \sum_{i=1}^{n}(-1)^{i+1} a_i = a_1 - a_2 + a_3 - a_4 + \cdots + (-1)^{n+1} a_n.$$

Because $\{a_n\}$ is decreasing, the amount by which S_n bounces up/down decreases. Moreover, the odd terms of S_n form a decreasing, bounded sequence, while the even terms of S_n form an increasing, bounded sequence. Since bounded, monotonic sequences converge (see Theorem 8.1.5) and the terms of $\{a_n\}$ approach 0, one can show the odd and even terms of S_n converge to the same common limit L, the sum of the series.

Example 8.5.1   Applying the Alternating Series Test

Determine if the Alternating Series Test applies to each of the following series.

1. $\displaystyle\sum_{n=1}^{\infty}(-1)^{n+1}\frac{1}{n}$ 2. $\displaystyle\sum_{n=1}^{\infty}(-1)^{n}\frac{\ln n}{n}$ 3. $\displaystyle\sum_{n=1}^{\infty}(-1)^{n+1}\frac{|\sin n|}{n^2}$

SOLUTION

1. This is the Alternating Harmonic Series as seen previously. The underlying sequence is $\{a_n\} = \{1/n\}$, which is positive, decreasing, and approaches 0 as $n \to \infty$. Therefore we can apply the Alternating Series Test and conclude this series converges.

   While the test does not state what the series converges to, we will see later that $\displaystyle\sum_{n=1}^{\infty}(-1)^{n+1}\frac{1}{n} = \ln 2$.

2. The underlying sequence is $\{a_n\} = \{\ln n/n\}$. This is positive and approaches 0 as $n \to \infty$ (use L'Hôpital's Rule). However, the sequence is not decreasing for all n. It is straightforward to compute $a_1 = 0$, $a_2 \approx 0.347$,

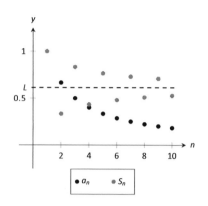

y

Figure 8.5.1:   Illustrating convergence with the Alternating Series Test.

Notes:

$a_3 \approx 0.366$, and $a_4 \approx 0.347$: the sequence is increasing for at least the first 3 terms.

We do not immediately conclude that we cannot apply the Alternating Series Test. Rather, consider the long–term behavior of $\{a_n\}$. Treating $a_n = a(n)$ as a continuous function of n defined on $[1, \infty)$, we can take its derivative:

$$a'(n) = \frac{1 - \ln n}{n^2}.$$

The derivative is negative for all $n \geq 3$ (actually, for all $n > e$), meaning $a(n) = a_n$ is decreasing on $[3, \infty)$. We can apply the Alternating Series Test to the series when we start with $n = 3$ and conclude that

$$\sum_{n=3}^{\infty} (-1)^n \frac{\ln n}{n}$$ converges; adding the terms with $n = 1$ and $n = 2$ do not

change the convergence (i.e., we apply Theorem 8.2.5).

The important lesson here is that as before, if a series fails to meet the criteria of the Alternating Series Test on only a finite number of terms, we can still apply the test.

3. The underlying sequence is $\{a_n\} = |\sin n|/n^2$. This sequence is positive and approaches 0 as $n \to \infty$. However, it is not a decreasing sequence; the value of $|\sin n|$ oscillates between 0 and 1 as $n \to \infty$. We cannot remove a finite number of terms to make $\{a_n\}$ decreasing, therefore we cannot apply the Alternating Series Test.

Keep in mind that this does not mean we conclude the series diverges; in fact, it does converge. We are just unable to conclude this based on Theorem 8.5.1.

Key Idea 8.2.1 gives the sum of some important series. Two of these are

$$\sum_{n=1}^{\infty} \frac{1}{n^2} = \frac{\pi^2}{6} \approx 1.64493 \quad \text{and} \quad \sum_{n=1}^{\infty} \frac{(-1)^{n+1}}{n^2} = \frac{\pi^2}{12} \approx 0.82247.$$

These two series converge to their sums at different rates. To be accurate to two places after the decimal, we need 202 terms of the first series though only 13 of the second. To get 3 places of accuracy, we need 1069 terms of the first series though only 33 of the second. Why is it that the second series converges so much faster than the first?

While there are many factors involved when studying rates of convergence, the alternating structure of an alternating series gives us a powerful tool when approximating the sum of a convergent series.

Notes:

Theorem 8.5.2   The Alternating Series Approximation Theorem

Let $\{a_n\}$ be a sequence that satisfies the hypotheses of the Alternating Series Test, and let S_n and L be the nth partial sums and sum, respectively, of either $\sum_{n=1}^{\infty}(-1)^n a_n$ or $\sum_{n=1}^{\infty}(-1)^{n+1} a_n$. Then

1. $|S_n - L| < a_{n+1}$, and

2. L is between S_n and S_{n+1}.

Part 1 of Theorem 8.5.2 states that the nth partial sum of a convergent alternating series will be within a_{n+1} of its total sum. Consider the alternating series we looked at before the statement of the theorem, $\sum_{n=1}^{\infty}\dfrac{(-1)^{n+1}}{n^2}$. Since $a_{14} = 1/14^2 \approx 0.0051$, we know that S_{13} is within 0.0051 of the total sum.

Moreover, Part 2 of the theorem states that since $S_{13} \approx 0.8252$ and $S_{14} \approx 0.8201$, we know the sum L lies between 0.8201 and 0.8252. One use of this is the knowledge that S_{14} is accurate to two places after the decimal.

Some alternating series converge slowly. In Example 8.5.1 we determined the series $\sum_{n=1}^{\infty}(-1)^{n+1}\dfrac{\ln n}{n}$ converged. With $n = 1001$, we find $\ln n/n \approx 0.0069$, meaning that $S_{1000} \approx 0.1633$ is accurate to one, maybe two, places after the decimal. Since $S_{1001} \approx 0.1564$, we know the sum L is $0.1564 \le L \le 0.1633$.

Example 8.5.2   Approximating the sum of convergent alternating series
Approximate the sum of the following series, accurate to within 0.001.

1. $\displaystyle\sum_{n=1}^{\infty}(-1)^{n+1}\dfrac{1}{n^3}$ 2. $\displaystyle\sum_{n=1}^{\infty}(-1)^{n+1}\dfrac{\ln n}{n}$.

SOLUTION

1. Using Theorem 8.5.2, we want to find n where $1/n^3 \le 0.001$:

$$\frac{1}{n^3} \le 0.001 = \frac{1}{1000}$$
$$n^3 \ge 1000$$
$$n \ge \sqrt[3]{1000}$$
$$n \ge 10.$$

Notes:

Let L be the sum of this series. By Part 1 of the theorem, $|S_9 - L| < a_{10} = 1/1000$. We can compute $S_9 = 0.902116$, which our theorem states is within 0.001 of the total sum.

We can use Part 2 of the theorem to obtain an even more accurate result. As we know the 10th term of the series is $-1/1000$, we can easily compute $S_{10} = 0.901116$. Part 2 of the theorem states that L is between S_9 and S_{10}, so $0.901116 < L < 0.902116$.

2. We want to find n where $\ln(n)/n < 0.001$. We start by solving $\ln(n)/n = 0.001$ for n. This cannot be solved algebraically, so we will use Newton's Method to approximate a solution.

    Let $f(x) = \ln(x)/x - 0.001$; we want to know where $f(x) = 0$. We make a guess that x must be "large," so our initial guess will be $x_1 = 1000$. Recall how Newton's Method works: given an approximate solution x_n, our next approximation x_{n+1} is given by

    $$x_{n+1} = x_n - \frac{f(x_n)}{f'(x_n)}.$$

    We find $f'(x) = (1 - \ln(x))/x^2$. This gives

    $$x_2 = 1000 - \frac{\ln(1000)/1000 - 0.001}{(1 - \ln(1000))/1000^2}$$
    $$= 2000.$$

    Using a computer, we find that Newton's Method seems to converge to a solution $x = 9118.01$ after 8 iterations. Taking the next integer higher, we have $n = 9119$, where $\ln(9119)/9119 = 0.000999903 < 0.001$.

    Again using a computer, we find $S_{9118} = -0.160369$. Part 1 of the theorem states that this is within 0.001 of the actual sum L. Already knowing the 9,119th term, we can compute $S_{9119} = -0.159369$, meaning $-0.159369 < L < -0.160369$.

Notice how the first series converged quite quickly, where we needed only 10 terms to reach the desired accuracy, whereas the second series took over 9,000 terms.

One of the famous results of mathematics is that the Harmonic Series, $\displaystyle\sum_{n=1}^{\infty} \frac{1}{n}$ diverges, yet the Alternating Harmonic Series, $\displaystyle\sum_{n=1}^{\infty} (-1)^{n+1} \frac{1}{n}$, converges. The

---

Notes:

notion that alternating the signs of the terms in a series can make a series converge leads us to the following definitions.

Note: In Definition 8.5.2, $\sum\limits_{n=1}^{\infty} a_n$ is not necessarily an alternating series; it just may have some negative terms.

Definition 8.5.2   Absolute and Conditional Convergence

1. A series $\sum\limits_{n=1}^{\infty} a_n$ **converges absolutely** if $\sum\limits_{n=1}^{\infty} |a_n|$ converges.

2. A series $\sum\limits_{n=1}^{\infty} a_n$ **converges conditionally** if $\sum\limits_{n=1}^{\infty} a_n$ converges but $\sum\limits_{n=1}^{\infty} |a_n|$ diverges.

Thus we say the Alternating Harmonic Series converges conditionally.

Example 8.5.3   Determining absolute and conditional convergence.
Determine if the following series converge absolutely, conditionally, or diverge.

1. $\sum\limits_{n=1}^{\infty}(-1)^n\dfrac{n+3}{n^2+2n+5}$ 2. $\sum\limits_{n=1}^{\infty}(-1)^n\dfrac{n^2+2n+5}{2^n}$ 3. $\sum\limits_{n=3}^{\infty}(-1)^n\dfrac{3n-3}{5n-10}$

SOLUTION

1. We can show the series

$$\sum_{n=1}^{\infty}\left|(-1)^n\frac{n+3}{n^2+2n+5}\right| = \sum_{n=1}^{\infty}\frac{n+3}{n^2+2n+5}$$

diverges using the Limit Comparison Test, comparing with $1/n$.

The series $\sum\limits_{n=1}^{\infty}(-1)^n\dfrac{n+3}{n^2+2n+5}$ converges using the Alternating Series Test; we conclude it converges conditionally.

2. We can show the series

$$\sum_{n=1}^{\infty}\left|(-1)^n\frac{n^2+2n+5}{2^n}\right| = \sum_{n=1}^{\infty}\frac{n^2+2n+5}{2^n}$$

converges using the Ratio Test.

Notes:

Therefore we conclude $\displaystyle\sum_{n=1}^{\infty}(-1)^n \frac{n^2 + 2n + 5}{2^n}$ converges absolutely.

3. The series

$$\sum_{n=3}^{\infty}\left|(-1)^n \frac{3n - 3}{5n - 10}\right| = \sum_{n=3}^{\infty}\frac{3n - 3}{5n - 10}$$

diverges using the nth Term Test, so it does not converge absolutely.

The series $\displaystyle\sum_{n=3}^{\infty}(-1)^n \frac{3n - 3}{5n - 10}$ fails the conditions of the Alternating Series Test as $(3n - 3)/(5n - 10)$ does not approach 0 as $n \to \infty$. We can state further that this series diverges; as $n \to \infty$, the series effectively adds and subtracts $3/5$ over and over. This causes the sequence of partial sums to oscillate and not converge.

Therefore the series $\displaystyle\sum_{n=1}^{\infty}(-1)^n \frac{3n - 3}{5n - 10}$ diverges.

Knowing that a series converges absolutely allows us to make two important statements, given in the following theorem. The first is that absolute convergence is "stronger" than regular convergence. That is, just because $\displaystyle\sum_{n=1}^{\infty} a_n$ converges, we cannot conclude that $\displaystyle\sum_{n=1}^{\infty} |a_n|$ will converge, but knowing a series converges absolutely tells us that $\displaystyle\sum_{n=1}^{\infty} a_n$ will converge.

One reason this is important is that our convergence tests all require that the underlying sequence of terms be positive. By taking the absolute value of the terms of a series where not all terms are positive, we are often able to apply an appropriate test and determine absolute convergence. This, in turn, determines that the series we are given also converges.

The second statement relates to **rearrangements** of series. When dealing with a finite set of numbers, the sum of the numbers does not depend on the order which they are added. (So $1 + 2 + 3 = 3 + 1 + 2$.) One may be surprised to find out that when dealing with an infinite set of numbers, the same statement does not always hold true: some infinite lists of numbers may be rearranged in different orders to achieve different sums. The theorem states that the terms of an absolutely convergent series can be rearranged in any way without affecting the sum.

Notes:

> **Theorem 8.5.3 Absolute Convergence Theorem**
>
> Let $\displaystyle\sum_{n=1}^{\infty} a_n$ be a series that converges absolutely.
>
> 1. $\displaystyle\sum_{n=1}^{\infty} a_n$ converges.
>
> 2. Let $\{b_n\}$ be any rearrangement of the sequence $\{a_n\}$. Then
> $$\sum_{n=1}^{\infty} b_n = \sum_{n=1}^{\infty} a_n.$$

In Example 8.5.3, we determined the series in part 2 converges absolutely. Theorem 8.5.3 tells us the series converges (which we could also determine using the Alternating Series Test).

The theorem states that rearranging the terms of an absolutely convergent series does not affect its sum. This implies that perhaps the sum of a conditionally convergent series can change based on the arrangement of terms. Indeed, it can. The Riemann Rearrangement Theorem (named after Bernhard Riemann) states that any conditionally convergent series can have its terms rearranged so that the sum is any desired value, including ∞!

As an example, consider the Alternating Harmonic Series once more. We have stated that

$$\sum_{n=1}^{\infty}(-1)^{n+1}\frac{1}{n} = 1 - \frac{1}{2} + \frac{1}{3} - \frac{1}{4} + \frac{1}{5} - \frac{1}{6} + \frac{1}{7}\cdots = \ln 2,$$

(see Key Idea 8.2.1 or Example 8.5.1).

Consider the rearrangement where every positive term is followed by two negative terms:

$$1 - \frac{1}{2} - \frac{1}{4} + \frac{1}{3} - \frac{1}{6} - \frac{1}{8} + \frac{1}{5} - \frac{1}{10} - \frac{1}{12}\cdots$$

(Convince yourself that these are exactly the same numbers as appear in the Alternating Harmonic Series, just in a different order.) Now group some terms

---

Notes:

and simplify:

$$\left(1 - \frac{1}{2}\right) - \frac{1}{4} + \left(\frac{1}{3} - \frac{1}{6}\right) - \frac{1}{8} + \left(\frac{1}{5} - \frac{1}{10}\right) - \frac{1}{12} + \cdots =$$

$$\frac{1}{2} - \frac{1}{4} + \frac{1}{6} - \frac{1}{8} + \frac{1}{10} - \frac{1}{12} + \cdots =$$

$$\frac{1}{2}\left(1 - \frac{1}{2} + \frac{1}{3} - \frac{1}{4} + \frac{1}{5} - \frac{1}{6} + \cdots\right) = \frac{1}{2}\ln 2.$$

By rearranging the terms of the series, we have arrived at a different sum! (One could *try* to argue that the Alternating Harmonic Series does not actually converge to ln 2, because rearranging the terms of the series *shouldn't* change the sum. However, the Alternating Series Test proves this series converges to L, for some number L, and if the rearrangement does not change the sum, then $L = L/2$, implying $L = 0$. But the Alternating Series Approximation Theorem quickly shows that $L > 0$. The only conclusion is that the rearrangement *did* change the sum.) This is an incredible result.

We end here our study of tests to determine convergence. The end of this text contains a table summarizing the tests that one may find useful.

While series are worthy of study in and of themselves, our ultimate goal within calculus is the study of Power Series, which we will consider in the next section. We will use power series to create functions where the output is the result of an infinite summation.

Notes:

Exercises 8.5

Terms and Concepts

1. Why is $\displaystyle\sum_{n=1}^{\infty} \sin n$ not an alternating series?

2. A series $\displaystyle\sum_{n=1}^{\infty} (-1)^n a_n$ converges when $\{a_n\}$ is _____, _____ and $\displaystyle\lim_{n\to\infty} a_n =$_____.

3. Give an example of a series where $\displaystyle\sum_{n=0}^{\infty} a_n$ converges but $\displaystyle\sum_{n=0}^{\infty} |a_n|$ does not.

4. The sum of a _____ convergent series can be changed by rearranging the order of its terms.

Problems

In Exercises 5 – 20, an alternating series $\displaystyle\sum_{n=i}^{\infty} a_n$ is given.

(a) Determine if the series converges or diverges.

(b) Determine if $\displaystyle\sum_{n=0}^{\infty} |a_n|$ converges or diverges.

(c) If $\displaystyle\sum_{n=0}^{\infty} a_n$ converges, determine if the convergence is conditional or absolute.

5. $\displaystyle\sum_{n=1}^{\infty} \frac{(-1)^{n+1}}{n^2}$

6. $\displaystyle\sum_{n=1}^{\infty} \frac{(-1)^{n+1}}{\sqrt{n!}}$

7. $\displaystyle\sum_{n=0}^{\infty} (-1)^n \frac{n+5}{3n-5}$

8. $\displaystyle\sum_{n=1}^{\infty} (-1)^n \frac{2^n}{n^2}$

9. $\displaystyle\sum_{n=0}^{\infty} (-1)^{n+1} \frac{3n+5}{n^2-3n+1}$

10. $\displaystyle\sum_{n=1}^{\infty} \frac{(-1)^n}{\ln n + 1}$

11. $\displaystyle\sum_{n=2}^{\infty} (-1)^n \frac{n}{\ln n}$

12. $\displaystyle\sum_{n=1}^{\infty} \frac{(-1)^{n+1}}{1+3+5+\cdots+(2n-1)}$

13. $\displaystyle\sum_{n=1}^{\infty} \cos(\pi n)$

14. $\displaystyle\sum_{n=2}^{\infty} \frac{\sin\left((n+1/2)\pi\right)}{n \ln n}$

15. $\displaystyle\sum_{n=0}^{\infty} \left(-\frac{2}{3}\right)^n$

16. $\displaystyle\sum_{n=0}^{\infty} (-e)^{-n}$

17. $\displaystyle\sum_{n=0}^{\infty} \frac{(-1)^n n^2}{n!}$

18. $\displaystyle\sum_{n=0}^{\infty} (-1)^n 2^{-n^2}$

19. $\displaystyle\sum_{n=1}^{\infty} \frac{(-1)^n}{\sqrt{n}}$

20. $\displaystyle\sum_{n=1}^{\infty} \frac{(-1000)^n}{n!}$

Let S_n be the nth partial sum of a series. In Exercises 21 – 24, a convergent alternating series is given and a value of n. Compute S_n and S_{n+1} and use these values to find bounds on the sum of the series.

21. $\displaystyle\sum_{n=1}^{\infty} \frac{(-1)^n}{\ln(n+1)}, \quad n = 5$

22. $\displaystyle\sum_{n=1}^{\infty} \frac{(-1)^{n+1}}{n^4}, \quad n = 4$

23. $\displaystyle\sum_{n=0}^{\infty} \frac{(-1)^n}{n!}, \quad n = 6$

24. $\displaystyle\sum_{n=0}^{\infty} \left(-\frac{1}{2}\right)^n, \quad n = 9$

In Exercises 25 – 28, a convergent alternating series is given along with its sum and a value of ε. Use Theorem 8.5.2 to find n such that the nth partial sum of the series is within ε of the sum of the series.

25. $\displaystyle\sum_{n=1}^{\infty} \frac{(-1)^{n+1}}{n^4} = \frac{7\pi^4}{720}, \quad \varepsilon = 0.001$

26. $\displaystyle\sum_{n=0}^{\infty} \frac{(-1)^n}{n!} = \frac{1}{e}, \quad \varepsilon = 0.0001$

27. $\displaystyle\sum_{n=0}^{\infty} \frac{(-1)^n}{2n+1} = \frac{\pi}{4}, \quad \varepsilon = 0.001$

28. $\displaystyle\sum_{n=0}^{\infty} \frac{(-1)^n}{(2n)!} = \cos 1, \quad \varepsilon = 10^{-8}$

8.6   Power Series

So far, our study of series has examined the question of "Is the sum of these infinite terms finite?," i.e., "Does the series converge?" We now approach series from a different perspective: as a function. Given a value of x, we evaluate $f(x)$ by finding the sum of a particular series that depends on x (assuming the series converges). We start this new approach to series with a definition.

Definition 8.6.1   Power Series

Let $\{a_n\}$ be a sequence, let x be a variable, and let c be a real number.

1. The **power series in x** is the series

$$\sum_{n=0}^{\infty} a_n x^n = a_0 + a_1 x + a_2 x^2 + a_3 x^3 + \dots$$

2. The **power series in x centered at c** is the series

$$\sum_{n=0}^{\infty} a_n (x - c)^n = a_0 + a_1(x - c) + a_2(x - c)^2 + a_3(x - c)^3 + \dots$$

Example 8.6.1   Examples of power series

Write out the first five terms of the following power series:

1. $\displaystyle\sum_{n=0}^{\infty} x^n$ 2. $\displaystyle\sum_{n=1}^{\infty} (-1)^{n+1} \frac{(x+1)^n}{n}$ 3. $\displaystyle\sum_{n=0}^{\infty} (-1)^{n+1} \frac{(x - \pi)^{2n}}{(2n)!}$.

SOLUTION

1. One of the conventions we adopt is that $x^0 = 1$ regardless of the value of x. Therefore

$$\sum_{n=0}^{\infty} x^n = 1 + x + x^2 + x^3 + x^4 + \dots$$

This is a geometric series in x.

2. This series is centered at $c = -1$. Note how this series starts with $n = 1$. We could rewrite this series starting at $n = 0$ with the understanding that $a_0 = 0$, and hence the first term is 0.

$$\sum_{n=1}^{\infty} (-1)^{n+1} \frac{(x+1)^n}{n} = (x+1) - \frac{(x+1)^2}{2} + \frac{(x+1)^3}{3} - \frac{(x+1)^4}{4} + \frac{(x+1)^5}{5} \dots$$

Notes:

3. This series is centered at $c = \pi$. Recall that $0! = 1$.

$$\sum_{n=0}^{\infty}(-1)^{n+1}\frac{(x-\pi)^{2n}}{(2n)!} = -1 + \frac{(x-\pi)^2}{2} - \frac{(x-\pi)^4}{24} + \frac{(x-\pi)^6}{6!} - \frac{(x-\pi)^8}{8!} \cdots$$

We introduced power series as a type of function, where a value of x is given and the sum of a series is returned. Of course, not every series converges. For instance, in part 1 of Example 8.6.1, we recognized the series $\sum_{n=0}^{\infty} x^n$ as a geometric series in x. Theorem 8.2.1 states that this series converges only when $|x| < 1$.

This raises the question: "For what values of x will a given power series converge?," which leads us to a theorem and definition.

Theorem 8.6.1   **Convergence of Power Series**

Let a power series $\sum_{n=0}^{\infty} a_n(x-c)^n$ be given. Then one of the following is true:

1. The series converges only at $x = c$.

2. There is an $R > 0$ such that the series converges for all x in $(c - R, c + R)$ and diverges for all $x < c - R$ and $x > c + R$.

3. The series converges for all x.

The value of R is important when understanding a power series, hence it is given a name in the following definition. Also, note that part 2 of Theorem 8.6.1 makes a statement about the interval $(c - R, c + R)$, but the not the endpoints of that interval. A series may/may not converge at these endpoints.

Notes:

Definition 8.6.2   Radius and Interval of Convergence

1. The number R given in Theorem 8.6.1 is the **radius of convergence** of a given series. When a series converges for only $x = c$, we say the radius of convergence is 0, i.e., $R = 0$. When a series converges for all x, we say the series has an infinite radius of convergence, i.e., $R = \infty$.

2. The **interval of convergence** is the set of all values of x for which the series converges.

To find the values of x for which a given series converges, we will use the convergence tests we studied previously (especially the Ratio Test). However, the tests all required that the terms of a series be positive. The following theorem gives us a work–around to this problem.

Theorem 8.6.2   The Radius of Convergence of a Series and Absolute Convergence

The series $\displaystyle\sum_{n=0}^{\infty} a_n(x - c)^n$ and $\displaystyle\sum_{n=0}^{\infty} \left| a_n(x - c)^n \right|$ have the same radius of convergence R.

Theorem 8.6.2 allows us to find the radius of convergence R of a series by applying the Ratio Test (or any applicable test) to the absolute value of the terms of the series. We practice this in the following example.

Example 8.6.2   Determining the radius and interval of convergence.
Find the radius and interval of convergence for each of the following series:

1. $\displaystyle\sum_{n=0}^{\infty} \frac{x^n}{n!}$ 2. $\displaystyle\sum_{n=1}^{\infty} (-1)^{n+1}\frac{x^n}{n}$ 3. $\displaystyle\sum_{n=0}^{\infty} 2^n(x - 3)^n$ 4. $\displaystyle\sum_{n=0}^{\infty} n!x^n$

SOLUTION

Notes:

1. We apply the Ratio Test to the series $\sum\limits_{n=0}^{\infty}\left|\dfrac{x^n}{n!}\right|$:

$$\lim_{n\to\infty}\frac{\left|x^{n+1}/(n+1)!\right|}{\left|x^n/n!\right|} = \lim_{n\to\infty}\left|\frac{x^{n+1}}{x^n}\cdot\frac{n!}{(n+1)!}\right|$$

$$= \lim_{n\to\infty}\left|\frac{x}{n+1}\right|$$

$$= 0 \text{ for all } x.$$

The Ratio Test shows us that regardless of the choice of x, the series converges. Therefore the radius of convergence is $R = \infty$, and the interval of convergence is $(-\infty, \infty)$.

2. We apply the Ratio Test to the series $\sum\limits_{n=1}^{\infty}\left|(-1)^{n+1}\dfrac{x^n}{n}\right| = \sum\limits_{n=1}^{\infty}\left|\dfrac{x^n}{n}\right|$:

$$\lim_{n\to\infty}\frac{\left|x^{n+1}/(n+1)\right|}{\left|x^n/n\right|} = \lim_{n\to\infty}\left|\frac{x^{n+1}}{x^n}\cdot\frac{n}{n+1}\right|$$

$$= \lim_{n\to\infty}|x|\frac{n}{n+1}$$

$$= |x|.$$

The Ratio Test states a series converges if the limit of $|a_{n+1}/a_n| = L < 1$. We found the limit above to be $|x|$; therefore, the power series converges when $|x| < 1$, or when x is in $(-1, 1)$. Thus the radius of convergence is $R = 1$.

To determine the interval of convergence, we need to check the endpoints of $(-1, 1)$. When $x = -1$, we have the opposite of the Harmonic Series:

$$\sum_{n=1}^{\infty}(-1)^{n+1}\frac{(-1)^n}{n} = \sum_{n=1}^{\infty}\frac{-1}{n}$$

$$= -\infty.$$

The series diverges when $x = -1$.

When $x = 1$, we have the series $\sum\limits_{n=1}^{\infty}(-1)^{n+1}\dfrac{(1)^n}{n}$, which is the Alternating Harmonic Series, which converges. Therefore the interval of convergence is $(-1, 1]$.

Notes:

3. We apply the Ratio Test to the series $\displaystyle\sum_{n=0}^{\infty} \left|2^n(x-3)^n\right|$:

$$\lim_{n\to\infty} \frac{\left|2^{n+1}(x-3)^{n+1}\right|}{\left|2^n(x-3)^n\right|} = \lim_{n\to\infty} \left|\frac{2^{n+1}}{2^n} \cdot \frac{(x-3)^{n+1}}{(x-3)^n}\right|$$

$$= \lim_{n\to\infty} \left|2(x-3)\right|.$$

According to the Ratio Test, the series converges when $\left|2(x-3)\right| < 1 \implies$ $\left|x-3\right| < 1/2$. The series is centered at 3, and x must be within $1/2$ of 3 in order for the series to converge. Therefore the radius of convergence is $R = 1/2$, and we know that the series converges absolutely for all x in $(3-1/2, 3+1/2) = (2.5, 3.5)$.

We check for convergence at the endpoints to find the interval of convergence. When $x = 2.5$, we have:

$$\sum_{n=0}^{\infty} 2^n(2.5-3)^n = \sum_{n=0}^{\infty} 2^n(-1/2)^n$$

$$= \sum_{n=0}^{\infty} (-1)^n,$$

which diverges. A similar process shows that the series also diverges at $x = 3.5$. Therefore the interval of convergence is $(2.5, 3.5)$.

4. We apply the Ratio Test to $\displaystyle\sum_{n=0}^{\infty} \left|n!x^n\right|$:

$$\lim_{n\to\infty} \frac{\left|(n+1)!x^{n+1}\right|}{\left|n!x^n\right|} = \lim_{n\to\infty} \left|(n+1)x\right|$$

$$= \infty \text{ for all } x, \text{ except } x = 0.$$

The Ratio Test shows that the series diverges for all x except $x = 0$. Therefore the radius of convergence is $R = 0$.

We can use a power series to define a function:

$$f(x) = \sum_{n=0}^{\infty} a_n x^n$$

where the domain of f is a subset of the interval of convergence of the power series. One can apply calculus techniques to such functions; in particular, we can find derivatives and antiderivatives.

Notes:

Theorem 8.6.3   **Derivatives and Indefinite Integrals of Power Series Functions**

Let $f(x) = \sum_{n=0}^{\infty} a_n(x - c)^n$ be a function defined by a power series, with radius of convergence R.

1. $f(x)$ is continuous and differentiable on $(c - R, c + R)$.

2. $f'(x) = \sum_{n=1}^{\infty} a_n \cdot n \cdot (x - c)^{n-1}$, with radius of convergence R.

3. $\int f(x)\,dx = C + \sum_{n=0}^{\infty} a_n \frac{(x - c)^{n+1}}{n + 1}$, with radius of convergence R.

A few notes about Theorem 8.6.3:

1. The theorem states that differentiation and integration do not change the radius of convergence. It does not state anything about the *interval* of convergence. They are not always the same.

2. Notice how the summation for $f'(x)$ starts with $n = 1$. This is because the constant term a_0 of $f(x)$ goes to 0.

3. Differentiation and integration are simply calculated term–by–term using the Power Rules.

Example 8.6.3   **Derivatives and indefinite integrals of power series**

Let $f(x) = \sum_{n=0}^{\infty} x^n$. Find $f'(x)$ and $F(x) = \int f(x)\,dx$, along with their respective intervals of convergence.

    SOLUTION   We find the derivative and indefinite integral of $f(x)$, following Theorem 8.6.3.

1. $f'(x) = \sum_{n=1}^{\infty} nx^{n-1} = 1 + 2x + 3x^2 + 4x^3 + \cdots$.

   In Example 8.6.1, we recognized that $\sum_{n=0}^{\infty} x^n$ is a geometric series in x. We know that such a geometric series converges when $|x| < 1$; that is, the interval of convergence is $(-1, 1)$.

Notes:

To determine the interval of convergence of $f'(x)$, we consider the end-points of $(-1, 1)$:

$$f'(-1) = 1 - 2 + 3 - 4 + \cdots, \quad \text{which diverges.}$$

$$f'(1) = 1 + 2 + 3 + 4 + \cdots, \quad \text{which diverges.}$$

Therefore, the interval of convergence of $f'(x)$ is $(-1, 1)$.

2. $F(x) = \displaystyle\int f(x)\, dx = C + \sum_{n=0}^{\infty} \frac{x^{n+1}}{n+1} = C + x + \frac{x^2}{2} + \frac{x^3}{3} + \cdots$

To find the interval of convergence of $F(x)$, we again consider the end-points of $(-1, 1)$:

$$F(-1) = C - 1 + 1/2 - 1/3 + 1/4 + \cdots$$

The value of C is irrelevant; notice that the rest of the series is an Alternating Series that whose terms converge to 0. By the Alternating Series Test, this series converges. (In fact, we can recognize that the terms of the series after C are the opposite of the Alternating Harmonic Series. We can thus say that $F(-1) = C - \ln 2$.)

$$F(1) = C + 1 + 1/2 + 1/3 + 1/4 + \cdots$$

Notice that this summation is $C +$ the Harmonic Series, which diverges. Since F converges for $x = -1$ and diverges for $x = 1$, the interval of convergence of $F(x)$ is $[-1, 1)$.

The previous example showed how to take the derivative and indefinite integral of a power series without motivation for why we care about such operations. We may care for the sheer mathematical enjoyment "that we can", which is motivation enough for many. However, we would be remiss to not recognize that we can learn a great deal from taking derivatives and indefinite integrals.

Recall that $f(x) = \displaystyle\sum_{n=0}^{\infty} x^n$ in Example 8.6.3 is a geometric series. According to Theorem 8.2.1, this series converges to $1/(1-x)$ when $|x| < 1$. Thus we can say

$$f(x) = \sum_{n=0}^{\infty} x^n = \frac{1}{1-x}, \quad \text{on} \quad (-1, 1).$$

Integrating the power series, (as done in Example 8.6.3,) we find

$$F(x) = C_1 + \sum_{n=0}^{\infty} \frac{x^{n+1}}{n+1}, \tag{8.4}$$

Notes:

while integrating the function $f(x) = 1/(1-x)$ gives

$$F(x) = -\ln|1-x| + C_2. \tag{8.5}$$

Equating Equations (8.4) and (8.5), we have

$$F(x) = C_1 + \sum_{n=0}^{\infty} \frac{x^{n+1}}{n+1} = -\ln|1-x| + C_2.$$

Letting $x = 0$, we have $F(0) = C_1 = C_2$. This implies that we can drop the constants and conclude

$$\sum_{n=0}^{\infty} \frac{x^{n+1}}{n+1} = -\ln|1-x|.$$

We established in Example 8.6.3 that the series on the left converges at $x = -1$; substituting $x = -1$ on both sides of the above equality gives

$$-1 + \frac{1}{2} - \frac{1}{3} + \frac{1}{4} - \frac{1}{5} + \cdots = -\ln 2.$$

On the left we have the opposite of the Alternating Harmonic Series; on the right, we have $-\ln 2$. We conclude that

$$1 - \frac{1}{2} + \frac{1}{3} - \frac{1}{4} + \cdots = \ln 2.$$

Important: We stated in Key Idea 8.2.1 (in Section 8.2) that the Alternating Harmonic Series converges to $\ln 2$, and referred to this fact again in Example 8.5.1 of Section 8.5. However, we never gave an argument for why this was the case. The work above finally shows how we conclude that the Alternating Harmonic Series converges to $\ln 2$.

We use this type of analysis in the next example.

Example 8.6.4   Analyzing power series functions

Let $f(x) = \sum_{n=0}^{\infty} \frac{x^n}{n!}$. Find $f'(x)$ and $\int f(x)\, dx$, and use these to analyze the behavior of $f(x)$.

SOLUTION   We start by making two notes: first, in Example 8.6.2, we found the interval of convergence of this power series is $(-\infty, \infty)$. Second, we will find it useful later to have a few terms of the series written out:

$$\sum_{n=0}^{\infty} \frac{x^n}{n!} = 1 + x + \frac{x^2}{2} + \frac{x^3}{6} + \frac{x^4}{24} + \cdots \tag{8.6}$$

Notes:

We now find the derivative:

$$f'(x) = \sum_{n=1}^{\infty} n \frac{x^{n-1}}{n!}$$

$$= \sum_{n=1}^{\infty} \frac{x^{n-1}}{(n-1)!} = 1 + x + \frac{x^2}{2!} + \cdots .$$

Since the series starts at $n = 1$ and each term refers to $(n-1)$, we can re-index the series starting with $n = 0$:

$$= \sum_{n=0}^{\infty} \frac{x^n}{n!}$$

$$= f(x).$$

We found the derivative of $f(x)$ is $f(x)$. The only functions for which this is true are of the form $y = ce^x$ for some constant c. As $f(0) = 1$ (see Equation (8.6)), c must be 1. Therefore we conclude that

$$f(x) = \sum_{n=0}^{\infty} \frac{x^n}{n!} = e^x$$

for all x.

We can also find $\int f(x)\, dx$:

$$\int f(x)\, dx = C + \sum_{n=0}^{\infty} \frac{x^{n+1}}{n!(n+1)}$$

$$= C + \sum_{n=0}^{\infty} \frac{x^{n+1}}{(n+1)!}$$

We write out a few terms of this last series:

$$C + \sum_{n=0}^{\infty} \frac{x^{n+1}}{(n+1)!} = C + x + \frac{x^2}{2} + \frac{x^3}{6} + \frac{x^4}{24} + \cdots$$

The integral of $f(x)$ differs from $f(x)$ only by a constant, again indicating that $f(x) = e^x$.

Example 8.6.4 and the work following Example 8.6.3 established relationships between a power series function and "regular" functions that we have dealt with in the past. In general, given a power series function, it is difficult (if

Notes:

not impossible) to express the function in terms of elementary functions. We chose examples where things worked out nicely.

In this section's last example, we show how to solve a simple differential equation with a power series.

Example 8.6.5   **Solving a differential equation with a power series.**
Give the first 4 terms of the power series solution to $y' = 2y$, where $y(0) = 1$.

SOLUTION   The differential equation $y' = 2y$ describes a function $y = f(x)$ where the derivative of y is twice y and $y(0) = 1$. This is a rather simple differential equation; with a bit of thought one should realize that if $y = Ce^{2x}$, then $y' = 2Ce^{2x}$, and hence $y' = 2y$. By letting $C = 1$ we satisfy the initial condition of $y(0) = 1$.

Let's ignore the fact that we already know the solution and find a power series function that satisfies the equation. The solution we seek will have the form

$$f(x) = \sum_{n=0}^{\infty} a_n x^n = a_0 + a_1 x + a_2 x^2 + a_3 x^3 + \cdots$$

for unknown coefficients a_n. We can find $f'(x)$ using Theorem 8.6.3:

$$f'(x) = \sum_{n=1}^{\infty} a_n \cdot n \cdot x^{n-1} = a_1 + 2a_2 x + 3a_3 x^2 + 4a_4 x^3 \cdots .$$

Since $f'(x) = 2f(x)$, we have

$$a_1 + 2a_2 x + 3a_3 x^2 + 4a_4 x^3 \cdots = 2\left(a_0 + a_1 x + a_2 x^2 + a_3 x^3 + \cdots\right)$$
$$= 2a_0 + 2a_1 x + 2a_2 x^2 + 2a_3 x^3 + \cdots$$

The coefficients of like powers of x must be equal, so we find that

$$a_1 = 2a_0, \quad 2a_2 = 2a_1, \quad 3a_3 = 2a_2, \quad 4a_4 = 2a_3, \quad \text{etc.}$$

The initial condition $y(0) = f(0) = 1$ indicates that $a_0 = 1$; with this, we can find the values of the other coefficients:

$$a_0 = 1 \text{ and } a_1 = 2a_0 \Rightarrow a_1 = 2;$$
$$a_1 = 2 \text{ and } 2a_2 = 2a_1 \Rightarrow a_2 = 4/2 = 2;$$
$$a_2 = 2 \text{ and } 3a_3 = 2a_2 \Rightarrow a_3 = 8/(2 \cdot 3) = 4/3;$$
$$a_3 = 4/3 \text{ and } 4a_4 = 2a_3 \Rightarrow a_4 = 16/(2 \cdot 3 \cdot 4) = 2/3.$$

Thus the first 5 terms of the power series solution to the differential equation $y' = 2y$ is

$$f(x) = 1 + 2x + 2x^2 + \frac{4}{3}x^3 + \frac{2}{3}x^4 + \cdots$$

Notes:

469

In Section 8.8, as we study Taylor Series, we will learn how to recognize this series as describing $y = e^{2x}$.

Our last example illustrates that it can be difficult to recognize an elementary function by its power series expansion. It is far easier to start with a known function, expressed in terms of elementary functions, and represent it as a power series function. One may wonder why we would bother doing so, as the latter function probably seems more complicated. In the next two sections, we show both *how* to do this and *why* such a process can be beneficial.

Notes:

Exercises 8.6

Terms and Concepts

1. We adopt the convention that $x^0 =$ _____, regardless of the value of x.

2. What is the difference between the radius of convergence and the interval of convergence?

3. If the radius of convergence of $\sum_{n=0}^{\infty} a_n x^n$ is 5, what is the radius of convergence of $\sum_{n=1}^{\infty} n \cdot a_n x^{n-1}$?

4. If the radius of convergence of $\sum_{n=0}^{\infty} a_n x^n$ is 5, what is the radius of convergence of $\sum_{n=0}^{\infty} (-1)^n a_n x^n$?

Problems

In Exercises 5 – 8, write out the sum of the first 5 terms of the given power series.

5. $\sum_{n=0}^{\infty} 2^n x^n$

6. $\sum_{n=1}^{\infty} \frac{1}{n^2} x^n$

7. $\sum_{n=0}^{\infty} \frac{1}{n!} x^n$

8. $\sum_{n=0}^{\infty} \frac{(-1)^n}{(2n)!} x^{2n}$

In Exercises 9 – 24, a power series is given.

   (a) Find the radius of convergence.

   (b) Find the interval of convergence.

9. $\sum_{n=0}^{\infty} \frac{(-1)^{n+1}}{n!} x^n$

10. $\sum_{n=0}^{\infty} n x^n$

11. $\sum_{n=1}^{\infty} \frac{(-1)^n (x-3)^n}{n}$

12. $\sum_{n=0}^{\infty} \frac{(x+4)^n}{n!}$

13. $\sum_{n=0}^{\infty} \frac{x^n}{2^n}$

14. $\sum_{n=0}^{\infty} \frac{(-1)^n (x-5)^n}{10^n}$

15. $\sum_{n=0}^{\infty} 5^n (x-1)^n$

16. $\sum_{n=0}^{\infty} (-2)^n x^n$

17. $\sum_{n=0}^{\infty} \sqrt{n} x^n$

18. $\sum_{n=0}^{\infty} \frac{n}{3^n} x^n$

19. $\sum_{n=0}^{\infty} \frac{3^n}{n!} (x-5)^n$

20. $\sum_{n=0}^{\infty} (-1)^n n! (x-10)^n$

21. $\sum_{n=1}^{\infty} \frac{x^n}{n^2}$

22. $\sum_{n=1}^{\infty} \frac{(x+2)^n}{n^3}$

23. $\sum_{n=0}^{\infty} n! \left(\frac{x}{10}\right)^n$

24. $\sum_{n=0}^{\infty} n^2 \left(\frac{x+4}{4}\right)^n$

In Exercises 25 – 30, a function $f(x) = \sum_{n=0}^{\infty} a_n x^n$ is given.

   (a) Give a power series for $f'(x)$ and its interval of convergence.

   (b) Give a power series for $\int f(x)\, dx$ and its interval of convergence.

25. $\sum_{n=0}^{\infty} n x^n$

26. $\sum_{n=1}^{\infty} \frac{x^n}{n}$

27. $\sum_{n=0}^{\infty} \left(\frac{x}{2}\right)^n$

28. $\displaystyle\sum_{n=0}^{\infty} (-3x)^n$

29. $\displaystyle\sum_{n=0}^{\infty} \frac{(-1)^n x^{2n}}{(2n)!}$

30. $\displaystyle\sum_{n=0}^{\infty} \frac{(-1)^n x^n}{n!}$

In Exercises 31 – 36, give the first 5 terms of the series that is a solution to the given differential equation.

31. $y' = 3y, \quad y(0) = 1$

32. $y' = 5y, \quad y(0) = 5$

33. $y' = y^2, \quad y(0) = 1$

34. $y' = y + 1, \quad y(0) = 1$

35. $y'' = -y, \quad y(0) = 0, y'(0) = 1$

36. $y'' = 2y, \quad y(0) = 1, y'(0) = 1$

8.7 Taylor Polynomials

Consider a function $y = f(x)$ and a point $(c, f(c))$. The derivative, $f'(c)$, gives the instantaneous rate of change of f at $x = c$. Of all lines that pass through the point $(c, f(c))$, the line that best approximates f at this point is the tangent line; that is, the line whose slope (rate of change) is $f'(c)$.

In Figure 8.7.1, we see a function $y = f(x)$ graphed. The table below the graph shows that $f(0) = 2$ and $f'(0) = 1$; therefore, the tangent line to f at $x = 0$ is $p_1(x) = 1(x - 0) + 2 = x + 2$. The tangent line is also given in the figure. Note that "near" $x = 0$, $p_1(x) \approx f(x)$; that is, the tangent line approximates f well.

One shortcoming of this approximation is that the tangent line only matches the slope of f; it does not, for instance, match the concavity of f. We can find a polynomial, $p_2(x)$, that does match the concavity without much difficulty, though. The table in Figure 8.7.1 gives the following information:

$$f(0) = 2 \qquad f'(0) = 1 \qquad f''(0) = 2.$$

Therefore, we want our polynomial $p_2(x)$ to have these same properties. That is, we need

$$p_2(0) = 2 \qquad p_2'(0) = 1 \qquad p_2''(0) = 2.$$

This is simply an initial–value problem. We can solve this using the techniques first described in Section 5.1. To keep $p_2(x)$ as simple as possible, we'll assume that not only $p_2''(0) = 2$, but that $p_2''(x) = 2$. That is, the second derivative of p_2 is constant.

If $p_2''(x) = 2$, then $p_2'(x) = 2x + C$ for some constant C. Since we have determined that $p_2'(0) = 1$, we find that $C = 1$ and so $p_2'(x) = 2x + 1$. Finally, we can compute $p_2(x) = x^2 + x + C$. Using our initial values, we know $p_2(0) = 2$ so $C = 2$. We conclude that $p_2(x) = x^2 + x + 2$. This function is plotted with f in Figure 8.7.2.

We can repeat this approximation process by creating polynomials of higher degree that match more of the derivatives of f at $x = 0$. In general, a polynomial of degree n can be created to match the first n derivatives of f. Figure 8.7.2 also shows $p_4(x) = -x^4/2 - x^3/6 + x^2 + x + 2$, whose first four derivatives at 0 match those of f. (Using the table in Figure 8.7.1, start with $p_4^{(4)}(x) = -12$ and solve the related initial–value problem.)

As we use more and more derivatives, our polynomial approximation to f gets better and better. In this example, the interval on which the approximation is "good" gets bigger and bigger. Figure 8.7.3 shows $p_{13}(x)$; we can visually affirm that this polynomial approximates f very well on $[-2, 3]$. (The polynomial $p_{13}(x)$ is not particularly "nice". It is

$$\frac{16901x^{13}}{6227020800} + \frac{13x^{12}}{1209600} - \frac{1321x^{11}}{39916800} - \frac{779x^{10}}{1814400} - \frac{359x^9}{362880} + \frac{x^8}{240} + \frac{139x^7}{5040} + \frac{11x^6}{360} - \frac{19x^5}{120} - \frac{x^4}{2} - \frac{x^3}{6} + x^2 + x + 2.)$$

Notes:

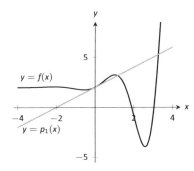

Figure 8.7.1: Plotting $y = f(x)$ and a table of derivatives of f evaluated at 0.

$f(0) = 2$	$f'''(0) = -1$
$f'(0) = 1$	$f^{(4)}(0) = -12$
$f''(0) = 2$	$f^{(5)}(0) = -19$

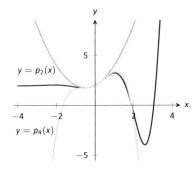

Figure 8.7.2: Plotting f, p_2 and p_4.

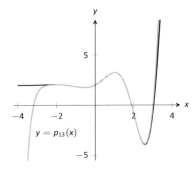

Figure 8.7.3: Plotting f and p_{13}.

The polynomials we have created are examples of *Taylor polynomials*, named after the British mathematician Brook Taylor who made important discoveries about such functions. While we created the above Taylor polynomials by solving initial–value problems, it can be shown that Taylor polynomials follow a general pattern that make their formation much more direct. This is described in the following definition.

Definition 8.7.1   Taylor Polynomial, Maclaurin Polynomial

Let f be a function whose first n derivatives exist at $x = c$.

1. The **Taylor polynomial of degree n of f at $x = c$** is

$$p_n(x) = f(c) + f'(c)(x-c) + \frac{f''(c)}{2!}(x-c)^2 + \frac{f'''(c)}{3!}(x-c)^3 + \cdots + \frac{f^{(n)}(c)}{n!}(x-c)^n.$$

2. A special case of the Taylor polynomial is the Maclaurin polynomial, where $c = 0$. That is, the **Maclaurin polynomial of degree n of f** is

$$p_n(x) = f(0) + f'(0)x + \frac{f''(0)}{2!}x^2 + \frac{f'''(0)}{3!}x^3 + \cdots + \frac{f^{(n)}(0)}{n!}x^n.$$

We will practice creating Taylor and Maclaurin polynomials in the following examples.

Example 8.7.1   Finding and using Maclaurin polynomials

1. Find the nth Maclaurin polynomial for $f(x) = e^x$.

2. Use $p_5(x)$ to approximate the value of e.

SOLUTION

1. We start with creating a table of the derivatives of e^x evaluated at $x = 0$. In this particular case, this is relatively simple, as shown in Figure 8.7.4. By the definition of the Maclaurin series, we have

$$p_n(x) = f(0) + f'(0)x + \frac{f''(0)}{2!}x^2 + \frac{f'''(0)}{3!}x^3 + \cdots + \frac{f^{(n)}(0)}{n!}x^n$$

$$= 1 + x + \frac{1}{2}x^2 + \frac{1}{6}x^3 + \frac{1}{24}x^4 + \cdots + \frac{1}{n!}x^n.$$

$$
\begin{aligned}
f(x) &= e^x &\Rightarrow\quad f(0) &= 1 \\
f'(x) &= e^x &\Rightarrow\quad f'(0) &= 1 \\
f''(x) &= e^x &\Rightarrow\quad f''(0) &= 1 \\
&\vdots & &\vdots \\
f^{(n)}(x) &= e^x &\Rightarrow\quad f^{(n)}(0) &= 1
\end{aligned}
$$

Figure 8.7.4: The derivatives of $f(x) = e^x$ evaluated at $x = 0$.

Notes:

2. Using our answer from part 1, we have

$$p_5 = 1 + x + \frac{1}{2}x^2 + \frac{1}{6}x^3 + \frac{1}{24}x^4 + \frac{1}{120}x^5.$$

To approximate the value of e, note that $e = e^1 = f(1) \approx p_5(1)$. It is very straightforward to evaluate $p_5(1)$:

$$p_5(1) = 1 + 1 + \frac{1}{2} + \frac{1}{6} + \frac{1}{24} + \frac{1}{120} = \frac{163}{60} \approx 2.71667.$$

A plot of $f(x) = e^x$ and $p_5(x)$ is given in Figure 8.7.5.

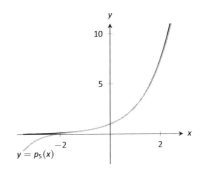

Figure 8.7.5: A plot of $f(x) = e^x$ and its 5th degree Maclaurin polynomial $p_5(x)$.

Example 8.7.2   Finding and using Taylor polynomials

1. Find the nth Taylor polynomial of $y = \ln x$ at $x = 1$.

2. Use $p_6(x)$ to approximate the value of $\ln 1.5$.

3. Use $p_6(x)$ to approximate the value of $\ln 2$.

SOLUTION

1. We begin by creating a table of derivatives of $\ln x$ evaluated at $x = 1$. While this is not as straightforward as it was in the previous example, a pattern does emerge, as shown in Figure 8.7.6.

   Using Definition 8.7.1, we have

$$p_n(x) = f(c) + f'(c)(x - c) + \frac{f''(c)}{2!}(x - c)^2 + \frac{f'''(c)}{3!}(x - c)^3 + \cdots + \frac{f^{(n)}(c)}{n!}(x - c)^n$$

$$= 0 + (x - 1) - \frac{1}{2}(x - 1)^2 + \frac{1}{3}(x - 1)^3 - \frac{1}{4}(x - 1)^4 + \cdots + \frac{(-1)^{n+1}}{n}(x - 1)^n.$$

$$\begin{aligned}
f(x) &= \ln x & \Rightarrow & \quad f(1) = 0 \\
f'(x) &= 1/x & \Rightarrow & \quad f'(1) = 1 \\
f''(x) &= -1/x^2 & \Rightarrow & \quad f''(1) = -1 \\
f'''(x) &= 2/x^3 & \Rightarrow & \quad f'''(1) = 2 \\
f^{(4)}(x) &= -6/x^4 & \Rightarrow & \quad f^{(4)}(1) = -6 \\
&\vdots & & \quad \vdots \\
f^{(n)}(x) &= & \Rightarrow & \quad f^{(n)}(1) = \\
\frac{(-1)^{n+1}(n-1)!}{x^n} & & & \quad (-1)^{n+1}(n-1)!
\end{aligned}$$

Figure 8.7.6: Derivatives of $\ln x$ evaluated at $x = 1$.

Note how the coefficients of the $(x - 1)$ terms turn out to be "nice."

2. We can compute $p_6(x)$ using our work above:

$$p_6(x) = (x-1) - \frac{1}{2}(x-1)^2 + \frac{1}{3}(x-1)^3 - \frac{1}{4}(x-1)^4 + \frac{1}{5}(x-1)^5 - \frac{1}{6}(x-1)^6.$$

Since $p_6(x)$ approximates $\ln x$ well near $x = 1$, we approximate $\ln 1.5 \approx p_6(1.5)$:

Notes:

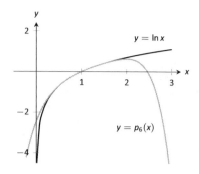

Figure 8.7.7: A plot of $y = \ln x$ and its 6th degree Taylor polynomial at $x = 1$.

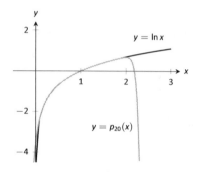

Figure 8.7.8: A plot of $y = \ln x$ and its 20th degree Taylor polynomial at $x = 1$.

$$p_6(1.5) = (1.5 - 1) - \frac{1}{2}(1.5 - 1)^2 + \frac{1}{3}(1.5 - 1)^3 - \frac{1}{4}(1.5 - 1)^4 + \cdots$$
$$\cdots + \frac{1}{5}(1.5 - 1)^5 - \frac{1}{6}(1.5 - 1)^6$$
$$= \frac{259}{640}$$
$$\approx 0.404688.$$

This is a good approximation as a calculator shows that $\ln 1.5 \approx 0.4055$. Figure 8.7.7 plots $y = \ln x$ with $y = p_6(x)$. We can see that $\ln 1.5 \approx p_6(1.5)$.

3. We approximate $\ln 2$ with $p_6(2)$:

$$p_6(2) = (2 - 1) - \frac{1}{2}(2 - 1)^2 + \frac{1}{3}(2 - 1)^3 - \frac{1}{4}(2 - 1)^4 + \cdots$$
$$\cdots + \frac{1}{5}(2 - 1)^5 - \frac{1}{6}(2 - 1)^6$$
$$= 1 - \frac{1}{2} + \frac{1}{3} - \frac{1}{4} + \frac{1}{5} - \frac{1}{6}$$
$$= \frac{37}{60}$$
$$\approx 0.616667.$$

This approximation is not terribly impressive: a hand held calculator shows that $\ln 2 \approx 0.693147$. The graph in Figure 8.7.7 shows that $p_6(x)$ provides less accurate approximations of $\ln x$ as x gets close to 0 or 2.

Surprisingly enough, even the 20th degree Taylor polynomial fails to approximate $\ln x$ for $x > 2$, as shown in Figure 8.7.8. We'll soon discuss why this is.

Taylor polynomials are used to approximate functions $f(x)$ in mainly two situations:

1. When $f(x)$ is known, but perhaps "hard" to compute directly. For instance, we can define $y = \cos x$ as either the ratio of sides of a right triangle ("adjacent over hypotenuse") or with the unit circle. However, neither of these provides a convenient way of computing $\cos 2$. A Taylor polynomial of sufficiently high degree can provide a reasonable method of computing such values using only operations usually hard–wired into a computer ($+$, $-$, $\times$ and $\div$).

Notes:

2. When $f(x)$ is not known, but information about its derivatives is known. This occurs more often than one might think, especially in the study of differential equations.

In both situations, a critical piece of information to have is "How good is my approximation?" If we use a Taylor polynomial to compute $\cos 2$, how do we know how accurate the approximation is?

We had the same problem when studying Numerical Integration. Theorem 5.5.1 provided bounds on the error when using, say, Simpson's Rule to approximate a definite integral. These bounds allowed us to determine that, for instance, using 10 subintervals provided an approximation within $\pm.01$ of the exact value. The following theorem gives similar bounds for Taylor (and hence Maclaurin) polynomials.

Note: Even though Taylor polynomials *could* be used in calculators and computers to calculate values of trigonometric functions, in practice they generally aren't. Other more efficient and accurate methods have been developed, such as the CORDIC algorithm.

Theorem 8.7.1 **Taylor's Theorem**

1. Let f be a function whose $n + 1^{\text{th}}$ derivative exists on an interval I and let c be in I. Then, for each x in I, there exists z_x between x and c such that

$$f(x) = f(c) + f'(c)(x - c) + \frac{f''(c)}{2!}(x - c)^2 + \cdots + \frac{f^{(n)}(c)}{n!}(x - c)^n + R_n(x),$$

where $R_n(x) = \frac{f^{(n+1)}(z_x)}{(n + 1)!}(x - c)^{(n+1)}$.

2. $\left| R_n(x) \right| \leq \frac{\max \left| f^{(n+1)}(z) \right|}{(n + 1)!} \left| (x - c)^{(n+1)} \right|$, where z is in I.

The first part of Taylor's Theorem states that $f(x) = p_n(x) + R_n(x)$, where $p_n(x)$ is the nth order Taylor polynomial and $R_n(x)$ is the remainder, or error, in the Taylor approximation. The second part gives bounds on how big that error can be. If the $(n + 1)^{\text{th}}$ derivative is large on I, the error may be large; if x is far from c, the error may also be large. However, the $(n + 1)!$ term in the denominator tends to ensure that the error gets smaller as n increases.

The following example computes error estimates for the approximations of $\ln 1.5$ and $\ln 2$ made in Example 8.7.2.

Example 8.7.3 **Finding error bounds of a Taylor polynomial**
Use Theorem 8.7.1 to find error bounds when approximating $\ln 1.5$ and $\ln 2$ with $p_6(x)$, the Taylor polynomial of degree 6 of $f(x) = \ln x$ at $x = 1$, as calculated in Example 8.7.2.

Notes:

SOLUTION

1. We start with the approximation of $\ln 1.5$ with $p_6(1.5)$. The theorem references an open interval I that contains both x and c. The smaller the interval we use the better; it will give us a more accurate (and smaller!) approximation of the error. We let $I = (0.9, 1.6)$, as this interval contains both $c = 1$ and $x = 1.5$.

The theorem references $\max\left|f^{(n+1)}(z)\right|$. In our situation, this is asking "How big can the 7th derivative of $y = \ln x$ be on the interval $(0.9, 1.6)$?" The seventh derivative is $y = -6!/x^7$. The largest value it attains on I is about 1506. Thus we can bound the error as:

$$\left|R_6(1.5)\right| \le \frac{\max\left|f^{(7)}(z)\right|}{7!}\left|(1.5 - 1)^7\right|$$
$$\le \frac{1506}{5040}\cdot\frac{1}{2^7}$$
$$\approx 0.0023.$$

We computed $p_6(1.5) = 0.404688$; using a calculator, we find $\ln 1.5 \approx 0.405465$, so the actual error is about 0.000778, which is less than our bound of 0.0023. This affirms Taylor's Theorem; the theorem states that our approximation would be within about 2 thousandths of the actual value, whereas the approximation was actually closer.

2. We again find an interval I that contains both $c = 1$ and $x = 2$; we choose $I = (0.9, 2.1)$. The maximum value of the seventh derivative of f on this interval is again about 1506 (as the largest values come near $x = 0.9$). Thus

$$\left|R_6(2)\right| \le \frac{\max\left|f^{(7)}(z)\right|}{7!}\left|(2 - 1)^7\right|$$
$$\le \frac{1506}{5040}\cdot 1^7$$
$$\approx 0.30.$$

This bound is not as nearly as good as before. Using the degree 6 Taylor polynomial at $x = 1$ will bring us within 0.3 of the correct answer. As $p_6(2) \approx 0.61667$, our error estimate guarantees that the actual value of $\ln 2$ is somewhere between 0.31667 and 0.91667. These bounds are not particularly useful.

In reality, our approximation was only off by about 0.07. However, we are approximating ostensibly because we do not know the real answer. In order to be assured that we have a good approximation, we would have to resort to using a polynomial of higher degree.

Notes:

We practice again. This time, we use Taylor's theorem to find n that guarantees our approximation is within a certain amount.

Example 8.7.4 Finding sufficiently accurate Taylor polynomials

Find n such that the nth Taylor polynomial of $f(x) = \cos x$ at $x = 0$ approximates $\cos 2$ to within 0.001 of the actual answer. What is $p_n(2)$?

SOLUTION Following Taylor's theorem, we need bounds on the size of the derivatives of $f(x) = \cos x$. In the case of this trigonometric function, this is easy. All derivatives of cosine are $\pm \sin x$ or $\pm \cos x$. In all cases, these functions are never greater than 1 in absolute value. We want the error to be less than 0.001. To find the appropriate n, consider the following inequalities:

$$\frac{\max \left| f^{(n+1)}(z) \right|}{(n+1)!} \left| (2-0)^{(n+1)} \right| \le 0.001$$

$$\frac{1}{(n+1)!} \cdot 2^{(n+1)} \le 0.001$$

We find an n that satisfies this last inequality with trial–and–error. When $n = 8$, we have $\dfrac{2^{8+1}}{(8+1)!} \approx 0.0014$; when $n = 9$, we have $\dfrac{2^{9+1}}{(9+1)!} \approx 0.000282 < 0.001$. Thus we want to approximate $\cos 2$ with $p_9(2)$.

We now set out to compute $p_9(x)$. We again need a table of the derivatives of $f(x) = \cos x$ evaluated at $x = 0$. A table of these values is given in Figure 8.7.9. Notice how the derivatives, evaluated at $x = 0$, follow a certain pattern. All the odd powers of x in the Taylor polynomial will disappear as their coefficient is 0. While our error bounds state that we need $p_9(x)$, our work shows that this will be the same as $p_8(x)$.

Since we are forming our polynomial at $x = 0$, we are creating a Maclaurin polynomial, and:

$$p_8(x) = f(0) + f'(0)x + \frac{f''(0)}{2!}x^2 + \frac{f'''(0)}{3!}x^3 + \cdots + \frac{f^{(8)}(0)}{8!}x^8$$

$$= 1 - \frac{1}{2!}x^2 + \frac{1}{4!}x^4 - \frac{1}{6!}x^6 + \frac{1}{8!}x^8$$

We finally approximate $\cos 2$:

$$\cos 2 \approx p_8(2) = -\frac{131}{315} \approx -0.41587.$$

Our error bound guarantee that this approximation is within 0.001 of the correct answer. Technology shows us that our approximation is actually within about 0.0003 of the correct answer.

Figure 8.7.10 shows a graph of $y = p_8(x)$ and $y = \cos x$. Note how well the two functions agree on about $(-\pi, \pi)$.

$$
\begin{array}{lcl}
f(x) = \cos x & \Rightarrow & f(0) = 1 \\
f'(x) = -\sin x & \Rightarrow & f'(0) = 0 \\
f''(x) = -\cos x & \Rightarrow & f''(0) = -1 \\
f'''(x) = \sin x & \Rightarrow & f'''(0) = 0 \\
f^{(4)}(x) = \cos x & \Rightarrow & f^{(4)}(0) = 1 \\
f^{(5)}(x) = -\sin x & \Rightarrow & f^{(5)}(0) = 0 \\
f^{(6)}(x) = -\cos x & \Rightarrow & f^{(6)}(0) = -1 \\
f^{(7)}(x) = \sin x & \Rightarrow & f^{(7)}(0) = 0 \\
f^{(8)}(x) = \cos x & \Rightarrow & f^{(8)}(0) = 1 \\
f^{(9)}(x) = -\sin x & \Rightarrow & f^{(9)}(0) = 0 \\
\end{array}
$$

Figure 8.7.9: A table of the derivatives of $f(x) = \cos x$ evaluated at $x = 0$.

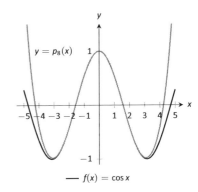

Figure 8.7.10: A graph of $f(x) = \cos x$ and its degree 8 Maclaurin polynomial.

Notes:

$$f(x) = \sqrt{x} \qquad \Rightarrow \qquad f(4) = 2$$

$$f'(x) = \frac{1}{2\sqrt{x}} \qquad \Rightarrow \qquad f'(4) = \frac{1}{4}$$

$$f''(x) = \frac{-1}{4x^{3/2}} \qquad \Rightarrow \qquad f''(4) = \frac{-1}{32}$$

$$f'''(x) = \frac{3}{8x^{5/2}} \qquad \Rightarrow \qquad f'''(4) = \frac{3}{256}$$

$$f^{(4)}(x) = \frac{-15}{16x^{7/2}} \qquad \Rightarrow \qquad f^{(4)}(4) = \frac{-15}{2048}$$

Figure 8.7.11: A table of the derivatives of $f(x) = \sqrt{x}$ evaluated at $x = 4$.

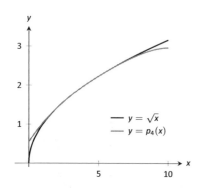

Figure 8.7.12: A graph of $f(x) = \sqrt{x}$ and its degree 4 Taylor polynomial at $x = 4$.

Example 8.7.5   Finding and using Taylor polynomials

1. Find the degree 4 Taylor polynomial, $p_4(x)$, for $f(x) = \sqrt{x}$ at $x = 4$.

2. Use $p_4(x)$ to approximate $\sqrt{3}$.

3. Find bounds on the error when approximating $\sqrt{3}$ with $p_4(3)$.

SOLUTION

1. We begin by evaluating the derivatives of f at $x = 4$. This is done in Figure 8.7.11. These values allow us to form the Taylor polynomial $p_4(x)$:

$$p_4(x) = 2 + \frac{1}{4}(x-4) + \frac{-1/32}{2!}(x-4)^2 + \frac{3/256}{3!}(x-4)^3 + \frac{-15/2048}{4!}(x-4)^4.$$

2. As $p_4(x) \approx \sqrt{x}$ near $x = 4$, we approximate $\sqrt{3}$ with $p_4(3) = 1.73212$.

3. To find a bound on the error, we need an open interval that contains $x = 3$ and $x = 4$. We set $I = (2.9, 4.1)$. The largest value the fifth derivative of $f(x) = \sqrt{x}$ takes on this interval is near $x = 2.9$, at about 0.0273. Thus

$$\left| R_4(3) \right| \le \frac{0.0273}{5!} \left| (3-4)^5 \right| \approx 0.00023.$$

This shows our approximation is accurate to at least the first 2 places after the decimal. (It turns out that our approximation is actually accurate to 4 places after the decimal.) A graph of $f(x) = \sqrt{x}$ and $p_4(x)$ is given in Figure 8.7.12. Note how the two functions are nearly indistinguishable on $(2, 7)$.

Our final example gives a brief introduction to using Taylor polynomials to solve differential equations.

Example 8.7.6   Approximating an unknown function
A function $y = f(x)$ is unknown save for the following two facts.

1. $y(0) = f(0) = 1$, and

2. $y' = y^2$

(This second fact says that amazingly, the derivative of the function is actually the function squared!)
Find the degree 3 Maclaurin polynomial $p_3(x)$ of $y = f(x)$.

Notes:

SOLUTION   One might initially think that not enough information is given to find $p_3(x)$. However, note how the second fact above actually lets us know what $y'(0)$ is:

$$y' = y^2 \Rightarrow y'(0) = y^2(0).$$

Since $y(0) = 1$, we conclude that $y'(0) = 1$.

Now we find information about y''. Starting with $y' = y^2$, take derivatives of both sides, *with respect to x*. That means we must use implicit differentiation.

$$y' = y^2$$
$$\frac{d}{dx}(y') = \frac{d}{dx}(y^2)$$
$$y'' = 2y \cdot y'.$$

Now evaluate both sides at $x = 0$:

$$y''(0) = 2y(0) \cdot y'(0)$$
$$y''(0) = 2$$

We repeat this once more to find $y'''(0)$. We again use implicit differentiation; this time the Product Rule is also required.

$$\frac{d}{dx}(y'') = \frac{d}{dx}(2yy')$$
$$y''' = 2y' \cdot y' + 2y \cdot y''.$$

Now evaluate both sides at $x = 0$:

$$y'''(0) = 2y'(0)^2 + 2y(0)y''(0)$$
$$y'''(0) = 2 + 4 = 6$$

In summary, we have:

$$y(0) = 1 \qquad y'(0) = 1 \qquad y''(0) = 2 \qquad y'''(0) = 6.$$

We can now form $p_3(x)$:

$$p_3(x) = 1 + x + \frac{2}{2!}x^2 + \frac{6}{3!}x^3$$
$$= 1 + x + x^2 + x^3.$$

It turns out that the differential equation we started with, $y' = y^2$, where $y(0) = 1$, can be solved without too much difficulty: $y = \dfrac{1}{1-x}$. Figure 8.7.13 shows this function plotted with $p_3(x)$. Note how similar they are near $x = 0$.

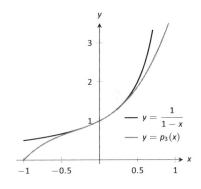

Figure 8.7.13: A graph of $y = -1/(x-1)$ and $y = p_3(x)$ from Example 8.7.6.

Notes:

It is beyond the scope of this text to pursue error analysis when using Taylor polynomials to approximate solutions to differential equations. This topic is often broached in introductory Differential Equations courses and usually covered in depth in Numerical Analysis courses. Such an analysis is very important; one needs to know how good their approximation is. We explored this example simply to demonstrate the usefulness of Taylor polynomials.

Most of this chapter has been devoted to the study of infinite series. This section has taken a step back from this study, focusing instead on finite summation of terms. In the next section, we explore **Taylor Series**, where we represent a function with an infinite series.

Notes:

Exercises 8.7

Terms and Concepts

1. What is the difference between a Taylor polynomial and a Maclaurin polynomial?

2. T/F: In general, $p_n(x)$ approximates $f(x)$ better and better as n gets larger.

3. For some function $f(x)$, the Maclaurin polynomial of degree 4 is $p_4(x) = 6 + 3x - 4x^2 + 5x^3 - 7x^4$. What is $p_2(x)$?

4. For some function $f(x)$, the Maclaurin polynomial of degree 4 is $p_4(x) = 6 + 3x - 4x^2 + 5x^3 - 7x^4$. What is $f'''(0)$?

Problems

In Exercises 5 – 12, find the Maclaurin polynomial of degree n for the given function.

5. $f(x) = e^{-x}, \quad n = 3$

6. $f(x) = \sin x, \quad n = 8$

7. $f(x) = x \cdot e^x, \quad n = 5$

8. $f(x) = \tan x, \quad n = 6$

9. $f(x) = e^{2x}, \quad n = 4$

10. $f(x) = \dfrac{1}{1-x}, \quad n = 4$

11. $f(x) = \dfrac{1}{1+x}, \quad n = 4$

12. $f(x) = \dfrac{1}{1+x}, \quad n = 7$

In Exercises 13 – 20, find the Taylor polynomial of degree n, at $x = c$, for the given function.

13. $f(x) = \sqrt{x}, \quad n = 4, \quad c = 1$

14. $f(x) = \ln(x + 1), \quad n = 4, \quad c = 1$

15. $f(x) = \cos x, \quad n = 6, \quad c = \pi/4$

16. $f(x) = \sin x, \quad n = 5, \quad c = \pi/6$

17. $f(x) = \dfrac{1}{x}, \quad n = 5, \quad c = 2$

18. $f(x) = \dfrac{1}{x^2}, \quad n = 8, \quad c = 1$

19. $f(x) = \dfrac{1}{x^2 + 1}, \quad n = 3, \quad c = -1$

20. $f(x) = x^2 \cos x, \quad n = 2, \quad c = \pi$

In Exercises 21 – 24, approximate the function value with the indicated Taylor polynomial and give approximate bounds on the error.

21. Approximate $\sin 0.1$ with the Maclaurin polynomial of degree 3.

22. Approximate $\cos 1$ with the Maclaurin polynomial of degree 4.

23. Approximate $\sqrt{10}$ with the Taylor polynomial of degree 2 centered at $x = 9$.

24. Approximate $\ln 1.5$ with the Taylor polynomial of degree 3 centered at $x = 1$.

Exercises 25 – 28 ask for an n to be found such that $p_n(x)$ approximates $f(x)$ within a certain bound of accuracy.

25. Find n such that the Maclaurin polynomial of degree n of $f(x) = e^x$ approximates e within 0.0001 of the actual value.

26. Find n such that the Taylor polynomial of degree n of $f(x) = \sqrt{x}$, centered at $x = 4$, approximates $\sqrt{3}$ within 0.0001 of the actual value.

27. Find n such that the Maclaurin polynomial of degree n of $f(x) = \cos x$ approximates $\cos \pi/3$ within 0.0001 of the actual value.

28. Find n such that the Maclaurin polynomial of degree n of $f(x) = \sin x$ approximates $\cos \pi$ within 0.0001 of the actual value.

In Exercises 29 – 34, find the nth term of the indicated Taylor polynomial.

29. Find a formula for the nth term of the Maclaurin polynomial for $f(x) = e^x$.

30. Find a formula for the nth term of the Maclaurin polynomial for $f(x) = \cos x$.

31. Find a formula for the nth term of the Maclaurin polynomial for $f(x) = \sin x$.

32. Find a formula for the nth term of the Maclaurin polynomial for $f(x) = \dfrac{1}{1-x}$.

33. Find a formula for the nth term of the Maclaurin polynomial for $f(x) = \dfrac{1}{1+x}$.

34. Find a formula for the nth term of the Taylor polynomial for $f(x) = \ln x$ centered at $x = 1$.

In Exercises 35 – 37, approximate the solution to the given differential equation with a degree 4 Maclaurin polynomial.

35. $y' = y, \qquad y(0) = 1$

36. $y' = 5y, \qquad y(0) = 3$

37. $y' = \dfrac{2}{y}, \qquad y(0) = 1$

8.8 Taylor Series

In Section 8.6, we showed how certain functions can be represented by a power series function. In Section 8.7, we showed how we can approximate functions with polynomials, given that enough derivative information is available. In this section we combine these concepts: if a function $f(x)$ is infinitely differentiable, we show how to represent it with a power series function.

Definition 8.8.1   Taylor and Maclaurin Series

Let $f(x)$ have derivatives of all orders at $x = c$.

1. The **Taylor Series of $f(x)$, centered at c** is

$$\sum_{n=0}^{\infty} \frac{f^{(n)}(c)}{n!}(x - c)^n.$$

2. Setting $c = 0$ gives the **Maclaurin Series of $f(x)$**:

$$\sum_{n=0}^{\infty} \frac{f^{(n)}(0)}{n!} x^n.$$

If $p_n(x)$ is the nth degree Taylor polynomial for $f(x)$ centered at $x = c$, we saw how $f(x)$ is *approximately equal* to $p_n(x)$ near $x = c$. We also saw how increasing the degree of the polynomial generally reduced the error.

We are now considering *series*, where we sum an infinite set of terms. Our ultimate hope is to see the error vanish and claim a function is *equal* to its Taylor series.

When creating the Taylor polynomial of degree n for a function $f(x)$ at $x = c$, we needed to evaluate f, and the first n derivatives of f, at $x = c$. When creating the Taylor series of f, it helps to find a pattern that describes the nth derivative of f at $x = c$. We demonstrate this in the next two examples.

Example 8.8.1   The Maclaurin series of $f(x) = \cos x$
Find the Maclaurin series of $f(x) = \cos x$.

SOLUTION   In Example 8.7.4 we found the 8th degree Maclaurin polynomial of $\cos x$. In doing so, we created the table shown in Figure 8.8.1. Notice how $f^{(n)}(0) = 0$ when n is odd, $f^{(n)}(0) = 1$ when n is divisible by 4, and $f^{(n)}(0) = -1$ when n is even but not divisible by 4. Thus the Maclaurin series

$$
\begin{aligned}
f(x) &= \cos x && \Rightarrow & f(0) &= 1 \\
f'(x) &= -\sin x && \Rightarrow & f'(0) &= 0 \\
f''(x) &= -\cos x && \Rightarrow & f''(0) &= -1 \\
f'''(x) &= \sin x && \Rightarrow & f'''(0) &= 0 \\
f^{(4)}(x) &= \cos x && \Rightarrow & f^{(4)}(0) &= 1 \\
f^{(5)}(x) &= -\sin x && \Rightarrow & f^{(5)}(0) &= 0 \\
f^{(6)}(x) &= -\cos x && \Rightarrow & f^{(6)}(0) &= -1 \\
f^{(7)}(x) &= \sin x && \Rightarrow & f^{(7)}(0) &= 0 \\
f^{(8)}(x) &= \cos x && \Rightarrow & f^{(8)}(0) &= 1 \\
f^{(9)}(x) &= -\sin x && \Rightarrow & f^{(9)}(0) &= 0
\end{aligned}
$$

Figure 8.8.1: A table of the derivatives of $f(x) = \cos x$ evaluated at $x = 0$.

Notes:

of $\cos x$ is

$$1 - \frac{x^2}{2} + \frac{x^4}{4!} - \frac{x^6}{6!} + \frac{x^8}{8!} - \cdots$$

We can go further and write this as a summation. Since we only need the terms where the power of x is even, we write the power series in terms of x^{2n}:

$$\sum_{n=0}^{\infty} (-1)^n \frac{x^{2n}}{(2n)!}.$$

This Maclaurin series is a special type of power series. As such, we should determine its interval of convergence. Applying the Ratio Test, we have

$$\lim_{n\to\infty} \left| (-1)^{n+1} \frac{x^{2(n+1)}}{(2(n+1))!} \right| \Bigg/ \left| (-1)^n \frac{x^{2n}}{(2n)!} \right| = \lim_{n\to\infty} \left| \frac{x^{2n+2}}{x^{2n}} \right| \frac{(2n)!}{(2n+2)!}$$

$$= \lim_{n\to\infty} \frac{x^2}{(2n+2)(2n+1)}.$$

For any fixed x, this limit is 0. Therefore this power series has an infinite radius of convergence, converging for all x. It is important to note what we have, and have not, determined: we have determined the Maclaurin series for $\cos x$ along with its interval of convergence. We *have not* shown that $\cos x$ is *equal* to this power series.

Example 8.8.2   The Taylor series of $f(x) = \ln x$ at $x = 1$
Find the Taylor series of $f(x) = \ln x$ centered at $x = 1$.

SOLUTION   Figure 8.8.2 shows the nth derivative of $\ln x$ evaluated at $x = 1$ for $n = 0, \ldots, 5$, along with an expression for the nth term:

$$f^{(n)}(1) = (-1)^{n+1}(n-1)! \quad \text{for } n \geq 1.$$

Remember that this is what distinguishes Taylor series from Taylor polynomials; we are very interested in finding a pattern for the nth term, not just finding a finite set of coefficients for a polynomial. Since $f(1) = \ln 1 = 0$, we skip the first term and start the summation with $n = 1$, giving the Taylor series for $\ln x$, centered at $x = 1$, as

$$\sum_{n=1}^{\infty} (-1)^{n+1}(n-1)! \frac{1}{n!} (x-1)^n = \sum_{n=1}^{\infty} (-1)^{n+1} \frac{(x-1)^n}{n}.$$

We now determine the interval of convergence, using the Ratio Test.

$$\lim_{n\to\infty} \left| (-1)^{n+2} \frac{(x-1)^{n+1}}{n+1} \right| \Bigg/ \left| (-1)^{n+1} \frac{(x-1)^n}{n} \right| = \lim_{n\to\infty} \left| \frac{(x-1)^{n+1}}{(x-1)^n} \right| \frac{n}{n+1}$$

$$= \left| (x-1) \right|.$$

$$f(x) = \ln x \qquad \Rightarrow \quad f(1) = 0$$
$$f'(x) = 1/x \qquad \Rightarrow \quad f'(1) = 1$$
$$f''(x) = -1/x^2 \qquad \Rightarrow \quad f''(1) = -1$$
$$f'''(x) = 2/x^3 \qquad \Rightarrow \quad f'''(1) = 2$$
$$f^{(4)}(x) = -6/x^4 \qquad \Rightarrow \quad f^{(4)}(1) = -6$$
$$f^{(5)}(x) = 24/x^5 \qquad \Rightarrow \quad f^{(5)}(1) = 24$$
$$\vdots \qquad\qquad\qquad \vdots$$
$$f^{(n)}(x) = \qquad\qquad \Rightarrow \quad f^{(n)}(1) =$$
$$\frac{(-1)^{n+1}(n-1)!}{x^n} \qquad\qquad (-1)^{n+1}(n-1)!$$

Figure 8.8.2: Derivatives of $\ln x$ evaluated at $x = 1$.

Notes:

By the Ratio Test, we have convergence when $\left|(x-1)\right| < 1$: the radius of convergence is 1, and we have convergence on $(0, 2)$. We now check the endpoints.

At $x = 0$, the series is

$$\sum_{n=1}^{\infty}(-1)^{n+1}\frac{(-1)^n}{n} = -\sum_{n=1}^{\infty}\frac{1}{n},$$

which diverges (it is the Harmonic Series times (-1).)

At $x = 2$, the series is

$$\sum_{n=1}^{\infty}(-1)^{n+1}\frac{(1)^n}{n} = \sum_{n=1}^{\infty}(-1)^{n+1}\frac{1}{n},$$

the Alternating Harmonic Series, which converges.

We have found the Taylor series of $\ln x$ centered at $x = 1$, and have determined the series converges on $(0, 2]$. We cannot (yet) say that $\ln x$ is equal to this Taylor series on $(0, 2]$.

It is important to note that Definition 8.8.1 defines a Taylor series given a function $f(x)$, but makes no claim about their equality. We will find that "most of the time" they are equal, but we need to consider the conditions that allow us to conclude this.

Theorem 8.7.1 states that the error between a function $f(x)$ and its nth–degree Taylor polynomial $p_n(x)$ is $R_n(x)$, where

$$\left|R_n(x)\right| \leq \frac{\max\left|f^{(n+1)}(z)\right|}{(n+1)!}\left|(x-c)^{(n+1)}\right|.$$

Note: It can be shown that $\ln x$ is equal to this Taylor series on $(0, 2]$. From the work in Example 8.8.2, this justifies our previous declaration that the Alternating Harmonic Series converges to $\ln 2$.

If $R_n(x)$ goes to 0 for each x in an interval I as n approaches infinity, we conclude that the function is equal to its Taylor series expansion.

Theorem 8.8.1   Function and Taylor Series Equality

Let $f(x)$ have derivatives of all orders at $x = c$, let $R_n(x)$ be as stated in Theorem 8.7.1, and let I be an interval on which the Taylor series of $f(x)$ converges. If $\lim_{n\to\infty} R_n(x) = 0$ for all x in I, then

$$f(x) = \sum_{n=0}^{\infty}\frac{f^{(n)}(c)}{n!}(x-c)^n \text{ on } I.$$

Notes:

We demonstrate the use of this theorem in an example.

Example 8.8.3   Establishing equality of a function and its Taylor series
Show that $f(x) = \cos x$ is equal to its Maclaurin series, as found in Example 8.8.1, for all x.

SOLUTION   Given a value x, the magnitude of the error term $R_n(x)$ is bounded by

$$\left|R_n(x)\right| \leq \frac{\max \left|f^{(n+1)}(z)\right|}{(n+1)!}\left|x^{n+1}\right|.$$

Since all derivatives of $\cos x$ are $\pm \sin x$ or $\pm \cos x$, whose magnitudes are bounded by 1, we can state

$$\left|R_n(x)\right| \leq \frac{1}{(n+1)!}\left|x^{n+1}\right|$$

which implies

$$-\frac{\left|x^{n+1}\right|}{(n+1)!} \leq R_n(x) \leq \frac{\left|x^{n+1}\right|}{(n+1)!}. \tag{8.7}$$

For any x, $\lim\limits_{n\to\infty} \dfrac{x^{n+1}}{(n+1)!} = 0$. Applying the Squeeze Theorem to Equation (8.7), we conclude that $\lim\limits_{n\to\infty} R_n(x) = 0$ for all x, and hence

$$\cos x = \sum_{n=0}^{\infty} (-1)^n \frac{x^{2n}}{(2n)!} \quad \text{for all } x.$$

It is natural to assume that a function is equal to its Taylor series on the series' interval of convergence, but this is not always the case. In order to properly establish equality, one must use Theorem 8.8.1. This is a bit disappointing, as we developed beautiful techniques for determining the interval of convergence of a power series, and proving that $R_n(x) \to 0$ can be difficult. For instance, it is not a simple task to show that $\ln x$ equals its Taylor series on $(0, 2]$ as found in Example 8.8.2; in the Exercises, the reader is only asked to show equality on $(1, 2)$, which is simpler.

There is good news. A function $f(x)$ that is equal to its Taylor series, centered at any point the domain of $f(x)$, is said to be an **analytic function**, and most, if not all, functions that we encounter within this course are analytic functions. Generally speaking, any function that one creates with elementary functions (polynomials, exponentials, trigonometric functions, etc.) that is not piecewise defined is probably analytic. For most functions, we assume the function is equal to its Taylor series on the series' interval of convergence and only use Theorem 8.8.1 when we suspect something may not work as expected.

Notes:

We develop the Taylor series for one more important function, then give a table of the Taylor series for a number of common functions.

Example 8.8.4   The Binomial Series

Find the Maclaurin series of $f(x) = (1 + x)^k$, $k \neq 0$.

SOLUTION   When k is a positive integer, the Maclaurin series is finite. For instance, when $k = 4$, we have

$$f(x) = (1 + x)^4 = 1 + 4x + 6x^2 + 4x^3 + x^4.$$

The coefficients of x when k is a positive integer are known as the *binomial coefficients*, giving the series we are developing its name.

When $k = 1/2$, we have $f(x) = \sqrt{1 + x}$. Knowing a series representation of this function would give a useful way of approximating $\sqrt{1.3}$, for instance.

To develop the Maclaurin series for $f(x) = (1 + x)^k$ for any value of $k \neq 0$, we consider the derivatives of f evaluated at $x = 0$:

$$f(x) = (1 + x)^k \qquad\qquad\qquad f(0) = 1$$
$$f'(x) = k(1 + x)^{k-1} \qquad\qquad\qquad f'(0) = k$$
$$f''(x) = k(k - 1)(1 + x)^{k-2} \qquad\qquad\qquad f''(0) = k(k - 1)$$
$$f'''(x) = k(k - 1)(k - 2)(1 + x)^{k-3} \qquad\qquad\qquad f'''(0) = k(k - 1)(k - 2)$$

$$\vdots \qquad\qquad\qquad\qquad\qquad\qquad \vdots$$

$$f^{(n)}(x) = k(k - 1)\cdots\big(k - (n - 1)\big)(1 + x)^{k-n} \qquad f^{(n)}(0) = k(k - 1)\cdots\big(k - (n - 1)\big)$$

Thus the Maclaurin series for $f(x) = (1 + x)^k$ is

$$1 + kx + \frac{k(k - 1)}{2!}x^2 + \frac{k(k - 1)(k - 2)}{3!}x^3 + \ldots + \frac{k(k - 1)\cdots\big(k - (n - 1)\big)}{n!}x^n + \ldots$$

It is important to determine the interval of convergence of this series. With

$$a_n = \frac{k(k - 1)\cdots\big(k - (n - 1)\big)}{n!}x^n,$$

we apply the Ratio Test:

$$\lim_{n\to\infty} \frac{|a_{n+1}|}{|a_n|} = \lim_{n\to\infty} \left|\frac{k(k - 1)\cdots(k - n)}{(n + 1)!}x^{n+1}\right| \Big/ \left|\frac{k(k - 1)\cdots\big(k - (n - 1)\big)}{n!}x^n\right|$$

$$= \lim_{n\to\infty} \left|\frac{k - n}{n + 1}x\right|$$

$$= |x|.$$

Notes:

The series converges absolutely when the limit of the Ratio Test is less than 1; therefore, we have absolute convergence when $|x| < 1$.

While outside the scope of this text, the interval of convergence depends on the value of k. When $k > 0$, the interval of convergence is $[-1, 1]$. When $-1 < k < 0$, the interval of convergence is $[-1, 1)$. If $k \leq -1$, the interval of convergence is $(-1, 1)$.

We learned that Taylor polynomials offer a way of approximating a "difficult to compute" function with a polynomial. Taylor series offer a way of exactly representing a function with a series. One probably can see the use of a good approximation; is there any use of representing a function exactly as a series?

While we should not overlook the mathematical beauty of Taylor series (which is reason enough to study them), there are practical uses as well. They provide a valuable tool for solving a variety of problems, including problems relating to integration and differential equations.

In Key Idea 8.8.1 (on the following page) we give a table of the Taylor series of a number of common functions. We then give a theorem about the "algebra of power series," that is, how we can combine power series to create power series of new functions. This allows us to find the Taylor series of functions like $f(x) = e^x \cos x$ by knowing the Taylor series of e^x and $\cos x$.

Before we investigate combining functions, consider the Taylor series for the arctangent function (see Key Idea 8.8.1). Knowing that $\tan^{-1}(1) = \pi/4$, we can use this series to approximate the value of π:

$$\frac{\pi}{4} = \tan^{-1}(1) = 1 - \frac{1}{3} + \frac{1}{5} - \frac{1}{7} + \frac{1}{9} - \cdots$$
$$\pi = 4\left(1 - \frac{1}{3} + \frac{1}{5} - \frac{1}{7} + \frac{1}{9} - \cdots\right)$$

Unfortunately, this particular expansion of π converges very slowly. The first 100 terms approximate π as 3.13159, which is not particularly good.

Notes:

Key Idea 8.8.1   Important Taylor Series Expansions

Function and Series	First Few Terms	Interval of Convergence
$e^x = \displaystyle\sum_{n=0}^{\infty} \frac{x^n}{n!}$	$1 + x + \dfrac{x^2}{2!} + \dfrac{x^3}{3!} + \cdots$	$(-\infty, \infty)$
$\sin x = \displaystyle\sum_{n=0}^{\infty} (-1)^n \frac{x^{2n+1}}{(2n+1)!}$	$x - \dfrac{x^3}{3!} + \dfrac{x^5}{5!} - \dfrac{x^7}{7!} + \cdots$	$(-\infty, \infty)$
$\cos x = \displaystyle\sum_{n=0}^{\infty} (-1)^n \frac{x^{2n}}{(2n)!}$	$1 - \dfrac{x^2}{2!} + \dfrac{x^4}{4!} - \dfrac{x^6}{6!} + \cdots$	$(-\infty, \infty)$
$\ln x = \displaystyle\sum_{n=1}^{\infty} (-1)^{n+1} \frac{(x-1)^n}{n}$	$(x-1) - \dfrac{(x-1)^2}{2} + \dfrac{(x-1)^3}{3} - \cdots$	$(0, 2]$
$\dfrac{1}{1-x} = \displaystyle\sum_{n=0}^{\infty} x^n$	$1 + x + x^2 + x^3 + \cdots$	$(-1, 1)$
$(1+x)^k = \displaystyle\sum_{n=0}^{\infty} \frac{k(k-1)\cdots(k-(n-1))}{n!} x^n$	$1 + kx + \dfrac{k(k-1)}{2!} x^2 + \cdots$	$(-1, 1)^a$
$\tan^{-1} x = \displaystyle\sum_{n=0}^{\infty} (-1)^n \frac{x^{2n+1}}{2n+1}$	$x - \dfrac{x^3}{3} + \dfrac{x^5}{5} - \dfrac{x^7}{7} + \cdots$	$[-1, 1]$

^{a}Convergence at $x = \pm 1$ depends on the value of k.

Theorem 8.8.2   Algebra of Power Series

Let $f(x) = \displaystyle\sum_{n=0}^{\infty} a_n x^n$ and $g(x) = \displaystyle\sum_{n=0}^{\infty} b_n x^n$ converge absolutely for $|x| < R$, and let $h(x)$ be continuous.

1. $f(x) \pm g(x) = \displaystyle\sum_{n=0}^{\infty} (a_n \pm b_n) x^n$ for $|x| < R$.

2. $f(x)g(x) = \left(\displaystyle\sum_{n=0}^{\infty} a_n x^n \right) \left(\displaystyle\sum_{n=0}^{\infty} b_n x^n \right) = \displaystyle\sum_{n=0}^{\infty} (a_0 b_n + a_1 b_{n-1} + \ldots a_n b_0) x^n$ for $|x| < R$.

3. $f(h(x)) = \displaystyle\sum_{n=0}^{\infty} a_n (h(x))^n$ for $|h(x)| < R$.

Notes:

Example 8.8.5   Combining Taylor series

Write out the first 3 terms of the Taylor Series for $f(x) = e^x \cos x$ using Key Idea 8.8.1 and Theorem 8.8.2.

SOLUTION   Key Idea 8.8.1 informs us that

$$e^x = 1 + x + \frac{x^2}{2!} + \frac{x^3}{3!} + \cdots \quad \text{and} \quad \cos x = 1 - \frac{x^2}{2!} + \frac{x^4}{4!} + \cdots .$$

Applying Theorem 8.8.2, we find that

$$e^x \cos x = \left(1 + x + \frac{x^2}{2!} + \frac{x^3}{3!} + \cdots\right)\left(1 - \frac{x^2}{2!} + \frac{x^4}{4!} + \cdots\right).$$

Distribute the right hand expression across the left:

$$= 1\left(1 - \frac{x^2}{2!} + \frac{x^4}{4!} + \cdots\right) + x\left(1 - \frac{x^2}{2!} + \frac{x^4}{4!} + \cdots\right) + \frac{x^2}{2!}\left(1 - \frac{x^2}{2!} + \frac{x^4}{4!} + \cdots\right)$$

$$+ \frac{x^3}{3!}\left(1 - \frac{x^2}{2!} + \frac{x^4}{4!} + \cdots\right) + \frac{x^4}{4!}\left(1 - \frac{x^2}{2!} + \frac{x^4}{4!} + \cdots\right) + \cdots$$

Distribute again and collect like terms.

$$= 1 + x - \frac{x^3}{3} - \frac{x^4}{6} - \frac{x^5}{30} + \frac{x^7}{630} + \cdots$$

While this process is a bit tedious, it is much faster than evaluating all the necessary derivatives of $e^x \cos x$ and computing the Taylor series directly.

Because the series for e^x and $\cos x$ both converge on $(-\infty, \infty)$, so does the series expansion for $e^x \cos x$.

Example 8.8.6   Creating new Taylor series

Use Theorem 8.8.2 to create series for $y = \sin(x^2)$ and $y = \ln(\sqrt{x})$.

SOLUTION   Given that

$$\sin x = \sum_{n=0}^{\infty} (-1)^n \frac{x^{2n+1}}{(2n+1)!} = x - \frac{x^3}{3!} + \frac{x^5}{5!} - \frac{x^7}{7!} + \cdots ,$$

we simply substitute x^2 for x in the series, giving

$$\sin(x^2) = \sum_{n=0}^{\infty} (-1)^n \frac{(x^2)^{2n+1}}{(2n+1)!} = x^2 - \frac{x^6}{3!} + \frac{x^{10}}{5!} - \frac{x^{14}}{7!} \cdots .$$

Notes:

Since the Taylor series for $\sin x$ has an infinite radius of convergence, so does the Taylor series for $\sin(x^2)$.

The Taylor expansion for $\ln x$ given in Key Idea 8.8.1 is centered at $x = 1$, so we will center the series for $\ln(\sqrt{x})$ at $x = 1$ as well. With

$$\ln x = \sum_{n=1}^{\infty} (-1)^{n+1} \frac{(x-1)^n}{n} = (x-1) - \frac{(x-1)^2}{2} + \frac{(x-1)^3}{3} - \cdots ,$$

we substitute $\sqrt{x}$ for x to obtain

$$\ln(\sqrt{x}) = \sum_{n=1}^{\infty} (-1)^{n+1} \frac{(\sqrt{x}-1)^n}{n} = (\sqrt{x}-1) - \frac{(\sqrt{x}-1)^2}{2} + \frac{(\sqrt{x}-1)^3}{3} - \cdots .$$

While this is not strictly a power series, it is a series that allows us to study the function $\ln(\sqrt{x})$. Since the interval of convergence of $\ln x$ is $(0, 2]$, and the range of $\sqrt{x}$ on $(0, 4]$ is $(0, 2]$, the interval of convergence of this series expansion of $\ln(\sqrt{x})$ is $(0, 4]$.

Note: In Example 8.8.6, one could create a series for $\ln(\sqrt{x})$ by simply recognizing that $\ln(\sqrt{x}) = \ln(x^{1/2}) = 1/2 \ln x$, and hence multiplying the Taylor series for $\ln x$ by $1/2$. This example was chosen to demonstrate other aspects of series, such as the fact that the interval of convergence changes.

Example 8.8.7 Using Taylor series to evaluate definite integrals

Use the Taylor series of e^{-x^2} to evaluate $\int_0^1 e^{-x^2} dx$.

SOLUTION   We learned, when studying Numerical Integration, that e^{-x^2} does not have an antiderivative expressible in terms of elementary functions. This means any definite integral of this function must have its value approximated, and not computed exactly.

We can quickly write out the Taylor series for e^{-x^2} using the Taylor series of e^x:

$$e^x = \sum_{n=0}^{\infty} \frac{x^n}{n!} = 1 + x + \frac{x^2}{2!} + \frac{x^3}{3!} + \cdots$$

and so

$$e^{-x^2} = \sum_{n=0}^{\infty} \frac{(-x^2)^n}{n!}$$
$$= \sum_{n=0}^{\infty} (-1)^n \frac{x^{2n}}{n!}$$
$$= 1 - x^2 + \frac{x^4}{2!} - \frac{x^6}{3!} + \cdots .$$

Notes:

We use Theorem 8.6.3 to integrate:

$$\int e^{-x^2}\, dx = C + x - \frac{x^3}{3} + \frac{x^5}{5 \cdot 2!} - \frac{x^7}{7 \cdot 3!} + \cdots + (-1)^n \frac{x^{2n+1}}{(2n+1)n!} + \cdots$$

This *is* the antiderivative of e^{-x^2}; while we can write it out as a series, we cannot write it out in terms of elementary functions. We can evaluate the definite integral $\int_0^1 e^{-x^2}\, dx$ using this antiderivative; substituting 1 and 0 for x and subtracting gives

$$\int_0^1 e^{-x^2}\, dx = 1 - \frac{1}{3} + \frac{1}{5 \cdot 2!} - \frac{1}{7 \cdot 3!} + \frac{1}{9 \cdot 4!} \cdots .$$

Summing the 5 terms shown above give the approximation of 0.74749. Since this is an alternating series, we can use the Alternating Series Approximation Theorem, (Theorem 8.5.2), to determine how accurate this approximation is. The next term of the series is $1/(11 \cdot 5!) \approx 0.00075758$. Thus we know our approximation is within 0.00075758 of the actual value of the integral. This is arguably much less work than using Simpson's Rule to approximate the value of the integral.

Example 8.8.8 Using Taylor series to solve differential equations
Solve the differential equation $y' = 2y$ in terms of a power series, and use the theory of Taylor series to recognize the solution in terms of an elementary function.

SOLUTION We found the first 5 terms of the power series solution to this differential equation in Example 8.6.5 in Section 8.6. These are:

$$a_0 = 1, \quad a_1 = 2, \quad a_2 = \frac{4}{2} = 2, \quad a_3 = \frac{8}{2 \cdot 3} = \frac{4}{3}, \quad a_4 = \frac{16}{2 \cdot 3 \cdot 4} = \frac{2}{3}.$$

We include the "unsimplified" expressions for the coefficients found in Example 8.6.5 as we are looking for a pattern. It can be shown that $a_n = 2^n/n!$. Thus the solution, written as a power series, is

$$y = \sum_{n=0}^{\infty} \frac{2^n}{n!} x^n = \sum_{n=0}^{\infty} \frac{(2x)^n}{n!}.$$

Using Key Idea 8.8.1 and Theorem 8.8.2, we recognize $f(x) = e^{2x}$:

$$e^x = \sum_{n=0}^{\infty} \frac{x^n}{n!} \quad \Rightarrow \quad e^{2x} = \sum_{n=0}^{\infty} \frac{(2x)^n}{n!}.$$

Notes:

Finding a pattern in the coefficients that match the series expansion of a known function, such as those shown in Key Idea 8.8.1, can be difficult. What if the coefficients in the previous example were given in their reduced form; how could we still recover the function $y = e^{2x}$?

Suppose that all we know is that

$$a_0 = 1, \quad a_1 = 2, \quad a_2 = 2, \quad a_3 = \frac{4}{3}, \quad a_4 = \frac{2}{3}.$$

Definition 8.8.1 states that each term of the Taylor expansion of a function includes an $n!$. This allows us to say that

$$a_2 = 2 = \frac{b_2}{2!}, \quad a_3 = \frac{4}{3} = \frac{b_3}{3!}, \quad \text{and} \quad a_4 = \frac{2}{3} = \frac{b_4}{4!}$$

for some values b_2, b_3 and b_4. Solving for these values, we see that $b_2 = 4$, $b_3 = 8$ and $b_4 = 16$. That is, we are recovering the pattern we had previously seen, allowing us to write

$$f(x) = \sum_{n=0}^{\infty} a_n x^n = \sum_{n=0}^{\infty} \frac{b_n}{n!} x^n$$
$$= 1 + 2x + \frac{4}{2!}x^2 + \frac{8}{3!}x^3 + \frac{16}{4!}x^4 + \cdots$$

From here it is easier to recognize that the series is describing an exponential function.

There are simpler, more direct ways of solving the differential equation $y' = 2y$. We applied power series techniques to this equation to demonstrate its utility, and went on to show how *sometimes* we are able to recover the solution in terms of elementary functions using the theory of Taylor series. Most differential equations faced in real scientific and engineering situations are much more complicated than this one, but power series can offer a valuable tool in finding, or at least approximating, the solution.

This chapter introduced sequences, which are ordered lists of numbers, followed by series, wherein we add up the terms of a sequence. We quickly saw that such sums do not always add up to "infinity," but rather converge. We studied tests for convergence, then ended the chapter with a formal way of defining functions based on series. Such "series–defined functions" are a valuable tool in solving a number of different problems throughout science and engineering.

Coming in the next chapters are new ways of defining curves in the plane apart from using functions of the form $y = f(x)$. Curves created by these new methods can be beautiful, useful, and important.

Notes:

Exercises 8.8

Terms and Concepts

1. What is the difference between a Taylor polynomial and a Taylor series?

2. What theorem must we use to show that a function is equal to its Taylor series?

Problems

Key Idea 8.8.1 gives the nth term of the Taylor series of common functions. In Exercises 3 – 6, verify the formula given in the Key Idea by finding the first few terms of the Taylor series of the given function and identifying a pattern.

3. $f(x) = e^x$; $c = 0$

4. $f(x) = \sin x$; $c = 0$

5. $f(x) = 1/(1 - x)$; $c = 0$

6. $f(x) = \tan^{-1} x$; $c = 0$

In Exercises 7 – 12, find a formula for the nth term of the Taylor series of $f(x)$, centered at c, by finding the coefficients of the first few powers of x and looking for a pattern. (The formulas for several of these are found in Key Idea 8.8.1; show work verifying these formula.)

7. $f(x) = \cos x$; $c = \pi/2$

8. $f(x) = 1/x$; $c = 1$

9. $f(x) = e^{-x}$; $c = 0$

10. $f(x) = \ln(1 + x)$; $c = 0$

11. $f(x) = x/(x + 1)$; $c = 1$

12. $f(x) = \sin x$; $c = \pi/4$

In Exercises 13 – 16, show that the Taylor series for $f(x)$, as given in Key Idea 8.8.1, is equal to $f(x)$ by applying Theorem 8.8.1; that is, show $\lim\limits_{n \to \infty} R_n(x) = 0$.

13. $f(x) = e^x$

14. $f(x) = \sin x$

15. $f(x) = \ln x$ (show equality only on $(1, 2)$)

16. $f(x) = 1/(1 - x)$ (show equality only on $(-1, 0)$)

In Exercises 17 – 20, use the Taylor series given in Key Idea 8.8.1 to verify the given identity.

17. $\cos(-x) = \cos x$

18. $\sin(-x) = -\sin x$

19. $\frac{d}{dx}\left(\sin x\right) = \cos x$

20. $\frac{d}{dx}\left(\cos x\right) = -\sin x$

In Exercises 21 – 24, write out the first 5 terms of the Binomial series with the given k-value.

21. $k = 1/2$

22. $k = -1/2$

23. $k = 1/3$

24. $k = 4$

In Exercises 25 – 30, use the Taylor series given in Key Idea 8.8.1 to create the Taylor series of the given functions.

25. $f(x) = \cos\left(x^2\right)$

26. $f(x) = e^{-x}$

27. $f(x) = \sin\left(2x + 3\right)$

28. $f(x) = \tan^{-1}\left(x/2\right)$

29. $f(x) = e^x \sin x$ (only find the first 4 terms)

30. $f(x) = (1 + x)^{1/2} \cos x$ (only find the first 4 terms)

In Exercises 31 – 32, approximate the value of the given definite integral by using the first 4 nonzero terms of the integrand's Taylor series.

31. $\displaystyle\int_0^{\sqrt{\pi}} \sin\left(x^2\right) dx$

32. $\displaystyle\int_0^{\pi^2/4} \cos\left(\sqrt{x}\right) dx$

9: CURVES IN THE PLANE

We have explored functions of the form $y = f(x)$ closely throughout this text. We have explored their limits, their derivatives and their antiderivatives; we have learned to identify key features of their graphs, such as relative maxima and minima, inflection points and asymptotes; we have found equations of their tangent lines, the areas between portions of their graphs and the x-axis, and the volumes of solids generated by revolving portions of their graphs about a horizontal or vertical axis.

Despite all this, the graphs created by functions of the form $y = f(x)$ are limited. Since each x-value can correspond to only 1 y-value, common shapes like circles cannot be fully described by a function in this form. Fittingly, the "vertical line test" excludes vertical lines from being functions of x, even though these lines are important in mathematics.

In this chapter we'll explore new ways of drawing curves in the plane. We'll still work within the framework of functions, as an input will still only correspond to one output. However, our new techniques of drawing curves will render the vertical line test pointless, and allow us to create important – and beautiful – new curves. Once these curves are defined, we'll apply the concepts of calculus to them, continuing to find equations of tangent lines and the areas of enclosed regions.

9.1 Conic Sections

The ancient Greeks recognized that interesting shapes can be formed by intersecting a plane with a *double napped* cone (i.e., two identical cones placed tip–to–tip as shown in the following figures). As these shapes are formed as sections of conics, they have earned the official name "conic sections."

The three "most interesting" conic sections are given in the top row of Figure 9.1.1. They are the parabola, the ellipse (which includes circles) and the hyperbola. In each of these cases, the plane does not intersect the tips of the cones (usually taken to be the origin).

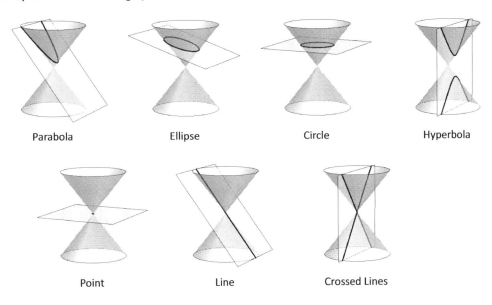

Figure 9.1.1: Conic Sections

When the plane does contain the origin, three **degenerate** cones can be formed as shown the bottom row of Figure 9.1.1: a point, a line, and crossed lines. We focus here on the nondegenerate cases.

While the above geometric constructs define the conics in an intuitive, visual way, these constructs are not very helpful when trying to analyze the shapes algebraically or consider them as the graph of a function. It can be shown that all conics can be defined by the general second–degree equation

$$Ax^2 + Bxy + Cy^2 + Dx + Ey + F = 0.$$

While this algebraic definition has its uses, most find another geometric perspective of the conics more beneficial.

Each nondegenerate conic can be defined as the **locus**, or set, of points that satisfy a certain distance property. These distance properties can be used to generate an algebraic formula, allowing us to study each conic as the graph of a function.

Parabolas

> **Definition 9.1.1   Parabola**
>
> A **parabola** is the locus of all points equidistant from a point (called a **focus**) and a line (called the **directrix**) that does not contain the focus.

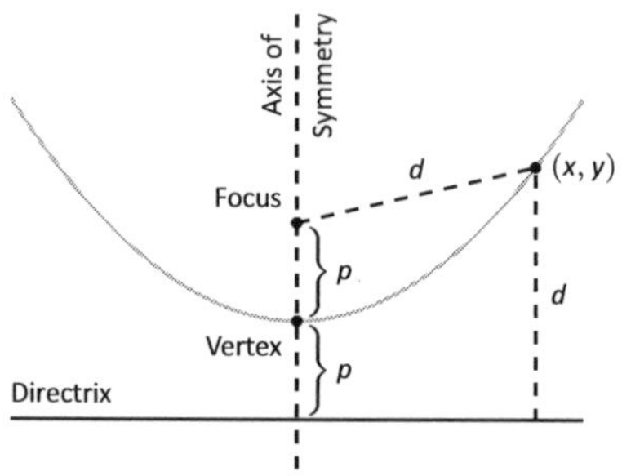

Figure 9.1.2: Illustrating the definition of the parabola and establishing an algebraic formula.

Figure 9.1.2 illustrates this definition. The point halfway between the focus and the directrix is the **vertex**. The line through the focus, perpendicular to the directrix, is the **axis of symmetry**, as the portion of the parabola on one side of this line is the mirror–image of the portion on the opposite side.

The definition leads us to an algebraic formula for the parabola. Let $P = (x, y)$ be a point on a parabola whose focus is at $F = (0, p)$ and whose directrix is at $y = -p$. (We'll assume for now that the focus lies on the y-axis; by placing the focus p units above the x-axis and the directrix p units below this axis, the vertex will be at $(0, 0)$.)

We use the Distance Formula to find the distance d_1 between F and P:

$$d_1 = \sqrt{(x - 0)^2 + (y - p)^2}.$$

The distance d_2 from P to the directrix is more straightforward:

$$d_2 = y - (-p) = y + p.$$

Notes:

These two distances are equal. Setting $d_1 = d_2$, we can solve for y in terms of x:

$$d_1 = d_2$$
$$\sqrt{x^2 + (y - p)^2} = y + p$$

Now square both sides.

$$x^2 + (y - p)^2 = (y + p)^2$$
$$x^2 + y^2 - 2yp + p^2 = y^2 + 2yp + p^2$$
$$x^2 = 4yp$$
$$y = \frac{1}{4p}x^2.$$

The geometric definition of the parabola has led us to the familiar quadratic function whose graph is a parabola with vertex at the origin. When we allow the vertex to not be at $(0, 0)$, we get the following standard form of the parabola.

Key Idea 9.1.1   General Equation of a Parabola

1. **Vertical Axis of Symmetry:** The equation of the parabola with vertex at (h, k) and directrix $y = k - p$ in standard form is

$$y = \frac{1}{4p}(x - h)^2 + k.$$

The focus is at $(h, k + p)$.

2. **Horizontal Axis of Symmetry:** The equation of the parabola with vertex at (h, k) and directrix $x = h - p$ in standard form is

$$x = \frac{1}{4p}(y - k)^2 + h.$$

The focus is at $(h + p, k)$.

Note: p is not necessarily a positive number.

Example 9.1.1   Finding the equation of a parabola

Give the equation of the parabola with focus at $(1, 2)$ and directrix at $y = 3$.

SOLUTION   The vertex is located halfway between the focus and directrix, so $(h, k) = (1, 2.5)$. This gives $p = -0.5$. Using Key Idea 9.1.1 we have the

---

Notes:

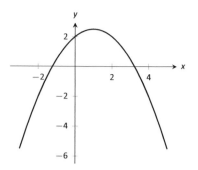

Figure 9.1.3: The parabola described in Example 9.1.1.

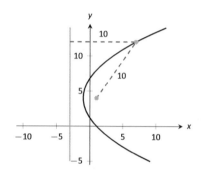

Figure 9.1.4: The parabola described in Example 9.1.2. The distances from a point on the parabola to the focus and directrix is given.

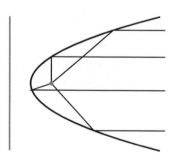

Figure 9.1.5: Illustrating the parabola's reflective property.

equation of the parabola as

$$y = \frac{1}{4(-0.5)}(x-1)^2 + 2.5 = -\frac{1}{2}(x-1)^2 + 2.5.$$

The parabola is sketched in Figure 9.1.3.

Example 9.1.2   Finding the focus and directrix of a parabola
Find the focus and directrix of the parabola $x = \frac{1}{8}y^2 - y + 1$. The point $(7, 12)$ lies on the graph of this parabola; verify that it is equidistant from the focus and directrix.

SOLUTION   We need to put the equation of the parabola in its general form. This requires us to complete the square:

$$\begin{aligned}
x &= \frac{1}{8}y^2 - y + 1 \\
&= \frac{1}{8}(y^2 - 8y + 8) \\
&= \frac{1}{8}(y^2 - 8y + 16 - 16 + 8) \\
&= \frac{1}{8}((y-4)^2 - 8) \\
&= \frac{1}{8}(y-4)^2 - 1.
\end{aligned}$$

Hence the vertex is located at $(-1, 4)$. We have $\frac{1}{8} = \frac{1}{4p}$, so $p = 2$. We conclude that the focus is located at $(1, 4)$ and the directrix is $x = -3$. The parabola is graphed in Figure 9.1.4, along with its focus and directrix.

The point $(7, 12)$ lies on the graph and is $7 - (-3) = 10$ units from the directrix. The distance from $(7, 12)$ to the focus is:

$$\sqrt{(7-1)^2 + (12-4)^2} = \sqrt{100} = 10.$$

Indeed, the point on the parabola is equidistant from the focus and directrix.

Reflective Property

One of the fascinating things about the nondegenerate conic sections is their reflective properties. Parabolas have the following reflective property:

> Any ray emanating from the focus that intersects the parabola reflects off along a line perpendicular to the directrix.

This is illustrated in Figure 9.1.5. The following theorem states this more rigorously.

Notes:

Theorem 9.1.1 **Reflective Property of the Parabola**

Let P be a point on a parabola. The tangent line to the parabola at P makes equal angles with the following two lines:

1. The line containing P and the focus F, and

2. The line perpendicular to the directrix through P.

Because of this reflective property, paraboloids (the 3D analogue of parabolas) make for useful flashlight reflectors as the light from the bulb, ideally located at the focus, is reflected along parallel rays. Satellite dishes also have paraboloid shapes. Signals coming from satellites effectively approach the dish along parallel rays. The dish then *focuses* these rays at the focus, where the sensor is located.

Ellipses

Definition 9.1.2 **Ellipse**

An **ellipse** is the locus of all points whose sum of distances from two fixed points, each a **focus** of the ellipse, is constant.

An easy way to visualize this construction of an ellipse is to pin both ends of a string to a board. The pins become the foci. Holding a pencil tight against the string places the pencil on the ellipse; the sum of distances from the pencil to the pins is constant: the length of the string. See Figure 9.1.6.

We can again find an algebraic equation for an ellipse using this geometric definition. Let the foci be located along the x-axis, c units from the origin. Let these foci be labeled as $F_1 = (-c, 0)$ and $F_2 = (c, 0)$. Let $P = (x, y)$ be a point on the ellipse. The sum of distances from F_1 to P (d_1) and from F_2 to P (d_2) is a constant d. That is, $d_1 + d_2 = d$. Using the Distance Formula, we have

$$\sqrt{(x + c)^2 + y^2} + \sqrt{(x - c)^2 + y^2} = d.$$

Using a fair amount of algebra can produce the following equation of an ellipse (note that the equation is an implicitly defined function; it has to be, as an ellipse fails the Vertical Line Test):

$$\frac{x^2}{\left(\frac{d}{2}\right)^2} + \frac{y^2}{\left(\frac{d}{2}\right)^2 - c^2} = 1.$$

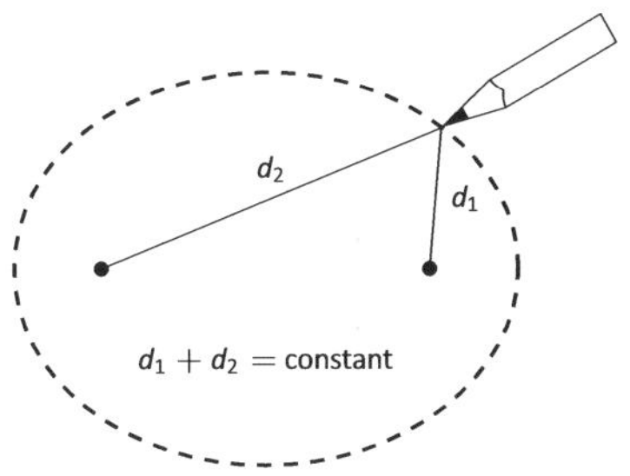

Figure 9.1.6: Illustrating the construction of an ellipse with pins, pencil and string.

Notes:

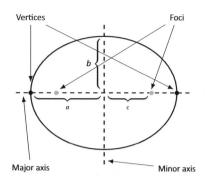

Figure 9.1.7: Labeling the significant features of an ellipse.

This is not particularly illuminating, but by making the substitution $a = d/2$ and $b = \sqrt{a^2 - c^2}$, we can rewrite the above equation as

$$\frac{x^2}{a^2} + \frac{y^2}{b^2} = 1.$$

This choice of a and b is not without reason; as shown in Figure 9.1.7, the values of a and b have geometric meaning in the graph of the ellipse.

In general, the two foci of an ellipse lie on the **major axis** of the ellipse, and the midpoint of the segment joining the two foci is the **center**. The major axis intersects the ellipse at two points, each of which is a **vertex**. The line segment through the center and perpendicular to the major axis is the **minor axis**. The "constant sum of distances" that defines the ellipse is the length of the major axis, i.e., $2a$.

Allowing for the shifting of the ellipse gives the following standard equations.

Key Idea 9.1.2   Standard Equation of the Ellipse

The equation of an ellipse centered at (h, k) with major axis of length $2a$ and minor axis of length $2b$ in standard form is:

1. **Horizontal major axis:** $\dfrac{(x - h)^2}{a^2} + \dfrac{(y - k)^2}{b^2} = 1$.

2. **Vertical major axis:** $\dfrac{(x - h)^2}{b^2} + \dfrac{(y - k)^2}{a^2} = 1$.

The foci lie along the major axis, c units from the center, where $c^2 = a^2 - b^2$.

Example 9.1.3   Finding the equation of an ellipse
Find the general equation of the ellipse graphed in Figure 9.1.8.

SOLUTION   The center is located at $(-3, 1)$. The distance from the center to a vertex is 5 units, hence $a = 5$. The minor axis seems to have length 4, so $b = 2$. Thus the equation of the ellipse is

$$\frac{(x + 3)^2}{4} + \frac{(y - 1)^2}{25} = 1.$$

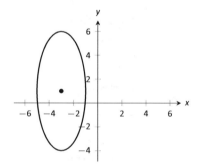

Figure 9.1.8: The ellipse used in Example 9.1.3.

Example 9.1.4   Graphing an ellipse
Graph the ellipse defined by $4x^2 + 9y^2 - 8x - 36y = -4$.

Notes:

SOLUTION It is simple to graph an ellipse once it is in standard form. In order to put the given equation in standard form, we must complete the square with both the x and y terms. We first rewrite the equation by regrouping:

$$4x^2 + 9y^2 - 8x - 36y = -4 \quad \Rightarrow \quad (4x^2 - 8x) + (9y^2 - 36y) = -4.$$

Now we complete the squares.

$$(4x^2 - 8x) + (9y^2 - 36y) = -4$$
$$4(x^2 - 2x) + 9(y^2 - 4y) = -4$$
$$4(x^2 - 2x + 1 - 1) + 9(y^2 - 4y + 4 - 4) = -4$$
$$4((x - 1)^2 - 1) + 9((y - 2)^2 - 4) = -4$$
$$4(x - 1)^2 - 4 + 9(y - 2)^2 - 36 = -4$$
$$4(x - 1)^2 + 9(y - 2)^2 = 36$$
$$\frac{(x - 1)^2}{9} + \frac{(y - 2)^2}{4} = 1.$$

We see the center of the ellipse is at $(1, 2)$. We have $a = 3$ and $b = 2$; the major axis is horizontal, so the vertices are located at $(-2, 2)$ and $(4, 2)$. We find $c = \sqrt{9 - 4} = \sqrt{5} \approx 2.24$. The foci are located along the major axis, approximately 2.24 units from the center, at $(1 \pm 2.24, 2)$. This is all graphed in Figure 9.1.9 .

Eccentricity

When $a = b$, we have a circle. The general equation becomes

$$\frac{(x - h)^2}{a^2} + \frac{(y - k)^2}{a^2} = 1 \quad \Rightarrow (x - h)^2 + (y - k)^2 = a^2,$$

the familiar equation of the circle centered at (h, k) with radius a. Since $a = b$, $c = \sqrt{a^2 - b^2} = 0$. The circle has "two" foci, but they lie on the same point, the center of the circle.

Consider Figure 9.1.10, where several ellipses are graphed with $a = 1$. In (a), we have $c = 0$ and the ellipse is a circle. As c grows, the resulting ellipses look less and less circular. A measure of this "noncircularness" is *eccentricity*.

Definition 9.1.3   **Eccentricity of an Ellipse**

The eccentricity e of an ellipse is $e = \dfrac{c}{a}$.

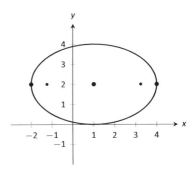

Figure 9.1.9: Graphing the ellipse in Example 9.1.4.

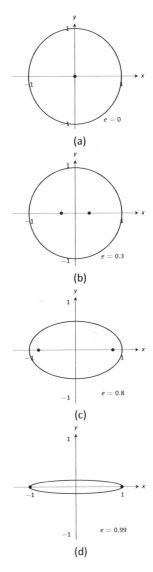

(a) $e = 0$

(b) $e = 0.3$

(c) $e = 0.8$

(d) $e = 0.99$

Figure 9.1.10: Understanding the eccentricity of an ellipse.

Notes:

The eccentricity of a circle is 0; that is, a circle has no "noncircularness." As c approaches a, e approaches 1, giving rise to a very noncircular ellipse, as seen in Figure 9.1.10 (d).

It was long assumed that planets had circular orbits. This is known to be incorrect; the orbits are elliptical. Earth has an eccentricity of 0.0167 – it has a nearly circular orbit. Mercury's orbit is the most eccentric, with $e = 0.2056$. (Pluto's eccentricity is greater, at $e = 0.248$, the greatest of all the currently known dwarf planets.) The planet with the most circular orbit is Venus, with $e = 0.0068$. The Earth's moon has an eccentricity of $e = 0.0549$, also very circular.

Reflective Property

The ellipse also possesses an interesting reflective property. Any ray emanating from one focus of an ellipse reflects off the ellipse along a line through the other focus, as illustrated in Figure 9.1.11. This property is given formally in the following theorem.

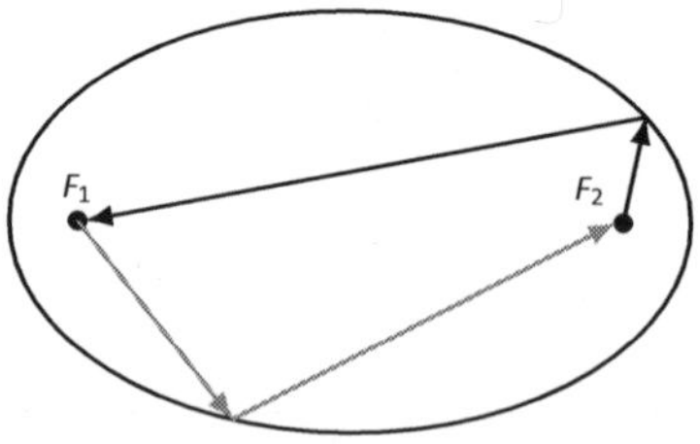

Figure 9.1.11: Illustrating the reflective property of an ellipse.

Theorem 9.1.2   Reflective Property of an Ellipse

Let P be a point on a ellipse with foci F_1 and F_2. The tangent line to the ellipse at P makes equal angles with the following two lines:

1. The line through F_1 and P, and

2. The line through F_2 and P.

This reflective property is useful in optics and is the basis of the phenomena experienced in whispering halls.

Hyperbolas

The definition of a hyperbola is very similar to the definition of an ellipse; we essentially just change the word "sum" to "difference."

Definition 9.1.4   Hyperbola

A **hyperbola** is the locus of all points where the absolute value of difference of distances from two fixed points, each a focus of the hyperbola, is constant.

Notes:

We do not have a convenient way of visualizing the construction of a hyperbola as we did for the ellipse. The geometric definition does allow us to find an algebraic expression that describes it. It will be useful to define some terms first.

The two foci lie on the **transverse axis** of the hyperbola; the midpoint of the line segment joining the foci is the **center** of the hyperbola. The transverse axis intersects the hyperbola at two points, each a **vertex** of the hyperbola. The line through the center and perpendicular to the transverse axis is the **conjugate axis.** This is illustrated in Figure 9.1.12. It is easy to show that the constant difference of distances used in the definition of the hyperbola is the distance between the vertices, i.e., $2a$.

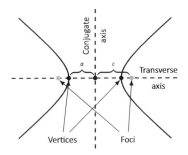

Figure 9.1.12: Labeling the significant features of a hyperbola.

Key Idea 9.1.3   Standard Equation of a Hyperbola

The equation of a hyperbola centered at (h, k) in standard form is:

1. **Horizontal Transverse Axis:** $\dfrac{(x-h)^2}{a^2} - \dfrac{(y-k)^2}{b^2} = 1.$

2. **Vertical Transverse Axis:** $\dfrac{(y-k)^2}{a^2} - \dfrac{(x-h)^2}{b^2} = 1.$

The vertices are located a units from the center and the foci are located c units from the center, where $c^2 = a^2 + b^2$.

Graphing Hyperbolas

Consider the hyperbola $\frac{x^2}{9} - \frac{y^2}{1} = 1$. Solving for y, we find $y = \pm\sqrt{x^2/9 - 1}$. As x grows large, the "-1" part of the equation for y becomes less significant and $y \approx \pm\sqrt{x^2/9} = \pm x/3$. That is, as x gets large, the graph of the hyperbola looks very much like the lines $y = \pm x/3$. These lines are asymptotes of the hyperbola, as shown in Figure 9.1.13.

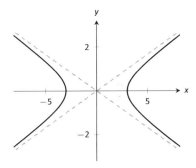

Figure 9.1.13: Graphing the hyperbola $\frac{x^2}{9} - \frac{y^2}{1} = 1$ along with its asymptotes, $y = \pm x/3$.

This is a valuable tool in sketching. Given the equation of a hyperbola in general form, draw a rectangle centered at (h, k) with sides of length $2a$ parallel to the transverse axis and sides of length $2b$ parallel to the conjugate axis. (See Figure 9.1.14 for an example with a horizontal transverse axis.) The diagonals of the rectangle lie on the asymptotes.

These lines pass through (h, k). When the transverse axis is horizontal, the slopes are $\pm b/a$; when the transverse axis is vertical, their slopes are $\pm a/b$. This gives equations:

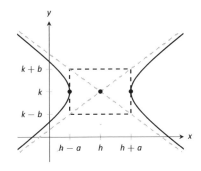

Figure 9.1.14: Using the asymptotes of a hyperbola as a graphing aid.

Notes:

	Horizontal Transverse Axis	Vertical Transverse Axis

$$y = \pm\frac{b}{a}(x - h) + k \qquad y = \pm\frac{a}{b}(x - h) + k.$$

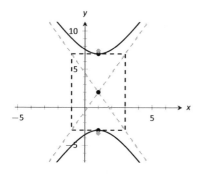

Figure 9.1.15: Graphing the hyperbola in Example 9.1.5.

Example 9.1.5   Graphing a hyperbola

Sketch the hyperbola given by $\dfrac{(y - 2)^2}{25} - \dfrac{(x - 1)^2}{4} = 1$.

SOLUTION   The hyperbola is centered at $(1, 2)$; $a = 5$ and $b = 2$. In Figure 9.1.15 we draw the prescribed rectangle centered at $(1, 2)$ along with the asymptotes defined by its diagonals. The hyperbola has a vertical transverse axis, so the vertices are located at $(1, 7)$ and $(1, -3)$. This is enough to make a good sketch.

We also find the location of the foci: as $c^2 = a^2 + b^2$, we have $c = \sqrt{29} \approx 5.4$. Thus the foci are located at $(1, 2 \pm 5.4)$ as shown in the figure.

Example 9.1.6   Graphing a hyperbola

Sketch the hyperbola given by $9x^2 - y^2 + 2y = 10$.

SOLUTION   We must complete the square to put the equation in general form. (We recognize this as a hyperbola since it is a general quadratic equation and the x^2 and y^2 terms have opposite signs.)

$$9x^2 - y^2 + 2y = 10$$
$$9x^2 - (y^2 - 2y) = 10$$
$$9x^2 - (y^2 - 2y + 1 - 1) = 10$$
$$9x^2 - ((y - 1)^2 - 1) = 10$$
$$9x^2 - (y - 1)^2 = 9$$
$$x^2 - \frac{(y - 1)^2}{9} = 1$$

We see the hyperbola is centered at $(0, 1)$, with a horizontal transverse axis, where $a = 1$ and $b = 3$. The appropriate rectangle is sketched in Figure 9.1.16 along with the asymptotes of the hyperbola. The vertices are located at $(\pm 1, 1)$. We have $c = \sqrt{10} \approx 3.2$, so the foci are located at $(\pm 3.2, 1)$ as shown in the figure.

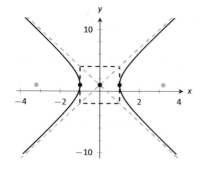

Figure 9.1.16: Graphing the hyperbola in Example 9.1.6.

Notes:

Eccentricity

Definition 9.1.5   Eccentricity of a Hyperbola

The eccentricity of a hyperbola is $e = \dfrac{c}{a}$.

Note that this is the definition of eccentricity as used for the ellipse. When c is close in value to a (i.e., $e \approx 1$), the hyperbola is very narrow (looking almost like crossed lines). Figure 9.1.17 shows hyperbolas centered at the origin with $a = 1$. The graph in (a) has $c = 1.05$, giving an eccentricity of $e = 1.05$, which is close to 1. As c grows larger, the hyperbola widens and begins to look like parallel lines, as shown in part (d) of the figure.

Reflective Property

Hyperbolas share a similar reflective property with ellipses. However, in the case of a hyperbola, a ray emanating from a focus that intersects the hyperbola reflects along a line containing the other focus, but moving *away* from that focus. This is illustrated in Figure 9.1.19 (on the next page). Hyperbolic mirrors are commonly used in telescopes because of this reflective property. It is stated formally in the following theorem.

Theorem 9.1.3   Reflective Property of Hyperbolas

Let P be a point on a hyperbola with foci F_1 and F_2. The tangent line to the hyperbola at P makes equal angles with the following two lines:

1. The line through F_1 and P, and

2. The line through F_2 and P.

Location Determination

Determining the location of a known event has many practical uses (locating the epicenter of an earthquake, an airplane crash site, the position of the person speaking in a large room, etc.).

To determine the location of an earthquake's epicenter, seismologists use *trilateration* (not to be confused with *triangulation*). A seismograph allows one

---

Notes:

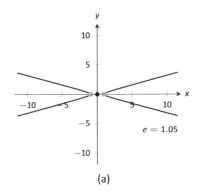

(a)

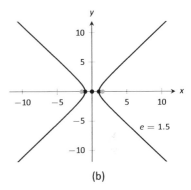

(b)

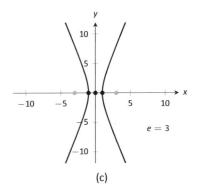

(c)

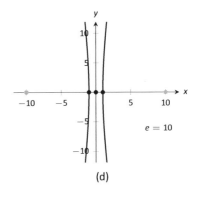

(d)

Figure 9.1.17: Understanding the eccentricity of a hyperbola.

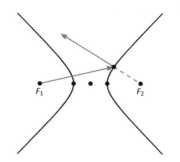

Figure 9.1.19: Illustrating the reflective property of a hyperbola.

to determine how far away the epicenter was; using three separate readings, the location of the epicenter can be approximated.

A key to this method is knowing distances. What if this information is not available? Consider three microphones at positions *A*, *B* and *C* which all record a noise (a person's voice, an explosion, etc.) created at unknown location *D*. The microphone does not "know" when the sound was *created*, only when the sound was *detected*. How can the location be determined in such a situation?

If each location has a clock set to the same time, hyperbolas can be used to determine the location. Suppose the microphone at position *A* records the sound at exactly 12:00, location *B* records the time exactly 1 second later, and location *C* records the noise exactly 2 seconds after that. We are interested in the *difference* of times. Since the speed of sound is approximately 340 m/s, we can conclude quickly that the sound was created 340 meters closer to position *A* than position *B*. If *A* and *B* are a known distance apart (as shown in Figure 9.1.18 (a)), then we can determine a hyperbola on which *D* must lie.

The "difference of distances" is 340; this is also the distance between vertices of the hyperbola. So we know $2a = 340$. Positions *A* and *B* lie on the foci, so $2c = 1000$. From this we can find $b \approx 470$ and can sketch the hyperbola, given in part (b) of the figure. We only care about the side closest to *A*. (Why?)

We can also find the hyperbola defined by positions *B* and *C*. In this case, $2a = 680$ as the sound traveled an extra 2 seconds to get to *C*. We still have $2c = 1000$, centering this hyperbola at $(-500, 500)$. We find $b \approx 367$. This hyperbola is sketched in part (c) of the figure. The intersection point of the two graphs is the location of the sound, at approximately $(188, -222.5)$.

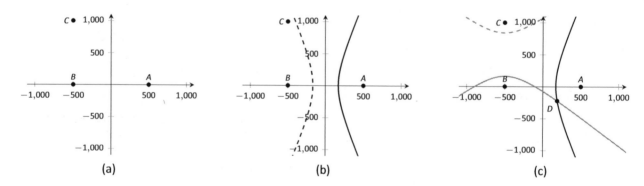

Figure 9.1.18: Using hyperbolas in location detection.

This chapter explores curves in the plane, in particular curves that cannot be described by functions of the form $y = f(x)$. In this section, we learned of ellipses and hyperbolas that are defined implicitly, not explicitly. In the following sections, we will learn completely new ways of describing curves in the plane, using *parametric equations* and *polar coordinates*, then study these curves using calculus techniques.

Notes:

Exercises 9.1

Terms and Concepts

1. What is the difference between degenerate and nondegenerate conics?

2. Use your own words to explain what the eccentricity of an ellipse measures.

3. What has the largest eccentricity: an ellipse or a hyperbola?

4. Explain why the following is true: "If the coefficient of the x^2 term in the equation of an ellipse in standard form is smaller than the coefficient of the y^2 term, then the ellipse has a horizontal major axis."

5. Explain how one can quickly look at the equation of a hyperbola in standard form and determine whether the transverse axis is horizontal or vertical.

6. Fill in the blank: It can be said that ellipses and hyperbolas share the *same* reflective property: "A ray emanating from one focus will reflect off the conic along a _____ that contains the other focus."

Problems

In Exercises 7 – 14, find the equation of the parabola defined by the given information. Sketch the parabola.

7. Focus: $(3, 2)$; directrix: $y = 1$

8. Focus: $(-1, -4)$; directrix: $y = 2$

9. Focus: $(1, 5)$; directrix: $x = 3$

10. Focus: $(1/4, 0)$; directrix: $x = -1/4$

11. Focus: $(1, 1)$; vertex: $(1, 2)$

12. Focus: $(-3, 0)$; vertex: $(0, 0)$

13. Vertex: $(0, 0)$; directrix: $y = -1/16$

14. Vertex: $(2, 3)$; directrix: $x = 4$

In Exercises 15 – 16, the equation of a parabola and a point on its graph are given. Find the focus and directrix of the parabola, and verify that the given point is equidistant from the focus and directrix.

15. $y = \frac{1}{4}x^2$, $P = (2, 1)$

16. $x = \frac{1}{8}(y - 2)^2 + 3$, $P = (11, 10)$

In Exercises 17 – 18, sketch the ellipse defined by the given equation. Label the center, foci and vertices.

17. $\dfrac{(x - 1)^2}{3} + \dfrac{(y - 2)^2}{5} = 1$

18. $\dfrac{1}{25}x^2 + \dfrac{1}{9}(y + 3)^2 = 1$

In Exercises 19 – 20, find the equation of the ellipse shown in the graph. Give the location of the foci and the eccentricity of the ellipse.

19.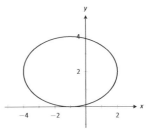

20.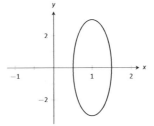

In Exercises 21 – 24, find the equation of the ellipse defined by the given information. Sketch the elllipse.

21. Foci: $(\pm 2, 0)$; vertices: $(\pm 3, 0)$

22. Foci: $(-1, 3)$ and $(5, 3)$; vertices: $(-3, 3)$ and $(7, 3)$

23. Foci: $(2, \pm 2)$; vertices: $(2, \pm 7)$

24. Focus: $(-1, 5)$; vertex: $(-1, -4)$; center: $(-1, 1)$

In Exercises 25 – 28, write the equation of the given ellipse in standard form.

25. $x^2 - 2x + 2y^2 - 8y = -7$

26. $5x^2 + 3y^2 = 15$

27. $3x^2 + 2y^2 - 12y + 6 = 0$

28. $x^2 + y^2 - 4x - 4y + 4 = 0$

In Exercises 29 – 32, find the equation of the hyperbola shown in the graph.

29.

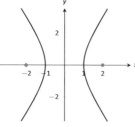

30.

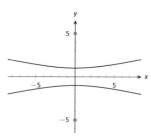

31.

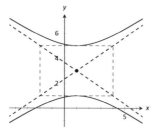

32.

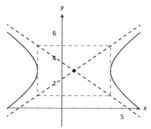

In Exercises 33 – 34, sketch the hyperbola defined by the given equation. Label the center and foci.

33. $\dfrac{(x-1)^2}{16} - \dfrac{(y+2)^2}{9} = 1$

34. $(y-4)^2 - \dfrac{(x+1)^2}{25} = 1$

In Exercises 35 – 38, find the equation of the hyperbola defined by the given information. Sketch the hyperbola.

35. Foci: $(\pm 3, 0)$; vertices: $(\pm 2, 0)$

36. Foci: $(0, \pm 3)$; vertices: $(0, \pm 2)$

37. Foci: $(-2, 3)$ and $(8, 3)$; vertices: $(-1, 3)$ and $(7, 3)$

38. Foci: $(3, -2)$ and $(3, 8)$; vertices: $(3, 0)$ and $(3, 6)$

In Exercises 39 – 42, write the equation of the hyperbola in standard form.

39. $3x^2 - 4y^2 = 12$

40. $3x^2 - y^2 + 2y = 10$

41. $x^2 - 10y^2 + 40y = 30$

42. $(4y - x)(4y + x) = 4$

43. Consider the ellipse given by $\dfrac{(x-1)^2}{4} + \dfrac{(y-3)^2}{12} = 1$.

   (a) Verify that the foci are located at $(1, 3 \pm 2\sqrt{2})$.

   (b) The points $P_1 = (2, 6)$ and $P_2 = (1 + \sqrt{2}, 3 + \sqrt{6}) \approx$ $(2.414, 5.449)$ lie on the ellipse. Verify that the sum of distances from each point to the foci is the same.

44. Johannes Kepler discovered that the planets of our solar system have elliptical orbits with the Sun at one focus. The Earth's elliptical orbit is used as a standard unit of distance; the distance from the center of Earth's elliptical orbit to one vertex is 1 Astronomical Unit, or A.U.

   The following table gives information about the orbits of three planets.

	Distance from center to vertex	eccentricity
Mercury	0.387 A.U.	0.2056
Earth	1 A.U.	0.0167
Mars	1.524 A.U.	0.0934

   (a) In an ellipse, knowing $c^2 = a^2 - b^2$ and $e = c/a$ allows us to find b in terms of a and e. Show $b = a\sqrt{1 - e^2}$.

   (b) For each planet, find equations of their elliptical orbit of the form $\dfrac{x^2}{a^2} + \dfrac{y^2}{b^2} = 1$. (This places the center at $(0, 0)$, but the Sun is in a different location for each planet.)

   (c) Shift the equations so that the Sun lies at the origin. Plot the three elliptical orbits.

45. A loud sound is recorded at three stations that lie on a line as shown in the figure below. Station A recorded the sound 1 second after Station B, and Station C recorded the sound 3 seconds after B. Using the speed of sound as 340m/s, determine the location of the sound's origination.

   A 1000m B   2000m   C

9.2 Parametric Equations

We are familiar with sketching shapes, such as parabolas, by following this basic procedure:

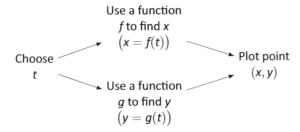

$$\text{Choose } x \xrightarrow{\text{Use a function } f \text{ to find } y \ (y=f(x))} \text{Plot point } (x,y)$$

The **rectangular equation** $y = f(x)$ works well for some shapes like a parabola with a vertical axis of symmetry, but in the previous section we encountered several shapes that could not be sketched in this manner. (To plot an ellipse using the above procedure, we need to plot the "top" and "bottom" separately.)

In this section we introduce a new sketching procedure:

$$\text{Choose } t \nearrow \text{Use a function } f \text{ to find } x \ (x=f(t)) \searrow \text{Plot point } (x,y)$$
$$\searrow \text{Use a function } g \text{ to find } y \ (y=g(t)) \nearrow$$

Here, x and y are found separately but then plotted together. This leads us to a definition.

Definition 9.2.1   Parametric Equations and Curves

Let f and g be continuous functions on an interval I. The set of all points $(x, y) = (f(t), g(t))$ in the Cartesian plane, as t varies over I, is the **graph** of the **parametric equations** $x = f(t)$ and $y = g(t)$, where t is the **parameter**. A **curve** is a graph along with the parametric equations that define it.

This is a formal definition of the word *curve*. When a curve lies in a plane (such as the Cartesian plane), it is often referred to as a **plane curve**. Examples will help us understand the concepts introduced in the definition.

Example 9.2.1   Plotting parametric functions

Plot the graph of the parametric equations $x = t^2$, $y = t + 1$ for t in $[-2, 2]$.

Notes:

t	x	y
−2	4	−1
−1	1	0
0	0	1
1	1	2
2	4	3

(a)

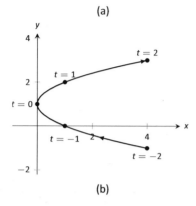

(b)

Figure 9.2.1: A table of values of the parametric equations in Example 9.2.1 along with a sketch of their graph.

t	x	y
0	1	2
$\pi/4$	1/2	$1 + \sqrt{2}/2$
$\pi/2$	0	1
$3\pi/4$	1/2	$1 - \sqrt{2}/2$
π	1	0

(a)

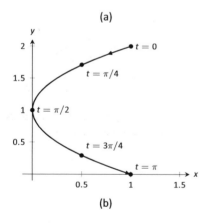

(b)

Figure 9.2.2: A table of values of the parametric equations in Example 9.2.2 along with a sketch of their graph.

SOLUTION   We plot the graphs of parametric equations in much the same manner as we plotted graphs of functions like $y = f(x)$: we make a table of values, plot points, then connect these points with a "reasonable" looking curve. Figure 9.2.1(a) shows such a table of values; note how we have 3 columns.

The points (x, y) from the table are plotted in Figure 9.2.1(b). The points have been connected with a smooth curve. Each point has been labeled with its corresponding t-value. These values, along with the two arrows along the curve, are used to indicate the **orientation** of the graph. This information helps us determine the direction in which the graph is "moving."

We often use the letter t as the parameter as we often regard t as representing *time*. Certainly there are many contexts in which the parameter is not time, but it can be helpful to think in terms of time as one makes sense of parametric plots and their orientation (for instance, "At time $t = 0$ the position is $(1, 2)$ and at time $t = 3$ the position is $(5, 1)$.").

Example 9.2.2   Plotting parametric functions

Sketch the graph of the parametric equations $x = \cos^2 t$, $y = \cos t + 1$ for t in $[0, \pi]$.

SOLUTION   We again start by making a table of values in Figure 9.2.2(a), then plot the points (x, y) on the Cartesian plane in Figure 9.2.2(b).

It is not difficult to show that the curves in Examples 9.2.1 and 9.2.2 are portions of the same parabola. While the *parabola* is the same, the *curves* are different. In Example 9.2.1, if we let t vary over all real numbers, we'd obtain the entire parabola. In this example, letting t vary over all real numbers would still produce the same graph; this portion of the parabola would be traced, and re–traced, infinitely many times. The orientation shown in Figure 9.2.2 shows the orientation on $[0, \pi]$, but this orientation is reversed on $[\pi, 2\pi]$.

These examples begin to illustrate the powerful nature of parametric equations. Their graphs are far more diverse than the graphs of functions produced by "$y = f(x)$" functions.

Technology Note: Most graphing utilities can graph functions given in parametric form. Often the word "parametric" is abbreviated as "PAR" or "PARAM" in the options. The user usually needs to determine the graphing window (i.e, the minimum and maximum x- and y-values), along with the values of t that are to be plotted. The user is often prompted to give a t minimum, a t maximum, and a "t-step" or "Δt." Graphing utilities effectively plot parametric functions just as we've shown here: they plots lots of points. A smaller t-step plots more points, making for a smoother graph (but may take longer). In Figure 9.2.1, the t-step is

Notes:

1; in Figure 9.2.2, the t-step is $\pi/4$.

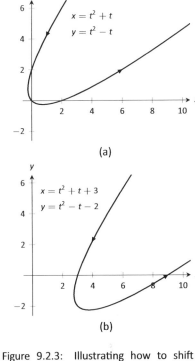

(a)

One nice feature of parametric equations is that their graphs are easy to shift. While this is not too difficult in the "$y = f(x)$" context, the resulting function can look rather messy. (Plus, to shift to the right by two, we replace x with $x - 2$, which is counter–intuitive.) The following example demonstrates this.

Example 9.2.3   Shifting the graph of parametric functions
Sketch the graph of the parametric equations $x = t^2 + t$, $y = t^2 - t$. Find new parametric equations that shift this graph to the right 3 places and down 2.

SOLUTION   The graph of the parametric equations is given in Figure 9.2.3 (a). It is a parabola with a axis of symmetry along the line $y = x$; the vertex is at $(0,0)$.

In order to shift the graph to the right 3 units, we need to increase the x-value by 3 for every point. The straightforward way to accomplish this is simply to add 3 to the function defining x: $x = t^2 + t + 3$. To shift the graph down by 2 units, we wish to decrease each y-value by 2, so we subtract 2 from the function defining y: $y = t^2 - t - 2$. Thus our parametric equations for the shifted graph are $x = t^2 + t + 3$, $y = t^2 - t - 2$. This is graphed in Figure 9.2.3 (b). Notice how the vertex is now at $(3, -2)$.

(b)

Figure 9.2.3: Illustrating how to shift graphs in Example 9.2.3.

Because the x- and y-values of a graph are determined independently, the graphs of parametric functions often possess features not seen on "$y = f(x)$" type graphs. The next example demonstrates how such graphs can arrive at the same point more than once.

Example 9.2.4   Graphs that cross themselves
Plot the parametric functions $x = t^3 - 5t^2 + 3t + 11$ and $y = t^2 - 2t + 3$ and determine the t-values where the graph crosses itself.

SOLUTION   Using the methods developed in this section, we again plot points and graph the parametric equations as shown in Figure 9.2.4. It appears that the graph crosses itself at the point $(2,6)$, but we'll need to analytically determine this.

We are looking for two different values, say, s and t, where $x(s) = x(t)$ and $y(s) = y(t)$. That is, the x-values are the same precisely when the y-values are the same. This gives us a system of 2 equations with 2 unknowns:

$$s^3 - 5s^2 + 3s + 11 = t^3 - 5t^2 + 3t + 11$$
$$s^2 - 2s + 3 = t^2 - 2t + 3$$

Solving this system is not trivial but involves only algebra. Using the quadratic

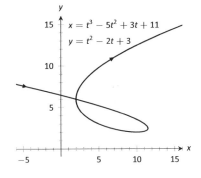

Figure 9.2.4: A graph of the parametric equations from Example 9.2.4.

Notes:

formula, one can solve for t in the second equation and find that $t = 1 \pm \sqrt{s^2 - 2s + 1}$. This can be substituted into the first equation, revealing that the graph crosses itself at $t = -1$ and $t = 3$. We confirm our result by computing $x(-1) = x(3) = 2$ and $y(-1) = y(3) = 6$.

Converting between rectangular and parametric equations

It is sometimes useful to rewrite equations in rectangular form (i.e., $y = f(x)$) into parametric form, and vice–versa. Converting from rectangular to parametric can be very simple: given $y = f(x)$, the parametric equations $x = t$, $y = f(t)$ produce the same graph. As an example, given $y = x^2$, the parametric equations $x = t$, $y = t^2$ produce the familiar parabola. However, other parametrizations can be used. The following example demonstrates one possible alternative.

Example 9.2.5   **Converting from rectangular to parametric**
Consider $y = x^2$. Find parametric equations $x = f(t)$, $y = g(t)$ for the parabola where $t = \frac{dy}{dx}$. That is, $t = a$ corresponds to the point on the graph whose tangent line has slope a.

SOLUTION   We start by computing $\frac{dy}{dx}$: $y' = 2x$. Thus we set $t = 2x$. We can solve for x and find $x = t/2$. Knowing that $y = x^2$, we have $y = t^2/4$. Thus parametric equations for the parabola $y = x^2$ are

$$x = t/2 \quad y = t^2/4.$$

To find the point where the tangent line has a slope of -2, we set $t = -2$. This gives the point $(-1, 1)$. We can verify that the slope of the line tangent to the curve at this point indeed has a slope of -2.

We sometimes choose the parameter to accurately model physical behavior.

Example 9.2.6   **Converting from rectangular to parametric**
An object is fired from a height of 0ft and lands 6 seconds later, 192ft away. Assuming ideal projectile motion, the height, in feet, of the object can be described by $h(x) = -x^2/64 + 3x$, where x is the distance in feet from the initial location. (Thus $h(0) = h(192) = 0$ft.) Find parametric equations $x = f(t)$, $y = g(t)$ for the path of the projectile where x is the horizontal distance the object has traveled at time t (in seconds) and y is the height at time t.

SOLUTION   Physics tells us that the horizontal motion of the projectile is linear; that is, the horizontal speed of the projectile is constant. Since the object travels 192ft in 6s, we deduce that the object is moving horizontally at a rate of 32ft/s, giving the equation $x = 32t$. As $y = -x^2/64 + 3x$, we find

Notes:

$y = -16t^2 + 96t$. We can quickly verify that $y'' = -32\text{ft/s}^2$, the acceleration due to gravity, and that the projectile reaches its maximum at $t = 3$, halfway along its path.

These parametric equations make certain determinations about the object's location easy: 2 seconds into the flight the object is at the point $\big(x(2), y(2)\big) = (64, 128)$. That is, it has traveled horizontally 64ft and is at a height of 128ft, as shown in Figure 9.2.5.

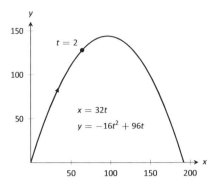

Figure 9.2.5: Graphing projectile motion in Example 9.2.6.

It is sometimes necessary to convert given parametric equations into rectangular form. This can be decidedly more difficult, as some "simple" looking parametric equations can have very "complicated" rectangular equations. This conversion is often referred to as "eliminating the parameter," as we are looking for a relationship between x and y that does not involve the parameter t.

Example 9.2.7   **Eliminating the parameter**
Find a rectangular equation for the curve described by

$$x = \frac{1}{t^2 + 1} \quad \text{and} \quad y = \frac{t^2}{t^2 + 1}.$$

SOLUTION   There is not a set way to eliminate a parameter. One method is to solve for t in one equation and then substitute that value in the second. We use that technique here, then show a second, simpler method.

Starting with $x = 1/(t^2 + 1)$, solve for t: $t = \pm\sqrt{1/x - 1}$. Substitute this value for t in the equation for y:

$$
\begin{aligned}
y &= \frac{t^2}{t^2 + 1} \\
&= \frac{1/x - 1}{1/x - 1 + 1} \\
&= \frac{1/x - 1}{1/x} \\
&= \left(\frac{1}{x} - 1\right) \cdot x \\
&= 1 - x.
\end{aligned}
$$

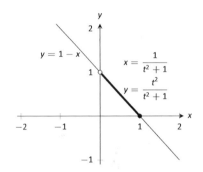

Figure 9.2.6: Graphing parametric and rectangular equations for a graph in Example 9.2.7.

Thus $y = 1 - x$. One may have recognized this earlier by manipulating the equation for y:

$$y = \frac{t^2}{t^2 + 1} = 1 - \frac{1}{t^2 + 1} = 1 - x.$$

Notes:

This is a shortcut that is very specific to this problem; sometimes shortcuts exist and are worth looking for.

We should be careful to limit the domain of the function $y = 1 - x$. The parametric equations limit x to values in $(0, 1]$, thus to produce the same graph we should limit the domain of $y = 1 - x$ to the same.

The graphs of these functions is given in Figure 9.2.6. The portion of the graph defined by the parametric equations is given in a thick line; the graph defined by $y = 1 - x$ with unrestricted domain is given in a thin line.

Example 9.2.8   Eliminating the parameter
Eliminate the parameter in $x = 4 \cos t + 3$, $y = 2 \sin t + 1$

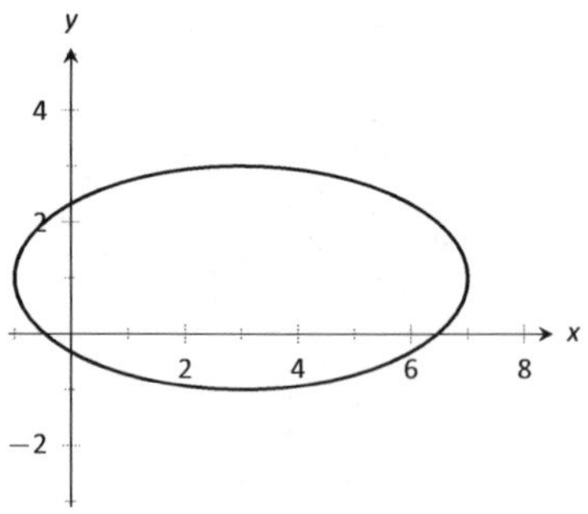

SOLUTION   We should not try to solve for t in this situation as the resulting algebra/trig would be messy. Rather, we solve for $\cos t$ and $\sin t$ in each equation, respectively. This gives

$$\cos t = \frac{x - 3}{4} \quad \text{and} \quad \sin t = \frac{y - 1}{2}.$$

Figure 9.2.7: Graphing the parametric equations $x = 4 \cos t + 3$, $y = 2 \sin t + 1$ in Example 9.2.8.

The Pythagorean Theorem gives $\cos^2 t + \sin^2 t = 1$, so:

$$\cos^2 t + \sin^2 t = 1$$
$$\left(\frac{x - 3}{4}\right)^2 + \left(\frac{y - 1}{2}\right)^2 = 1$$
$$\frac{(x - 3)^2}{16} + \frac{(y - 1)^2}{4} = 1$$

This final equation should look familiar – it is the equation of an ellipse! Figure 9.2.7 plots the parametric equations, demonstrating that the graph is indeed of an ellipse with a horizontal major axis and center at $(3, 1)$.

The Pythagorean Theorem can also be used to identify parametric equations for hyperbolas. We give the parametric equations for ellipses and hyperbolas in the following Key Idea.

Notes:

Key Idea 9.2.1   Parametric Equations of Ellipses and Hyperbolas

- The parametric equations

$$x = a\cos t + h, \quad y = b\sin t + k$$

  define an ellipse with horizontal axis of length $2a$ and vertical axis of length $2b$, centered at (h, k).

- The parametric equations

$$x = a\tan t + h, \quad y = \pm b\sec t + k$$

  define a hyperbola with vertical transverse axis centered at (h, k), and

$$x = \pm a\sec t + h, \quad y = b\tan t + k$$

  defines a hyperbola with horizontal transverse axis. Each has asymptotes at $y = \pm b/a(x - h) + k$.

Special Curves

Figure 9.2.8 gives a small gallery of "interesting" and "famous" curves along with parametric equations that produce them. Interested readers can begin learning more about these curves through internet searches.

One might note a feature shared by two of these graphs: "sharp corners," or **cusps**. We have seen graphs with cusps before and determined that such functions are not differentiable at these points. This leads us to a definition.

Definition 9.2.2   Smooth

A curve C defined by $x = f(t), y = g(t)$ is **smooth** on an interval I if f' and g' are continuous on I and not simultaneously 0 (except possibly at the endpoints of I). A curve is **piecewise smooth** on I if I can be partitioned into subintervals where C is smooth on each subinterval.

Consider the astroid, given by $x = \cos^3 t, y = \sin^3 t$. Taking derivatives, we have:

$$x' = -3\cos^2 t \sin t \quad \text{and} \quad y' = 3\sin^2 t \cos t.$$

It is clear that each is 0 when $t = 0, \pi/2, \pi, \ldots$. Thus the astroid is not smooth

Notes:

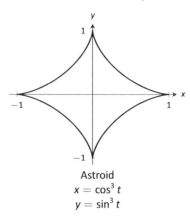

Astroid
$x = \cos^3 t$
$y = \sin^3 t$

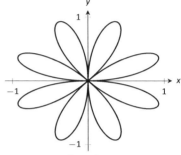

Rose Curve
$x = \cos(t)\sin(4t)$
$y = \sin(t)\sin(4t)$

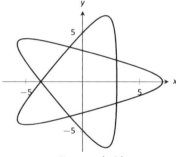

Hypotrochoid
$x = 2\cos(t) + 5\cos(2t/3)$
$y = 2\sin(t) - 5\sin(2t/3)$

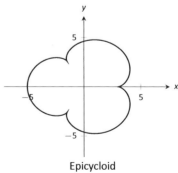

Epicycloid
$x = 4\cos(t) - \cos(4t)$
$y = 4\sin(t) - \sin(4t)$

Figure 9.2.8: A gallery of interesting planar curves.

at these points, corresponding to the cusps seen in the figure.

We demonstrate this once more.

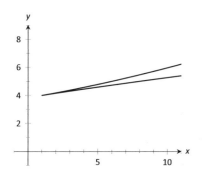

Figure 9.2.9: Graphing the curve in Example 9.2.9; note it is not smooth at $(1, 4)$.

Example 9.2.9   Determine where a curve is not smooth

Let a curve C be defined by the parametric equations $x = t^3 - 12t + 17$ and $y = t^2 - 4t + 8$. Determine the points, if any, where it is not smooth.

SOLUTION   We begin by taking derivatives.

$$x' = 3t^2 - 12, \quad y' = 2t - 4.$$

We set each equal to 0:

$$x' = 0 \Rightarrow 3t^2 - 12 = 0 \Rightarrow t = \pm 2$$
$$y' = 0 \Rightarrow 2t - 4 = 0 \Rightarrow t = 2$$

We see at $t = 2$ both x' and y' are 0; thus C is not smooth at $t = 2$, corresponding to the point $(1, 4)$. The curve is graphed in Figure 9.2.9, illustrating the cusp at $(1, 4)$.

If a curve is not smooth at $t = t_0$, it means that $x'(t_0) = y'(t_0) = 0$ as defined. This, in turn, means that rate of change of x (and y) is 0; that is, at that instant, neither x nor y is changing. If the parametric equations describe the path of some object, this means the object is at rest at t_0. An object at rest can make a "sharp" change in direction, whereas moving objects tend to change direction in a "smooth" fashion.

One should be careful to note that a "sharp corner" does not have to occur when a curve is not smooth. For instance, one can verify that $x = t^3$, $y = t^6$ produce the familiar $y = x^2$ parabola. However, in this parametrization, the curve is not smooth. A particle traveling along the parabola according to the given parametric equations comes to rest at $t = 0$, though no sharp point is created.

Our previous experience with cusps taught us that a function was not differentiable at a cusp. This can lead us to wonder about derivatives in the context of parametric equations and the application of other calculus concepts. Given a curve defined parametrically, how do we find the slopes of tangent lines? Can we determine concavity? We explore these concepts and more in the next section.

Notes:

Exercises 9.2

Terms and Concepts

1. T/F: When sketching the graph of parametric equations, the x and y values are found separately, then plotted together.

2. The direction in which a graph is "moving" is called the ____ of the graph.

3. An equation written as $y = f(x)$ is written in ____ form.

4. Create parametric equations $x = f(t)$, $y = g(t)$ and sketch their graph. Explain any interesting features of your graph based on the functions f and g.

Problems

In Exercises 5 – 8, sketch the graph of the given parametric equations by hand, making a table of points to plot. Be sure to indicate the orientation of the graph.

5. $x = t^2 + t$, $y = 1 - t^2$, $-3 \le t \le 3$

6. $x = 1$, $y = 5 \sin t$, $-\pi/2 \le t \le \pi/2$

7. $x = t^2$, $y = 2$, $-2 \le t \le 2$

8. $x = t^3 - t + 3$, $y = t^2 + 1$, $-2 \le t \le 2$

In Exercises 9 – 18, sketch the graph of the given parametric equations; using a graphing utility is advisable. Be sure to indicate the orientation of the graph.

9. $x = t^3 - 2t^2$, $y = t^2$, $-2 \le t \le 3$

10. $x = 1/t$, $y = \sin t$, $0 < t \le 10$

11. $x = 3 \cos t$, $y = 5 \sin t$, $0 \le t \le 2\pi$

12. $x = 3 \cos t + 2$, $y = 5 \sin t + 3$, $0 \le t \le 2\pi$

13. $x = \cos t$, $y = \cos(2t)$, $0 \le t \le \pi$

14. $x = \cos t$, $y = \sin(2t)$, $0 \le t \le 2\pi$

15. $x = 2 \sec t$, $y = 3 \tan t$, $-\pi/2 < t < \pi/2$

16. $x = \cosh t$, $y = \sinh t$, $-2 \le t \le 2$

17. $x = \cos t + \frac{1}{4} \cos(8t)$, $y = \sin t + \frac{1}{4} \sin(8t)$, $0 \le t \le 2\pi$

18. $x = \cos t + \frac{1}{4} \sin(8t)$, $y = \sin t + \frac{1}{4} \cos(8t)$, $0 \le t \le 2\pi$

In Exercises 19 – 20, four sets of parametric equations are given. Describe how their graphs are similar and different. Be sure to discuss orientation and ranges.

19. (a) $x = t$ $y = t^2$, $-\infty < t < \infty$

    (b) $x = \sin t$ $y = \sin^2 t$, $-\infty < t < \infty$

    (c) $x = e^t$ $y = e^{2t}$, $-\infty < t < \infty$

    (d) $x = -t$ $y = t^2$, $-\infty < t < \infty$

20. (a) $x = \cos t$ $y = \sin t$, $0 \le t \le 2\pi$

    (b) $x = \cos(t^2)$ $y = \sin(t^2)$, $0 \le t \le 2\pi$

    (c) $x = \cos(1/t)$ $y = \sin(1/t)$, $0 < t < 1$

    (d) $x = \cos(\cos t)$ $y = \sin(\cos t)$, $0 \le t \le 2\pi$

In Exercises 21 – 30, eliminate the parameter in the given parametric equations.

21. $x = 2t + 5$, $y = -3t + 1$

22. $x = \sec t$, $y = \tan t$

23. $x = 4 \sin t + 1$, $y = 3 \cos t - 2$

24. $x = t^2$, $y = t^3$

25. $x = \dfrac{1}{t+1}$, $y = \dfrac{3t+5}{t+1}$

26. $x = e^t$, $y = e^{3t} - 3$

27. $x = \ln t$, $y = t^2 - 1$

28. $x = \cot t$, $y = \csc t$

29. $x = \cosh t$, $y = \sinh t$

30. $x = \cos(2t)$, $y = \sin t$

In Exercises 31 – 34, eliminate the parameter in the given parametric equations. Describe the curve defined by the parametric equations based on its rectangular form.

31. $x = at + x_0$, $y = bt + y_0$

32. $x = r \cos t$, $y = r \sin t$

33. $x = a \cos t + h$, $y = b \sin t + k$

34. $x = a \sec t + h$, $y = b \tan t + k$

In Exercises 35 – 38, find parametric equations for the given rectangular equation using the parameter $t = \dfrac{dy}{dx}$. Verify that at $t = 1$, the point on the graph has a tangent line with slope of 1.

35. $y = 3x^2 - 11x + 2$

36. $y = e^x$

37. $y = \sin x$ on $[0, \pi]$

38. $y = \sqrt{x}$ on $[0, \infty)$

In Exercises 39 – 42, find the values of t where the graph of the parametric equations crosses itself.

39. $x = t^3 - t + 3, \quad y = t^2 - 3$

40. $x = t^3 - 4t^2 + t + 7, \quad y = t^2 - t$

41. $x = \cos t, \quad y = \sin(2t)$ on $[0, 2\pi]$

42. $x = \cos t \cos(3t), \quad y = \sin t \cos(3t)$ on $[0, \pi]$

In Exercises 43 – 46, find the value(s) of t where the curve defined by the parametric equations is not smooth.

43. $x = t^3 + t^2 - t, \quad y = t^2 + 2t + 3$

44. $x = t^2 - 4t, \quad y = t^3 - 2t^2 - 4t$

45. $x = \cos t, \quad y = 2\cos t$

46. $x = 2\cos t - \cos(2t), \quad y = 2\sin t - \sin(2t)$

In Exercises 47 – 55, find parametric equations that describe the given situation.

47. A projectile is fired from a height of 0ft, landing 16ft away in 4s.

48. A projectile is fired from a height of 0ft, landing 200ft away in 4s.

49. A projectile is fired from a height of 0ft, landing 200ft away in 20s.

50. A circle of radius 2, centered at the origin, that is traced clockwise once on $[0, 2\pi]$.

51. A circle of radius 3, centered at $(1, 1)$, that is traced once counter–clockwise on $[0, 1]$.

52. An ellipse centered at $(1, 3)$ with vertical major axis of length 6 and minor axis of length 2.

53. An ellipse with foci at $(\pm 1, 0)$ and vertices at $(\pm 5, 0)$.

54. A hyperbola with foci at $(5, -3)$ and $(-1, -3)$, and with vertices at $(1, -3)$ and $(3, -3)$.

55. A hyperbola with vertices at $(0, \pm 6)$ and asymptotes $y = \pm 3x$.

9.3 Calculus and Parametric Equations

The previous section defined curves based on parametric equations. In this section we'll employ the techniques of calculus to study these curves.

We are still interested in lines tangent to points on a curve. They describe how the y-values are changing with respect to the x-values, they are useful in making approximations, and they indicate instantaneous direction of travel.

The slope of the tangent line is still $\frac{dy}{dx}$, and the Chain Rule allows us to calculate this in the context of parametric equations. If $x = f(t)$ and $y = g(t)$, the Chain Rule states that

$$\frac{dy}{dt} = \frac{dy}{dx} \cdot \frac{dx}{dt}.$$

Solving for $\frac{dy}{dx}$, we get

$$\frac{dy}{dx} = \frac{dy}{dt} \bigg/ \frac{dx}{dt} = \frac{g'(t)}{f'(t)},$$

provided that $f'(t) \neq 0$. This is important so we label it a Key Idea.

Key Idea 9.3.1   Finding $\frac{dy}{dx}$ with Parametric Equations.

Let $x = f(t)$ and $y = g(t)$, where f and g are differentiable on some open interval I and $f'(t) \neq 0$ on I. Then

$$\frac{dy}{dx} = \frac{g'(t)}{f'(t)}.$$

We use this to define the tangent line.

Definition 9.3.1   Tangent and Normal Lines

Let a curve C be parametrized by $x = f(t)$ and $y = g(t)$, where f and g are differentiable functions on some interval I containing $t = t_0$. The **tangent line** to C at $t = t_0$ is the line through $\big(f(t_0), g(t_0)\big)$ with slope $m = g'(t_0)/f'(t_0)$, provided $f'(t_0) \neq 0$.

The **normal line** to C at $t = t_0$ is the line through $\big(f(t_0), g(t_0)\big)$ with slope $m = -f'(t_0)/g'(t_0)$, provided $g'(t_0) \neq 0$.

The definition leaves two special cases to consider. When the tangent line is horizontal, the normal line is undefined by the above definition as $g'(t_0) = 0$.

Notes:

Likewise, when the normal line is horizontal, the tangent line is undefined. It seems reasonable that these lines be defined (one can draw a line tangent to the "right side" of a circle, for instance), so we add the following to the above definition.

1. If the tangent line at $t = t_0$ has a slope of 0, the normal line to C at $t = t_0$ is the line $x = f(t_0)$.

2. If the normal line at $t = t_0$ has a slope of 0, the tangent line to C at $t = t_0$ is the line $x = f(t_0)$.

Example 9.3.1   Tangent and Normal Lines to Curves
Let $x = 5t^2 - 6t + 4$ and $y = t^2 + 6t - 1$, and let C be the curve defined by these equations.

1. Find the equations of the tangent and normal lines to C at $t = 3$.

2. Find where C has vertical and horizontal tangent lines.

SOLUTION

1. We start by computing $f'(t) = 10t - 6$ and $g'(t) = 2t + 6$. Thus

$$\frac{dy}{dx} = \frac{2t + 6}{10t - 6}.$$

Make note of something that might seem unusual: $\frac{dy}{dx}$ is a function of t, not x. Just as points on the curve are found in terms of t, so are the slopes of the tangent lines.

The point on C at $t = 3$ is $(31, 26)$. The slope of the tangent line is $m = 1/2$ and the slope of the normal line is $m = -2$. Thus,

- the equation of the tangent line is $y = \dfrac{1}{2}(x - 31) + 26$, and
- the equation of the normal line is $y = -2(x - 31) + 26$.

This is illustrated in Figure 9.3.1.

2. To find where C has a horizontal tangent line, we set $\frac{dy}{dx} = 0$ and solve for t. In this case, this amounts to setting $g'(t) = 0$ and solving for t (and making sure that $f'(t) \neq 0$).

$$g'(t) = 0 \quad \Rightarrow \quad 2t + 6 = 0 \quad \Rightarrow \quad t = -3.$$

The point on C corresponding to $t = -3$ is $(67, -10)$; the tangent line at that point is horizontal (hence with equation $y = -10$).

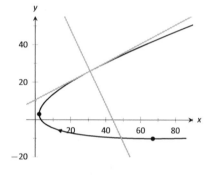

Figure 9.3.1: Graphing tangent and normal lines in Example 9.3.1.

Notes:

To find where C has a vertical tangent line, we find where it has a horizontal normal line, and set $-\frac{f'(t)}{g'(t)} = 0$. This amounts to setting $f'(t) = 0$ and solving for t (and making sure that $g'(t) \neq 0$).

$$f'(t) = 0 \quad \Rightarrow \quad 10t - 6 = 0 \quad \Rightarrow \quad t = 0.6.$$

The point on C corresponding to $t = 0.6$ is $(2.2, 2.96)$. The tangent line at that point is $x = 2.2$.

The points where the tangent lines are vertical and horizontal are indicated on the graph in Figure 9.3.1.

Example 9.3.2   Tangent and Normal Lines to a Circle

1. Find where the unit circle, defined by $x = \cos t$ and $y = \sin t$ on $[0, 2\pi]$, has vertical and horizontal tangent lines.

2. Find the equation of the normal line at $t = t_0$.

SOLUTION

1. We compute the derivative following Key Idea 9.3.1:
$$\frac{dy}{dx} = \frac{g'(t)}{f'(t)} = -\frac{\cos t}{\sin t}.$$

The derivative is 0 when $\cos t = 0$; that is, when $t = \pi/2,\ 3\pi/2$. These are the points $(0, 1)$ and $(0, -1)$ on the circle.

The normal line is horizontal (and hence, the tangent line is vertical) when $\sin t = 0$; that is, when $t = 0,\ \pi,\ 2\pi$, corresponding to the points $(-1, 0)$ and $(0, 1)$ on the circle. These results should make intuitive sense.

2. The slope of the normal line at $t = t_0$ is $m = \dfrac{\sin t_0}{\cos t_0} = \tan t_0$. This normal line goes through the point $(\cos t_0, \sin t_0)$, giving the line

$$y = \frac{\sin t_0}{\cos t_0}(x - \cos t_0) + \sin t_0$$
$$= (\tan t_0)x,$$

as long as $\cos t_0 \neq 0$. It is an important fact to recognize that the normal lines to a circle pass through its center, as illustrated in Figure 9.3.2. Stated in another way, any line that passes through the center of a circle intersects the circle at right angles.

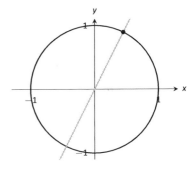

Figure 9.3.2: Illustrating how a circle's normal lines pass through its center.

Notes:

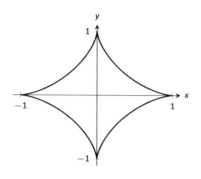

Figure 9.3.3: A graph of an astroid.

Example 9.3.3   **Tangent lines when $\frac{dy}{dx}$ is not defined**

Find the equation of the tangent line to the astroid $x = \cos^3 t$, $y = \sin^3 t$ at $t = 0$, shown in Figure 9.3.3.

SOLUTION   We start by finding $x'(t)$ and $y'(t)$:

$$x'(t) = -3\sin t \cos^2 t, \qquad y'(t) = 3\cos t \sin^2 t.$$

Note that both of these are 0 at $t = 0$; the curve is not smooth at $t = 0$ forming a cusp on the graph. Evaluating $\frac{dy}{dx}$ at this point returns the indeterminate form of "0/0".

We can, however, examine the slopes of tangent lines near $t = 0$, and take the limit as $t \to 0$.

$$\lim_{t \to 0} \frac{y'(t)}{x'(t)} = \lim_{t \to 0} \frac{3\cos t \sin^2 t}{-3\sin t \cos^2 t} \qquad \text{(We can cancel as } t \neq 0.\text{)}$$

$$= \lim_{t \to 0} -\frac{\sin t}{\cos t}$$

$$= 0.$$

We have accomplished something significant. When the derivative $\frac{dy}{dx}$ returns an indeterminate form at $t = t_0$, we can define its value by setting it to be $\lim\limits_{t \to t_0} \frac{dy}{dx}$, if that limit exists. This allows us to find slopes of tangent lines at cusps, which can be very beneficial.

We found the slope of the tangent line at $t = 0$ to be 0; therefore the tangent line is $y = 0$, the x-axis.

Concavity

We continue to analyze curves in the plane by considering their concavity; that is, we are interested in $\frac{d^2y}{dx^2}$, "the second derivative of y with respect to x." To find this, we need to find the derivative of $\frac{dy}{dx}$ with respect to x; that is,

$$\frac{d^2y}{dx^2} = \frac{d}{dx}\left[\frac{dy}{dx}\right],$$

but recall that $\frac{dy}{dx}$ is a function of t, not x, making this computation not straightforward.

To make the upcoming notation a bit simpler, let $h(t) = \frac{dy}{dx}$. We want $\frac{d}{dx}[h(t)]$; that is, we want $\frac{dh}{dx}$. We again appeal to the Chain Rule. Note:

$$\frac{dh}{dt} = \frac{dh}{dx} \cdot \frac{dx}{dt} \qquad \Rightarrow \qquad \frac{dh}{dx} = \frac{dh}{dt} \bigg/ \frac{dx}{dt}.$$

Notes:

In words, to find $\frac{d^2y}{dx^2}$, we first take the derivative of $\frac{dy}{dx}$ *with respect to t*, then divide by $x'(t)$. We restate this as a Key Idea.

Key Idea 9.3.2   **Finding $\frac{d^2y}{dx^2}$ with Parametric Equations**

Let $x = f(t)$ and $y = g(t)$ be twice differentiable functions on an open interval I, where $f'(t) \neq 0$ on I. Then

$$\frac{d^2y}{dx^2} = \frac{d}{dt}\left[\frac{dy}{dx}\right] \Big/ \frac{dx}{dt} = \frac{d}{dt}\left[\frac{dy}{dx}\right] \Big/ f'(t).$$

Examples will help us understand this Key Idea.

Example 9.3.4   Concavity of Plane Curves

Let $x = 5t^2 - 6t + 4$ and $y = t^2 + 6t - 1$ as in Example 9.3.1. Determine the t-intervals on which the graph is concave up/down.

SOLUTION   Concavity is determined by the second derivative of y with respect to x, $\frac{d^2y}{dx^2}$, so we compute that here following Key Idea 9.3.2.

In Example 9.3.1, we found $\frac{dy}{dx} = \frac{2t+6}{10t-6}$ and $f'(t) = 10t - 6$. So:

$$\frac{d^2y}{dx^2} = \frac{d}{dt}\left[\frac{2t+6}{10t-6}\right] \Big/ (10t-6)$$

$$= -\frac{72}{(10t-6)^2} \Big/ (10t-6)$$

$$= -\frac{72}{(10t-6)^3}$$

$$= -\frac{9}{(5t-3)^3}$$

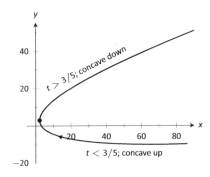

Figure 9.3.4: Graphing the parametric equations in Example 9.3.4 to demonstrate concavity.

The graph of the parametric functions is concave up when $\frac{d^2y}{dx^2} > 0$ and concave down when $\frac{d^2y}{dx^2} < 0$. We determine the intervals when the second derivative is greater/less than 0 by first finding when it is 0 or undefined.

As the numerator of $-\frac{9}{(5t-3)^3}$ is never 0, $\frac{d^2y}{dx^2} \neq 0$ for all t. It is undefined when $5t - 3 = 0$; that is, when $t = 3/5$. Following the work established in Section 3.4, we look at values of t greater/less than $3/5$ on a number line:

Notes:

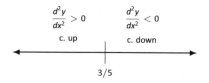

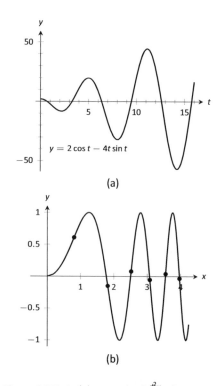

$y = 2\cos t - 4t\sin t$

(a)

(b)

Figure 9.3.5: In (a), a graph of $\frac{d^2y}{dx^2}$, showing where it is approximately 0. In (b), graph of the parametric equations in Example 9.3.5 along with the points of inflection.

Reviewing Example 9.3.1, we see that when $t = 3/5 = 0.6$, the graph of the parametric equations has a vertical tangent line. This point is also a point of inflection for the graph, illustrated in Figure 9.3.4.

Example 9.3.5   Concavity of Plane Curves
Find the points of inflection of the graph of the parametric equations $x = \sqrt{t}$, $y = \sin t$, for $0 \leq t \leq 16$.

SOLUTION   We need to compute $\frac{dy}{dx}$ and $\frac{d^2y}{dx^2}$.

$$\frac{dy}{dx} = \frac{y'(t)}{x'(t)} = \frac{\cos t}{1/(2\sqrt{t})} = 2\sqrt{t}\cos t.$$

$$\frac{d^2y}{dx^2} = \frac{\frac{d}{dt}\left[\frac{dy}{dx}\right]}{x'(t)} = \frac{\cos t/\sqrt{t} - 2\sqrt{t}\sin t}{1/(2\sqrt{t})} = 2\cos t - 4t\sin t.$$

The points of inflection are found by setting $\frac{d^2y}{dx^2} = 0$. This is not trivial, as equations that mix polynomials and trigonometric functions generally do not have "nice" solutions.

In Figure 9.3.5(a) we see a plot of the second derivative. It shows that it has zeros at approximately $t = 0.5$, 3.5, 6.5, 9.5, 12.5 and 16. These approximations are not very good, made only by looking at the graph. Newton's Method provides more accurate approximations. Accurate to 2 decimal places, we have:

$$t = 0.65, \ 3.29, \ 6.36, \ 9.48, \ 12.61 \text{ and } 15.74.$$

The corresponding points have been plotted on the graph of the parametric equations in Figure 9.3.5(b). Note how most occur near the x-axis, but not exactly on the axis.

Arc Length

We continue our study of the features of the graphs of parametric equations by computing their arc length.

Recall in Section 7.4 we found the arc length of the graph of a function, from $x = a$ to $x = b$, to be

$$L = \int_a^b \sqrt{1 + \left(\frac{dy}{dx}\right)^2}\, dx.$$

Notes:

We can use this equation and convert it to the parametric equation context. Letting $x = f(t)$ and $y = g(t)$, we know that $\frac{dy}{dx} = g'(t)/f'(t)$. It will also be useful to calculate the differential of x:

$$dx = f'(t)dt \qquad \Rightarrow \qquad dt = \frac{1}{f'(t)} \cdot dx.$$

Starting with the arc length formula above, consider:

$$L = \int_a^b \sqrt{1 + \left(\frac{dy}{dx}\right)^2}\, dx$$

$$= \int_a^b \sqrt{1 + \frac{g'(t)^2}{f'(t)^2}}\, dx.$$

Factor out the $f'(t)^2$:

$$= \int_a^b \sqrt{f'(t)^2 + g'(t)^2} \cdot \underbrace{\frac{1}{f'(t)}\, dx}_{=dt}$$

$$= \int_{t_1}^{t_2} \sqrt{f'(t)^2 + g'(t)^2}\, dt.$$

Note the new bounds (no longer "x" bounds, but "t" bounds). They are found by finding t_1 and t_2 such that $a = f(t_1)$ and $b = f(t_2)$. This formula is important, so we restate it as a theorem.

Theorem 9.3.1   Arc Length of Parametric Curves

Let $x = f(t)$ and $y = g(t)$ be parametric equations with f' and g' continuous on $[t_1, t_2]$, on which the graph traces itself only once. The arc length of the graph, from $t = t_1$ to $t = t_2$, is

$$L = \int_{t_1}^{t_2} \sqrt{f'(t)^2 + g'(t)^2}\, dt.$$

Note: Theorem 9.3.1 makes use of differentiability on closed intervals, just as was done in Section 7.4.

As before, these integrals are often not easy to compute. We start with a simple example, then give another where we approximate the solution.

Notes:

Example 9.3.6   **Arc Length of a Circle**

Find the arc length of the circle parametrized by $x = 3\cos t$, $y = 3\sin t$ on $[0, 3\pi/2]$.

SOLUTION   By direct application of Theorem 9.3.1, we have

$$L = \int_0^{3\pi/2} \sqrt{(-3\sin t)^2 + (3\cos t)^2}\, dt.$$

Apply the Pythagorean Theorem.

$$= \int_0^{3\pi/2} 3\, dt$$

$$= 3t\Big|_0^{3\pi/2} = 9\pi/2.$$

This should make sense; we know from geometry that the circumference of a circle with radius 3 is 6π; since we are finding the arc length of $3/4$ of a circle, the arc length is $3/4 \cdot 6\pi = 9\pi/2$.

Example 9.3.7   **Arc Length of a Parametric Curve**

The graph of the parametric equations $x = t(t^2 - 1)$, $y = t^2 - 1$ crosses itself as shown in Figure 9.3.6, forming a "teardrop." Find the arc length of the teardrop.

SOLUTION   We can see by the parametrizations of x and y that when $t = \pm 1$, $x = 0$ and $y = 0$. This means we'll integrate from $t = -1$ to $t = 1$. Applying Theorem 9.3.1, we have

$$L = \int_{-1}^1 \sqrt{(3t^2 - 1)^2 + (2t)^2}\, dt$$

$$= \int_{-1}^1 \sqrt{9t^4 - 2t^2 + 1}\, dt.$$

Unfortunately, the integrand does not have an antiderivative expressible by elementary functions. We turn to numerical integration to approximate its value. Using 4 subintervals, Simpson's Rule approximates the value of the integral as 2.65051. Using a computer, more subintervals are easy to employ, and $n = 20$ gives a value of 2.71559. Increasing n shows that this value is stable and a good approximation of the actual value.

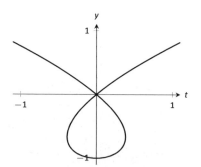

Figure 9.3.6: A graph of the parametric equations in Example 9.3.7, where the arc length of the teardrop is calculated.

Notes:

Surface Area of a Solid of Revolution

Related to the formula for finding arc length is the formula for finding surface area. We can adapt the formula found in Theorem 7.4.2 from Section 7.4 in a similar way as done to produce the formula for arc length done before.

Theorem 9.3.2   Surface Area of a Solid of Revolution

Consider the graph of the parametric equations $x = f(t)$ and $y = g(t)$, where f' and g' are continuous on an open interval I containing t_1 and t_2 on which the graph does not cross itself.

1. The surface area of the solid formed by revolving the graph about the x-axis is (where $g(t) \geq 0$ on $[t_1, t_2]$):

$$\text{Surface Area} = 2\pi \int_{t_1}^{t_2} g(t) \sqrt{f'(t)^2 + g'(t)^2} \, dt.$$

2. The surface area of the solid formed by revolving the graph about the y-axis is (where $f(t) \geq 0$ on $[t_1, t_2]$):

$$\text{Surface Area} = 2\pi \int_{t_1}^{t_2} f(t) \sqrt{f'(t)^2 + g'(t)^2} \, dt.$$

Example 9.3.8   Surface Area of a Solid of Revolution

Consider the teardrop shape formed by the parametric equations $x = t(t^2 - 1)$, $y = t^2 - 1$ as seen in Example 9.3.7. Find the surface area if this shape is rotated about the x-axis, as shown in Figure 9.3.7.

SOLUTION   The teardrop shape is formed between $t = -1$ and $t = 1$. Using Theorem 9.3.2, we see we need for $g(t) \geq 0$ on $[-1, 1]$, and this is not the case. To fix this, we simplify replace $g(t)$ with $-g(t)$, which flips the whole graph about the x-axis (and does not change the surface area of the resulting solid). The surface area is:

$$\text{Area } S = 2\pi \int_{-1}^{1} (1 - t^2) \sqrt{(3t^2 - 1)^2 + (2t)^2} \, dt$$

$$= 2\pi \int_{-1}^{1} (1 - t^2) \sqrt{9t^4 - 2t^2 + 1} \, dt.$$

Once again we arrive at an integral that we cannot compute in terms of ele-

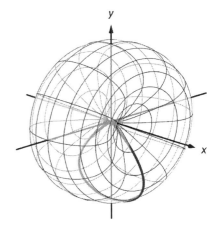

Figure 9.3.7: Rotating a teardrop shape about the x-axis in Example 9.3.8.

Notes:

mentary functions. Using Simpson's Rule with $n = 20$, we find the area to be $S = 9.44$. Using larger values of n shows this is accurate to 2 places after the decimal.

After defining a new way of creating curves in the plane, in this section we have applied calculus techniques to the parametric equation defining these curves to study their properties. In the next section, we define another way of forming curves in the plane. To do so, we create a new coordinate system, called *polar coordinates*, that identifies points in the plane in a manner different than from measuring distances from the y- and x- axes.

Notes:

Exercises 9.3

Terms and Concepts

1. T/F: Given parametric equations $x = f(t)$ and $y = g(t)$, $\frac{dy}{dx} = f'(t)/g'(t)$, as long as $g'(t) \neq 0$.

2. Given parametric equations $x = f(t)$ and $y = g(t)$, the derivative $\frac{dy}{dx}$ as given in Key Idea 9.3.1 is a function of _____ ?

3. T/F: Given parametric equations $x = f(t)$ and $y = g(t)$, to find $\frac{d^2y}{dx^2}$, one simply computes $\frac{d}{dt}\left(\frac{dy}{dx}\right)$.

4. T/F: If $\frac{dy}{dx} = 0$ at $t = t_0$, then the normal line to the curve at $t = t_0$ is a vertical line.

Problems

In Exercises 5 – 12, parametric equations for a curve are given.

(a) Find $\frac{dy}{dx}$.

(b) Find the equations of the tangent and normal line(s) at the point(s) given.

(c) Sketch the graph of the parametric functions along with the found tangent and normal lines.

5. $x = t, y = t^2$; $t = 1$

6. $x = \sqrt{t}, y = 5t + 2$; $t = 4$

7. $x = t^2 - t, y = t^2 + t$; $t = 1$

8. $x = t^2 - 1, y = t^3 - t$; $t = 0$ and $t = 1$

9. $x = \sec t, y = \tan t$ on $(-\pi/2, \pi/2)$; $t = \pi/4$

10. $x = \cos t, y = \sin(2t)$ on $[0, 2\pi]$; $t = \pi/4$

11. $x = \cos t \sin(2t), y = \sin t \sin(2t)$ on $[0, 2\pi]$; $t = 3\pi/4$

12. $x = e^{t/10} \cos t, y = e^{t/10} \sin t$; $t = \pi/2$

In Exercises 13 – 20, find t-values where the curve defined by the given parametric equations has a horizontal tangent line. Note: these are the same equations as in Exercises 5 – 12.

13. $x = t, y = t^2$

14. $x = \sqrt{t}, y = 5t + 2$

15. $x = t^2 - t, y = t^2 + t$

16. $x = t^2 - 1, y = t^3 - t$

17. $x = \sec t, y = \tan t$ on $(-\pi/2, \pi/2)$

18. $x = \cos t, y = \sin(2t)$ on $[0, 2\pi]$

19. $x = \cos t \sin(2t), y = \sin t \sin(2t)$ on $[0, 2\pi]$

20. $x = e^{t/10} \cos t, y = e^{t/10} \sin t$

In Exercises 21 – 24, find $t = t_0$ where the graph of the given parametric equations is not smooth, then find $\lim\limits_{t \to t_0} \dfrac{dy}{dx}$.

21. $x = \dfrac{1}{t^2 + 1}$, $y = t^3$

22. $x = -t^3 + 7t^2 - 16t + 13$, $y = t^3 - 5t^2 + 8t - 2$

23. $x = t^3 - 3t^2 + 3t - 1$, $y = t^2 - 2t + 1$

24. $x = \cos^2 t$, $y = 1 - \sin^2 t$

In Exercises 25 – 32, parametric equations for a curve are given. Find $\frac{d^2y}{dx^2}$, then determine the intervals on which the graph of the curve is concave up/down. Note: these are the same equations as in Exercises 5 – 12.

25. $x = t$, $y = t^2$

26. $x = \sqrt{t}$, $y = 5t + 2$

27. $x = t^2 - t$, $y = t^2 + t$

28. $x = t^2 - 1$, $y = t^3 - t$

29. $x = \sec t$, $y = \tan t$ on $(-\pi/2, \pi/2)$

30. $x = \cos t$, $y = \sin(2t)$ on $[0, 2\pi]$

31. $x = \cos t \sin(2t)$, $y = \sin t \sin(2t)$ on $[-\pi/2, \pi/2]$

32. $x = e^{t/10} \cos t$, $y = e^{t/10} \sin t$

In Exercises 33 – 36, find the arc length of the graph of the parametric equations on the given interval(s).

33. $x = -3 \sin(2t)$, $y = 3 \cos(2t)$ on $[0, \pi]$

34. $x = e^{t/10} \cos t$, $y = e^{t/10} \sin t$ on $[0, 2\pi]$ and $[2\pi, 4\pi]$

35. $x = 5t + 2$, $y = 1 - 3t$ on $[-1, 1]$

36. $x = 2t^{3/2}$, $y = 3t$ on $[0, 1]$

In Exercises 37 – 40, numerically approximate the given arc length.

37. Approximate the arc length of one petal of the rose curve $x = \cos t \cos(2t)$, $y = \sin t \cos(2t)$ using Simpson's Rule and $n = 4$.

38. Approximate the arc length of the "bow tie curve" $x = \cos t$, $y = \sin(2t)$ using Simpson's Rule and $n = 6$.

39. Approximate the arc length of the parabola $x = t^2 - t$, $y = t^2 + t$ on $[-1, 1]$ using Simpson's Rule and $n = 4$.

40. A common approximate of the circumference of an ellipse given by $x = a\cos t$, $y = b\sin t$ is $C \approx 2\pi\sqrt{\dfrac{a^2 + b^2}{2}}$. Use this formula to approximate the circumference of $x = 5\cos t$, $y = 3\sin t$ and compare this to the approximation given by Simpson's Rule and $n = 6$.

In Exercises 41 – 44, a solid of revolution is described. Find or approximate its surface area as specified.

41. Find the surface area of the sphere formed by rotating the circle $x = 2\cos t$, $y = 2\sin t$ about:

(a) the x-axis and

(b) the y-axis.

42. Find the surface area of the torus (or "donut") formed by rotating the circle $x = \cos t + 2$, $y = \sin t$ about the y-axis.

43. Approximate the surface area of the solid formed by rotating the "upper right half" of the bow tie curve $x = \cos t$, $y = \sin(2t)$ on $[0, \pi/2]$ about the x-axis, using Simpson's Rule and $n = 4$.

44. Approximate the surface area of the solid formed by rotating the one petal of the rose curve $x = \cos t\cos(2t)$, $y = \sin t\cos(2t)$ on $[0, \pi/4]$ about the x-axis, using Simpson's Rule and $n = 4$.

9.4 Introduction to Polar Coordinates

We are generally introduced to the idea of graphing curves by relating *x*-values to *y*-values through a function *f*. That is, we set $y = f(x)$, and plot lots of point pairs (x, y) to get a good notion of how the curve looks. This method is useful but has limitations, not least of which is that curves that "fail the vertical line test" cannot be graphed without using multiple functions.

The previous two sections introduced and studied a new way of plotting points in the x, y-plane. Using parametric equations, *x* and *y* values are computed independently and then plotted together. This method allows us to graph an extraordinary range of curves. This section introduces yet another way to plot points in the plane: using **polar coordinates**.

Polar Coordinates

Start with a point *O* in the plane called the **pole** (we will always identify this point with the origin). From the pole, draw a ray, called the **initial ray** (we will always draw this ray horizontally, identifying it with the positive *x*-axis). A point *P* in the plane is determined by the distance *r* that *P* is from *O*, and the angle θ formed between the initial ray and the segment $\overline{OP}$ (measured counter-clockwise). We record the distance and angle as an ordered pair (r, θ). To avoid confusion with rectangular coordinates, we will denote polar coordinates with the letter *P*, as in $P(r, \theta)$. This is illustrated in Figure 9.4.1

Practice will make this process more clear.

Example 9.4.1 **Plotting Polar Coordinates**
Plot the following polar coordinates:

$$A = P(1, \pi/4) \quad B = P(1.5, \pi) \quad C = P(2, -\pi/3) \quad D = P(-1, \pi/4)$$

SOLUTION To aid in the drawing, a polar grid is provided at the bottom of this page. To place the point *A*, go out 1 unit along the initial ray (putting you on the inner circle shown on the grid), then rotate counter-clockwise $\pi/4$ radians (or 45°). Alternately, one can consider the rotation first: think about the ray from *O* that forms an angle of $\pi/4$ with the initial ray, then move out 1 unit along this ray (again placing you on the inner circle of the grid).

To plot *B*, go out 1.5 units along the initial ray and rotate π radians (180°).

To plot *C*, go out 2 units along the initial ray then rotate *clockwise* $\pi/3$ radians, as the angle given is negative.

To plot *D*, move along the initial ray "−1" units − in other words, "back up" 1 unit, then rotate counter-clockwise by $\pi/4$. The results are given in Figure 9.4.2.

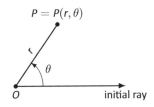

Figure 9.4.1: Illustrating polar coordinates.

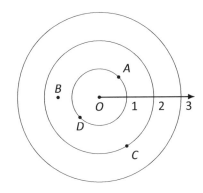

Figure 9.4.2: Plotting polar points in Example 9.4.1.

Notes:

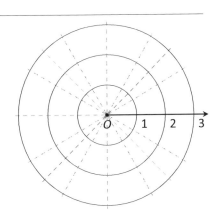

Consider the following two points: $A = P(1, \pi)$ and $B = P(-1, 0)$. To locate A, go out 1 unit on the initial ray then rotate π radians; to locate B, go out -1 units on the initial ray and don't rotate. One should see that A and B are located at the same point in the plane. We can also consider $C = P(1, 3\pi)$, or $D = P(1, -\pi)$; all four of these points share the same location.

This ability to identify a point in the plane with multiple polar coordinates is both a "blessing" and a "curse." We will see that it is beneficial as we can plot beautiful functions that intersect themselves (much like we saw with parametric functions). The unfortunate part of this is that it can be difficult to determine when this happens. We'll explore this more later in this section.

Polar to Rectangular Conversion

It is useful to recognize both the rectangular (or, Cartesian) coordinates of a point in the plane and its polar coordinates. Figure 9.4.3 shows a point P in the plane with rectangular coordinates (x, y) and polar coordinates $P(r, \theta)$. Using trigonometry, we can make the identities given in the following Key Idea.

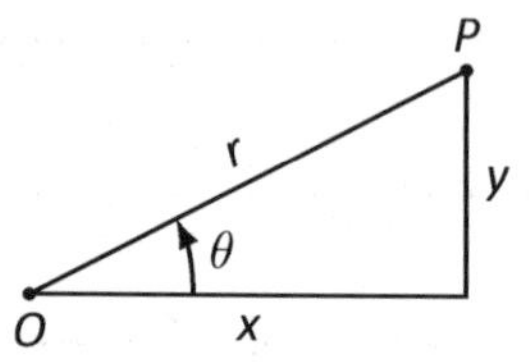

Figure 9.4.3: Converting between rectangular and polar coordinates.

Key Idea 9.4.1 **Converting Between Rectangular and Polar Coordinates**

Given the polar point $P(r, \theta)$, the rectangular coordinates are determined by
$$x = r\cos\theta \qquad y = r\sin\theta.$$

Given the rectangular coordinates (x, y), the polar coordinates are determined by
$$r^2 = x^2 + y^2 \qquad \tan\theta = \frac{y}{x}.$$

Example 9.4.2 **Converting Between Polar and Rectangular Coordinates**

1. Convert the polar coordinates $P(2, 2\pi/3)$ and $P(-1, 5\pi/4)$ to rectangular coordinates.

2. Convert the rectangular coordinates $(1, 2)$ and $(-1, 1)$ to polar coordinates.

SOLUTION

Notes:

1. (a) We start with $P(2, 2\pi/3)$. Using Key Idea 9.4.1, we have

   $$x = 2\cos(2\pi/3) = -1 \qquad y = 2\sin(2\pi/3) = \sqrt{3}.$$

   So the rectangular coordinates are $(-1, \sqrt{3}) \approx (-1, 1.732)$.

   (b) The polar point $P(-1, 5\pi/4)$ is converted to rectangular with:

   $$x = -1\cos(5\pi/4) = \sqrt{2}/2 \qquad y = -1\sin(5\pi/4) = \sqrt{2}/2.$$

   So the rectangular coordinates are $(\sqrt{2}/2, \sqrt{2}/2) \approx (0.707, 0.707)$.

   These points are plotted in Figure 9.4.4 (a). The rectangular coordinate system is drawn lightly under the polar coordinate system so that the relationship between the two can be seen.

2. (a) To convert the rectangular point $(1, 2)$ to polar coordinates, we use the Key Idea to form the following two equations:

   $$1^2 + 2^2 = r^2 \qquad \tan\theta = \frac{2}{1}.$$

   The first equation tells us that $r = \sqrt{5}$. Using the inverse tangent function, we find

   $$\tan\theta = 2 \quad \Rightarrow \quad \theta = \tan^{-1}2 \approx 1.11 \approx 63.43°.$$

   Thus polar coordinates of $(1, 2)$ are $P(\sqrt{5}, 1.11)$.

   (b) To convert $(-1, 1)$ to polar coordinates, we form the equations

   $$(-1)^2 + 1^2 = r^2 \qquad \tan\theta = \frac{1}{-1}.$$

   Thus $r = \sqrt{2}$. We need to be careful in computing θ: using the inverse tangent function, we have

   $$\tan\theta = -1 \quad \Rightarrow \quad \theta = \tan^{-1}(-1) = -\pi/4 = -45°.$$

   This is not the angle we desire. The range of $\tan^{-1}x$ is $(-\pi/2, \pi/2)$; that is, it returns angles that lie in the 1st and 4th quadrants. To find locations in the 2nd and 3rd quadrants, add π to the result of $\tan^{-1}x$. So $\pi + (-\pi/4)$ puts the angle at $3\pi/4$. Thus the polar point is $P(\sqrt{2}, 3\pi/4)$.

   An alternate method is to use the angle θ given by arctangent, but change the sign of r. Thus we could also refer to $(-1, 1)$ as $P(-\sqrt{2}, -\pi/4)$.

   These points are plotted in Figure 9.4.4 (b). The polar system is drawn lightly under the rectangular grid with rays to demonstrate the angles used.

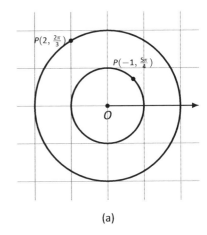

(a)

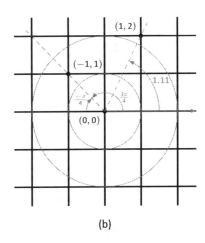

(b)

Figure 9.4.4: Plotting rectangular and polar points in Example 9.4.2.

Notes:

Polar Functions and Polar Graphs

Defining a new coordinate system allows us to create a new kind of function, a **polar function.** Rectangular coordinates lent themselves well to creating functions that related x and y, such as $y = x^2$. Polar coordinates allow us to create functions that relate r and θ. Normally these functions look like $r = f(\theta)$, although we can create functions of the form $\theta = f(r)$. The following examples introduce us to this concept.

Example 9.4.3   Introduction to Graphing Polar Functions
Describe the graphs of the following polar functions.

1. $r = 1.5$

2. $\theta = \pi/4$

SOLUTION

1. The equation $r = 1.5$ describes all points that are 1.5 units from the pole; as the angle is not specified, any θ is allowable. All points 1.5 units from the pole describes a circle of radius 1.5.

   We can consider the rectangular equivalent of this equation; using $r^2 = x^2 + y^2$, we see that $1.5^2 = x^2 + y^2$, which we recognize as the equation of a circle centered at $(0, 0)$ with radius 1.5. This is sketched in Figure 9.4.5.

2. The equation $\theta = \pi/4$ describes all points such that the line through them and the pole make an angle of $\pi/4$ with the initial ray. As the radius r is not specified, it can be any value (even negative). Thus $\theta = \pi/4$ describes the line through the pole that makes an angle of $\pi/4 = 45°$ with the initial ray.

   We can again consider the rectangular equivalent of this equation. Combine $\tan\theta = y/x$ and $\theta = \pi/4$:

   $$\tan\pi/4 = y/x \quad \Rightarrow x\tan\pi/4 = y \quad \Rightarrow y = x.$$

   This graph is also plotted in Figure 9.4.5.

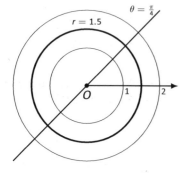

Figure 9.4.5: Plotting standard polar plots.

The basic rectangular equations of the form $x = h$ and $y = k$ create vertical and horizontal lines, respectively; the basic polar equations $r = h$ and $\theta = \alpha$ create circles and lines through the pole, respectively. With this as a foundation, we can create more complicated polar functions of the form $r = f(\theta)$. The input is an angle; the output is a length, how far in the direction of the angle to go out.

Notes:

We sketch these functions much like we sketch rectangular and parametric functions: we plot lots of points and "connect the dots" with curves. We demonstrate this in the following example.

Example 9.4.4   Sketching Polar Functions

Sketch the polar function $r = 1 + \cos\theta$ on $[0, 2\pi]$ by plotting points.

SOLUTION   A common question when sketching curves by plotting points is "Which points should I plot?" With rectangular equations, we often choose "easy" values – integers, then add more if needed. When plotting polar equations, start with the "common" angles – multiples of $\pi/6$ and $\pi/4$. Figure 9.4.6 gives a table of just a few values of θ in $[0, \pi]$.

Consider the point $P(2, 0)$ determined by the first line of the table. The angle is 0 radians – we do not rotate from the initial ray – then we go out 2 units from the pole. When $\theta = \pi/6$, $r = 1.866$ (actually, it is $1 + \sqrt{3}/2$); so rotate by $\pi/6$ radians and go out 1.866 units.

The graph shown uses more points, connected with straight lines. (The points on the graph that correspond to points in the table are signified with larger dots.) Such a sketch is likely good enough to give one an idea of what the graph looks like.

Technology Note: Plotting functions in this way can be tedious, just as it was with rectangular functions. To obtain very accurate graphs, technology is a great aid. Most graphing calculators can plot polar functions; in the menu, set the plotting mode to something like `polar` or `POL`, depending on one's calculator. As with plotting parametric functions, the viewing "window" no longer determines the x-values that are plotted, so additional information needs to be provided. Often with the "window" settings are the settings for the beginning and ending θ values (often called θ_{min} and θ_{max}) as well as the θ_{step} – that is, how far apart the θ values are spaced. The smaller the θ_{step} value, the more accurate the graph (which also increases plotting time). Using technology, we graphed the polar function $r = 1 + \cos\theta$ from Example 9.4.4 in Figure 9.4.7.

Example 9.4.5   Sketching Polar Functions

Sketch the polar function $r = \cos(2\theta)$ on $[0, 2\pi]$ by plotting points.

SOLUTION   We start by making a table of $\cos(2\theta)$ evaluated at common angles θ, as shown in Figure 9.4.8. These points are then plotted in Figure 9.4.9 (a). This particular graph "moves" around quite a bit and one can easily forget which points should be connected to each other. To help us with this, we numbered each point in the table and on the graph.

θ	$r = 1 + \cos\theta$
0	2
$\pi/6$	1.86603
$\pi/2$	1
$4\pi/3$	0.5
$7\pi/4$	1.70711

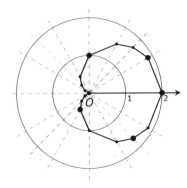

Figure 9.4.6: Graphing a polar function in Example 9.4.4 by plotting points.

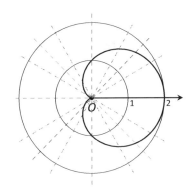

Figure 9.4.7: Using technology to graph a polar function.

Notes:

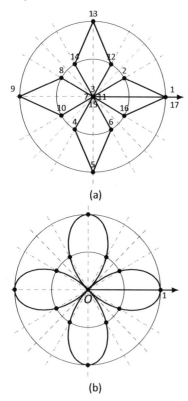

(a)

(b)

Figure 9.4.9: Polar plots from Example 9.4.5.

Pt.	θ	$\cos(2\theta)$		Pt.	θ	$\cos(2\theta)$
1	0	1.		10	$7\pi/6$	0.5
2	$\pi/6$	0.5		11	$5\pi/4$	0.
3	$\pi/4$	0.		12	$4\pi/3$	-0.5
4	$\pi/3$	-0.5		13	$3\pi/2$	$-1.$
5	$\pi/2$	$-1.$		14	$5\pi/3$	-0.5
6	$2\pi/3$	-0.5		15	$7\pi/4$	0.
7	$3\pi/4$	0.		16	$11\pi/6$	0.5
8	$5\pi/6$	0.5		17	2π	1.
9	π	1.				

Figure 9.4.8: Tables of points for plotting a polar curve.

Using more points (and the aid of technology) a smoother plot can be made as shown in Figure 9.4.9 (b). This plot is an example of a *rose curve*.

It is sometimes desirable to refer to a graph via a polar equation, and other times by a rectangular equation. Therefore it is necessary to be able to convert between polar and rectangular functions, which we practice in the following example. We will make frequent use of the identities found in Key Idea 9.4.1.

Example 9.4.6   Converting between rectangular and polar equations.

Convert from rectangular to polar.

1. $y = x^2$

2. $xy = 1$

Convert from polar to rectangular.

3. $r = \dfrac{2}{\sin\theta - \cos\theta}$

4. $r = 2\cos\theta$

SOLUTION

1. Replace y with $r\sin\theta$ and replace x with $r\cos\theta$, giving:

$$y = x^2$$
$$r\sin\theta = r^2\cos^2\theta$$
$$\frac{\sin\theta}{\cos^2\theta} = r$$

We have found that $r = \sin\theta/\cos^2\theta = \tan\theta\sec\theta$. The domain of this polar function is $(-\pi/2, \pi/2)$; plot a few points to see how the familiar parabola is traced out by the polar equation.

Notes:

2. We again replace x and y using the standard identities and work to solve for r:

$$xy = 1$$
$$r\cos\theta \cdot r\sin\theta = 1$$
$$r^2 = \frac{1}{\cos\theta\sin\theta}$$
$$r = \frac{1}{\sqrt{\cos\theta\sin\theta}}$$

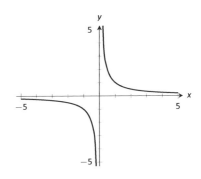

Figure 9.4.10: Graphing $xy = 1$ from Example 9.4.6.

This function is valid only when the product of $\cos\theta\sin\theta$ is positive. This occurs in the first and third quadrants, meaning the domain of this polar function is $(0, \pi/2) \cup (\pi, 3\pi/2)$.

We can rewrite the original rectangular equation $xy = 1$ as $y = 1/x$. This is graphed in Figure 9.4.10; note how it only exists in the first and third quadrants.

3. There is no set way to convert from polar to rectangular; in general, we look to form the products $r\cos\theta$ and $r\sin\theta$, and then replace these with x and y, respectively. We start in this problem by multiplying both sides by $\sin\theta - \cos\theta$:

$$r = \frac{2}{\sin\theta - \cos\theta}$$
$$r(\sin\theta - \cos\theta) = 2$$
$$r\sin\theta - r\cos\theta = 2. \qquad \text{Now replace with } y \text{ and } x:$$
$$y - x = 2$$
$$y = x + 2.$$

The original polar equation, $r = 2/(\sin\theta - \cos\theta)$ does not easily reveal that its graph is simply a line. However, our conversion shows that it is. The upcoming gallery of polar curves gives the general equations of lines in polar form.

4. By multiplying both sides by r, we obtain both an r^2 term and an $r\cos\theta$ term, which we replace with $x^2 + y^2$ and x, respectively.

$$r = 2\cos\theta$$
$$r^2 = 2r\cos\theta$$
$$x^2 + y^2 = 2x.$$

Notes:

We recognize this as a circle; by completing the square we can find its radius and center.

$$x^2 - 2x + y^2 = 0$$
$$(x - 1)^2 + y^2 = 1.$$

The circle is centered at $(1, 0)$ and has radius 1. The upcoming gallery of polar curves gives the equations of *some* circles in polar form; circles with arbitrary centers have a complicated polar equation that we do not consider here.

Some curves have very simple polar equations but rather complicated rectangular ones. For instance, the equation $r = 1 + \cos\theta$ describes a *cardioid* (a shape important to the sensitivity of microphones, among other things; one is graphed in the gallery in the Limaçon section). It's rectangular form is not nearly as simple; it is the implicit equation $x^4 + y^4 + 2x^2y^2 - 2xy^2 - 2x^3 - y^2 = 0$. The conversion is not "hard," but takes several steps, and is left as a problem in the Exercise section.

Gallery of Polar Curves

There are a number of basic and "classic" polar curves, famous for their beauty and/or applicability to the sciences. This section ends with a small gallery of some of these graphs. We encourage the reader to understand how these graphs are formed, and to investigate with technology other types of polar functions.

Lines

| **Through the origin:** | **Horizontal line:** | **Vertical line:** | **Not through origin:** |

$\theta = \alpha$

$r = a\csc\theta$

$r = a\sec\theta$

$r = \dfrac{b}{\sin\theta - m\cos\theta}$

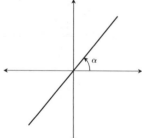

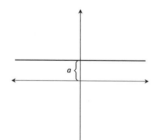

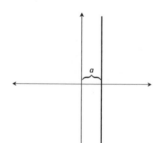

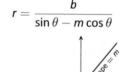

Notes:

Circles

Centered on *x*-axis:
$r = a \cos \theta$

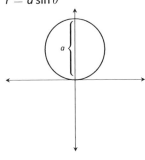

Centered on *y*-axis:
$r = a \sin \theta$

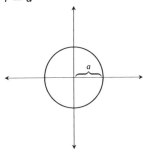

Centered on origin:
$r = a$

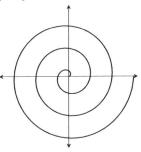

Spiral

Archimedean spiral
$r = \theta$

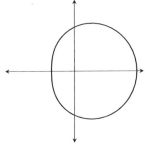

Limaçons

Symmetric about *x*-axis: $r = a \pm b \cos \theta$;   Symmetric about *y*-axis: $r = a \pm b \sin \theta$; $a, b > 0$

With inner loop:

$\dfrac{a}{b} < 1$

Cardioid:

$\dfrac{a}{b} = 1$

Dimpled:

$1 < \dfrac{a}{b} < 2$

Convex:

$\dfrac{a}{b} > 2$

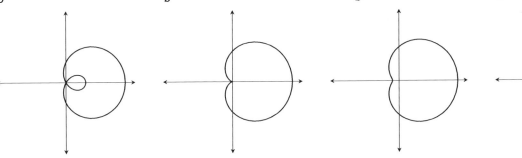

Rose Curves

Symmetric about *x*-axis: $r = a \cos(n\theta)$;   Symmetric about *y*-axis: $r = a \sin(n\theta)$
Curve contains 2*n* petals when *n* is even and *n* petals when *n* is odd.

$r = a \cos(2\theta)$

$r = a \sin(2\theta)$

$r = a \cos(3\theta)$

$r = a \sin(3\theta)$

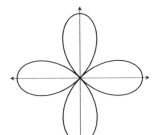
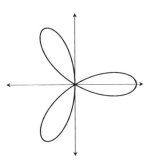
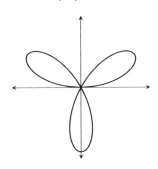

Special Curves

Rose curves

$r = a \sin(\theta/5)$

$r = a \sin(2\theta/5)$

Lemniscate:

$r^2 = a^2 \cos(2\theta)$

Eight Curve:

$r^2 = a^2 \sec^4 \theta \cos(2\theta)$

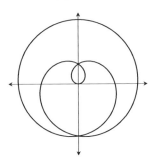
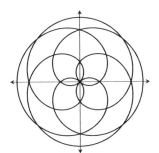
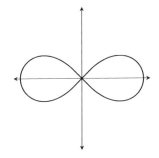
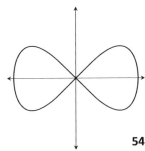

Earlier we discussed how each point in the plane does not have a unique representation in polar form. This can be a "good" thing, as it allows for the beautiful and interesting curves seen in the preceding gallery. However, it can also be a "bad" thing, as it can be difficult to determine where two curves intersect.

Example 9.4.7   Finding points of intersection with polar curves
Determine where the graphs of the polar equations $r = 1 + 3\cos\theta$ and $r = \cos\theta$ intersect.

SOLUTION   As technology is generally readily available, it is usually a good idea to start with a graph. We have graphed the two functions in Figure 9.4.11(a); to better discern the intersection points, part (b) of the figure zooms in around the origin. We start by setting the two functions equal to each other and solving for θ:

$$1 + 3\cos\theta = \cos\theta$$
$$2\cos\theta = -1$$
$$\cos\theta = -\frac{1}{2}$$
$$\theta = \frac{2\pi}{3}, \frac{4\pi}{3}.$$

(There are, of course, infinite solutions to the equation $\cos\theta = -1/2$; as the limaçon is traced out once on $[0, 2\pi]$, we restrict our solutions to this interval.)

We need to analyze this solution. When $\theta = 2\pi/3$ we obtain the point of intersection that lies in the 4th quadrant. When $\theta = 4\pi/3$, we get the point of intersection that lies in the 2nd quadrant. There is more to say about this second intersection point, however. The circle defined by $r = \cos\theta$ is traced out once on $[0, \pi]$, meaning that this point of intersection occurs while tracing out the circle a second time. It seems strange to pass by the point once and then recognize it as a point of intersection only when arriving there a "second time." The first time the circle arrives at this point is when $\theta = \pi/3$. It is key to understand that these two points are the same: $(\cos\pi/3, \pi/3)$ and $(\cos 4\pi/3, 4\pi/3)$.

To summarize what we have done so far, we have found two points of intersection: when $\theta = 2\pi/3$ and when $\theta = 4\pi/3$. When referencing the circle $r = \cos\theta$, the latter point is better referenced as when $\theta = \pi/3$.

There is yet another point of intersection: the pole (or, the origin). We did not recognize this intersection point using our work above as each graph arrives at the pole at a different θ value.

A graph intersects the pole when $r = 0$. Considering the circle $r = \cos\theta$, $r = 0$ when $\theta = \pi/2$ (and odd multiples thereof, as the circle is repeatedly

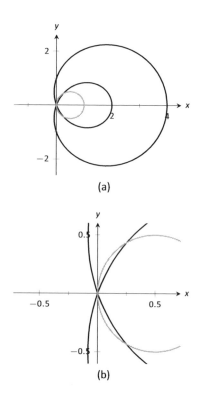

Figure 9.4.11: Graphs to help determine the points of intersection of the polar functions given in Example 9.4.7.

Notes:

traced). The limaçon intersects the pole when $1 + 3\cos\theta = 0$; this occurs when $\cos\theta = -1/3$, or for $\theta = \cos^{-1}(-1/3)$. This is a nonstandard angle, approximately $\theta = 1.9106 = 109.47°$. The limaçon intersects the pole twice in $[0, 2\pi]$; the other angle at which the limaçon is at the pole is the reflection of the first angle across the x-axis. That is, $\theta = 4.3726 = 250.53°$.

If all one is concerned with is the (x, y) coordinates at which the graphs intersect, much of the above work is extraneous. We know they intersect at $(0, 0)$; we might not care at what θ value. Likewise, using $\theta = 2\pi/3$ and $\theta = 4\pi/3$ can give us the needed rectangular coordinates. However, in the next section we apply calculus concepts to polar functions. When computing the area of a region bounded by polar curves, understanding the nuances of the points of intersection becomes important.

Notes:

Exercises 9.4

Terms and Concepts

1. In your own words, describe how to plot the polar point $P(r, \theta)$.

2. T/F: When plotting a point with polar coordinate $P(r, \theta)$, r must be positive.

3. T/F: Every point in the Cartesian plane can be represented by a polar coordinate.

4. T/F: Every point in the Cartesian plane can be represented uniquely by a polar coordinate.

Problems

5. Plot the points with the given polar coordinates.

   (a) $A = P(2, 0)$ (c) $C = P(-2, \pi/2)$
   (b) $B = P(1, \pi)$ (d) $D = P(1, \pi/4)$

6. Plot the points with the given polar coordinates.

   (a) $A = P(2, 3\pi)$ (c) $C = P(1, 2)$
   (b) $B = P(1, -\pi)$ (d) $D = P(1/2, 5\pi/6)$

7. For each of the given points give two sets of polar coordinates that identify it, where $0 \le \theta \le 2\pi$.

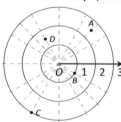

8. For each of the given points give two sets of polar coordinates that identify it, where $-\pi \le \theta \le \pi$.

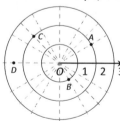

9. Convert each of the following polar coordinates to rectangular, and each of the following rectangular coordinates to polar.

   (a) $A = P(2, \pi/4)$ (c) $C = (2, -1)$
   (b) $B = P(2, -\pi/4)$ (d) $D = (-2, 1)$

10. Convert each of the following polar coordinates to rectangular, and each of the following rectangular coordinates to polar.

    (a) $A = P(3, \pi)$ (c) $C = (0, 4)$
    (b) $B = P(1, 2\pi/3)$ (d) $D = (1, -\sqrt{3})$

In Exercises 11 – 30, graph the polar function on the given interval.

11. $r = 2, \quad 0 \le \theta \le \pi/2$

12. $\theta = \pi/6, \quad -1 \le r \le 2$

13. $r = 1 - \cos\theta, \quad [0, 2\pi]$

14. $r = 2 + \sin\theta, \quad [0, 2\pi]$

15. $r = 2 - \sin\theta, \quad [0, 2\pi]$

16. $r = 1 - 2\sin\theta, \quad [0, 2\pi]$

17. $r = 1 + 2\sin\theta, \quad [0, 2\pi]$

18. $r = \cos(2\theta), \quad [0, 2\pi]$

19. $r = \sin(3\theta), \quad [0, \pi]$

20. $r = \cos(\theta/3), \quad [0, 3\pi]$

21. $r = \cos(2\theta/3), \quad [0, 6\pi]$

22. $r = \theta/2, \quad [0, 4\pi]$

23. $r = 3\sin(\theta), \quad [0, \pi]$

24. $r = 2\cos(\theta), \quad [0, \pi/2]$

25. $r = \cos\theta \sin\theta, \quad [0, 2\pi]$

26. $r = \theta^2 - (\pi/2)^2, \quad [-\pi, \pi]$

27. $r = \dfrac{3}{5\sin\theta - \cos\theta}, \quad [0, 2\pi]$

28. $r = \dfrac{-2}{3\cos\theta - 2\sin\theta}, \quad [0, 2\pi]$

29. $r = 3\sec\theta, \quad (-\pi/2, \pi/2)$

30. $r = 3\csc\theta, \quad (0, \pi)$

In Exercises 31 – 40, convert the polar equation to a rectangular equation.

31. $r = 6\cos\theta$

32. $r = -4\sin\theta$

33. $r = \cos\theta + \sin\theta$

34. $r = \dfrac{7}{5\sin\theta - 2\cos\theta}$

35. $r = \dfrac{3}{\cos\theta}$

36. $r = \dfrac{4}{\sin\theta}$

37. $r = \tan\theta$

38. $r = \cot\theta$

39. $r = 2$

40. $\theta = \pi/6$

In Exercises 41 – 48, convert the rectangular equation to a polar equation.

41. $y = x$

42. $y = 4x + 7$

43. $x = 5$

44. $y = 5$

45. $x = y^2$

46. $x^2 y = 1$

47. $x^2 + y^2 = 7$

48. $(x + 1)^2 + y^2 = 1$

In Exercises 49 – 56, find the points of intersection of the polar graphs.

49. $r = \sin(2\theta)$ and $r = \cos\theta$ on $[0, \pi]$

50. $r = \cos(2\theta)$ and $r = \cos\theta$ on $[0, \pi]$

51. $r = 2\cos\theta$ and $r = 2\sin\theta$ on $[0, \pi]$

52. $r = \sin\theta$ and $r = \sqrt{3} + 3\sin\theta$ on $[0, 2\pi]$

53. $r = \sin(3\theta)$ and $r = \cos(3\theta)$ on $[0, \pi]$

54. $r = 3\cos\theta$ and $r = 1 + \cos\theta$ on $[-\pi, \pi]$

55. $r = 1$ and $r = 2\sin(2\theta)$ on $[0, 2\pi]$

56. $r = 1 - \cos\theta$ and $r = 1 + \sin\theta$ on $[0, 2\pi]$

57. Pick a integer value for n, where $n \neq 2, 3$, and use technology to plot $r = \sin\left(\dfrac{m}{n}\theta\right)$ for three different integer values of m. Sketch these and determine a minimal interval on which the entire graph is shown.

58. Create your own polar function, $r = f(\theta)$ and sketch it. Describe why the graph looks as it does.

9.5 Calculus and Polar Functions

The previous section defined polar coordinates, leading to polar functions. We investigated plotting these functions and solving a fundamental question about their graphs, namely, where do two polar graphs intersect?

We now turn our attention to answering other questions, whose solutions require the use of calculus. A basis for much of what is done in this section is the ability to turn a polar function $r = f(\theta)$ into a set of parametric equations. Using the identities $x = r\cos\theta$ and $y = r\sin\theta$, we can create the parametric equations $x = f(\theta)\cos\theta$, $y = f(\theta)\sin\theta$ and apply the concepts of Section 9.3.

Polar Functions and $\dfrac{dy}{dx}$

We are interested in the lines tangent to a given graph, regardless of whether that graph is produced by rectangular, parametric, or polar equations. In each of these contexts, the slope of the tangent line is $\frac{dy}{dx}$. Given $r = f(\theta)$, we are generally *not* concerned with $r' = f'(\theta)$; that describes how fast r changes with respect to θ. Instead, we will use $x = f(\theta)\cos\theta$, $y = f(\theta)\sin\theta$ to compute $\frac{dy}{dx}$.

Using Key Idea 9.3.1 we have

$$\frac{dy}{dx} = \frac{dy}{d\theta} \Big/ \frac{dx}{d\theta}.$$

Each of the two derivatives on the right hand side of the equality requires the use of the Product Rule. We state the important result as a Key Idea.

Key Idea 9.5.1   **Finding $\frac{dy}{dx}$ with Polar Functions**

Let $r = f(\theta)$ be a polar function. With $x = f(\theta)\cos\theta$ and $y = f(\theta)\sin\theta$,

$$\frac{dy}{dx} = \frac{f'(\theta)\sin\theta + f(\theta)\cos\theta}{f'(\theta)\cos\theta - f(\theta)\sin\theta}.$$

Example 9.5.1   **Finding $\frac{dy}{dx}$ with polar functions.**
Consider the limaçon $r = 1 + 2\sin\theta$ on $[0, 2\pi]$.

1. Find the equations of the tangent and normal lines to the graph at $\theta = \pi/4$.

2. Find where the graph has vertical and horizontal tangent lines.

Notes:

SOLUTION

1. We start by computing $\frac{dy}{dx}$. With $f'(\theta) = 2\cos\theta$, we have

$$\frac{dy}{dx} = \frac{2\cos\theta\sin\theta + \cos\theta(1 + 2\sin\theta)}{2\cos^2\theta - \sin\theta(1 + 2\sin\theta)}$$

$$= \frac{\cos\theta(4\sin\theta + 1)}{2(\cos^2\theta - \sin^2\theta) - \sin\theta}.$$

When $\theta = \pi/4$, $\frac{dy}{dx} = -2\sqrt{2} - 1$ (this requires a bit of simplification). In rectangular coordinates, the point on the graph at $\theta = \pi/4$ is $(1 + \sqrt{2}/2, 1 + \sqrt{2}/2)$. Thus the rectangular equation of the line tangent to the limaçon at $\theta = \pi/4$ is

$$y = (-2\sqrt{2} - 1)\big(x - (1 + \sqrt{2}/2)\big) + 1 + \sqrt{2}/2 \approx -3.83x + 8.24.$$

The limaçon and the tangent line are graphed in Figure 9.5.1.

The normal line has the opposite–reciprocal slope as the tangent line, so its equation is

$$y \approx \frac{1}{3.83}x + 1.26.$$

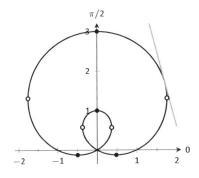

Figure 9.5.1: The limaçon in Example 9.5.1 with its tangent line at $\theta = \pi/4$ and points of vertical and horizontal tangency.

2. To find the horizontal lines of tangency, we find where $\frac{dy}{dx} = 0$; thus we find where the numerator of our equation for $\frac{dy}{dx}$ is 0.

$$\cos\theta(4\sin\theta + 1) = 0 \quad \Rightarrow \quad \cos\theta = 0 \quad \text{or} \quad 4\sin\theta + 1 = 0.$$

On $[0, 2\pi]$, $\cos\theta = 0$ when $\theta = \pi/2, 3\pi/2$.

Setting $4\sin\theta + 1 = 0$ gives $\theta = \sin^{-1}(-1/4) \approx -0.2527 = -14.48°$. We want the results in $[0, 2\pi]$; we also recognize there are two solutions, one in the 3rd quadrant and one in the 4th. Using reference angles, we have our two solutions as $\theta = 3.39$ and 6.03 radians. The four points we obtained where the limaçon has a horizontal tangent line are given in Figure 9.5.1 with black–filled dots.

To find the vertical lines of tangency, we set the denominator of $\frac{dy}{dx} = 0$.

$$2(\cos^2\theta - \sin^2\theta) - \sin\theta = 0.$$

Convert the $\cos^2\theta$ term to $1 - \sin^2\theta$:

$$2(1 - \sin^2\theta - \sin^2\theta) - \sin\theta = 0$$
$$4\sin^2\theta + \sin\theta - 2 = 0.$$

Notes:

Recognize this as a quadratic in the variable $\sin\theta$. Using the quadratic formula, we have

$$\sin\theta = \frac{-1 \pm \sqrt{33}}{8}.$$

We solve $\sin\theta = \frac{-1+\sqrt{33}}{8}$ and $\sin\theta = \frac{-1-\sqrt{33}}{8}$:

$$\sin\theta = \frac{-1+\sqrt{33}}{8} \qquad\qquad \sin\theta = \frac{-1-\sqrt{33}}{8}$$

$$\theta = \sin^{-1}\left(\frac{-1+\sqrt{33}}{8}\right) \qquad \theta = \sin^{-1}\left(\frac{-1-\sqrt{33}}{8}\right)$$

$$\theta = 0.6349 \qquad\qquad\qquad \theta = -1.0030$$

In each of the solutions above, we only get one of the possible two solutions as $\sin^{-1}x$ only returns solutions in $[-\pi/2, \pi/2]$, the 4th and 1st quadrants. Again using reference angles, we have:

$$\sin\theta = \frac{-1+\sqrt{33}}{8} \quad\Rightarrow\quad \theta = 0.6349,\ 2.5067 \text{ radians}$$

and

$$\sin\theta = \frac{-1-\sqrt{33}}{8} \quad\Rightarrow\quad \theta = 4.1446,\ 5.2802 \text{ radians}.$$

These points are also shown in Figure 9.5.1 with white–filled dots.

When the graph of the polar function $r = f(\theta)$ intersects the pole, it means that $f(\alpha) = 0$ for some angle α. Thus the formula for $\frac{dy}{dx}$ in such instances is very simple, reducing simply to

$$\frac{dy}{dx} = \tan\alpha.$$

This equation makes an interesting point. It tells us the slope of the tangent line at the pole is $\tan\alpha$; some of our previous work (see, for instance, Example 9.4.3) shows us that the line through the pole with slope $\tan\alpha$ has polar equation $\theta = \alpha$. Thus when a polar graph touches the pole at $\theta = \alpha$, the equation of the tangent line at the pole is $\theta = \alpha$.

Example 9.5.2   Finding tangent lines at the pole.

Let $r = 1 + 2\sin\theta$, a limaçon. Find the equations of the lines tangent to the graph at the pole.

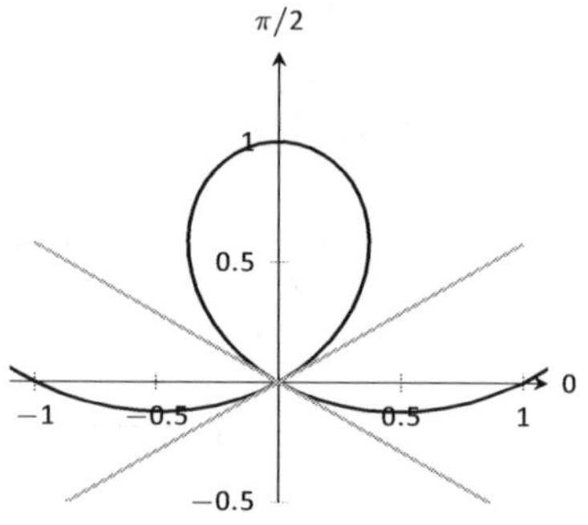

Figure 9.5.2: Graphing the tangent lines at the pole in Example 9.5.2.

Notes:

SOLUTION   We need to know when $r = 0$.

$$1 + 2\sin\theta = 0$$
$$\sin\theta = -1/2$$
$$\theta = \frac{7\pi}{6}, \frac{11\pi}{6}.$$

Thus the equations of the tangent lines, in polar, are $\theta = 7\pi/6$ and $\theta = 11\pi/6$. In rectangular form, the tangent lines are $y = \tan(7\pi/6)x$ and $y = \tan(11\pi/6)x$. The full limaçon can be seen in Figure 9.5.1; we zoom in on the tangent lines in Figure 9.5.2.

Area

When using rectangular coordinates, the equations $x = h$ and $y = k$ defined vertical and horizontal lines, respectively, and combinations of these lines create rectangles (hence the name "rectangular coordinates"). It is then somewhat natural to use rectangles to approximate area as we did when learning about the definite integral.

When using polar coordinates, the equations $\theta = \alpha$ and $r = c$ form lines through the origin and circles centered at the origin, respectively, and combinations of these curves form sectors of circles. It is then somewhat natural to calculate the area of regions defined by polar functions by first approximating with sectors of circles.

Consider Figure 9.5.3 (a) where a region defined by $r = f(\theta)$ on $[\alpha, \beta]$ is given. (Note how the "sides" of the region are the lines $\theta = \alpha$ and $\theta = \beta$ whereas in rectangular coordinates the "sides" of regions were often the vertical lines $x = a$ and $x = b$.)

Partition the interval $[\alpha, \beta]$ into n equally spaced subintervals as $\alpha = \theta_1 < \theta_2 < \cdots < \theta_{n+1} = \beta$ The length of each subinterval is $\Delta\theta = (\beta - \alpha)/n$, representing a small change in angle. The area of the region defined by the ith subinterval $[\theta_i, \theta_{i+1}]$ can be approximated with a sector of a circle with radius $f(c_i)$, for some c_i in $[\theta_i, \theta_{i+1}]$. The area of this sector is $\frac{1}{2}f(c_i)^2\Delta\theta$. This is shown in part (b) of the figure, where $[\alpha, \beta]$ has been divided into 4 subintervals. We approximate the area of the whole region by summing the areas of all sectors:

$$\text{Area} \approx \sum_{i=1}^{n} \frac{1}{2}f(c_i)^2\Delta\theta.$$

This is a Riemann sum. By taking the limit of the sum as $n \to \infty$, we find the

Note: Recall that the area of a sector of a circle with radius r subtended by an angle θ is $A = \frac{1}{2}\theta r^2$.

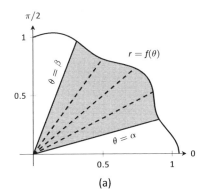

(a)

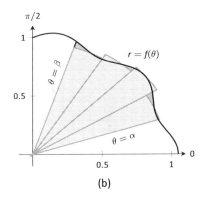

(b)

Figure 9.5.3: Computing the area of a polar region.

Notes:

exact area of the region in the form of a definite integral.

Theorem 9.5.1   Area of a Polar Region

Let f be continuous and non-negative on $[\alpha, \beta]$, where $0 \leq \beta - \alpha \leq 2\pi$. The area A of the region bounded by the curve $r = f(\theta)$ and the lines $\theta = \alpha$ and $\theta = \beta$ is

$$A = \frac{1}{2} \int_{\alpha}^{\beta} f(\theta)^2 \, d\theta = \frac{1}{2} \int_{\alpha}^{\beta} r^2 \, d\theta$$

Note: Example 9.5.3 requires the use of the integral $\int \cos^2 \theta \, d\theta$. This is handled well by using the power reducing formula as found at the end of this text. Due to the nature of the area formula, integrating $\cos^2 \theta$ and $\sin^2 \theta$ is required often. We offer here these indefinite integrals as a time–saving measure.

$$\int \cos^2 \theta \, d\theta = \frac{1}{2}\theta + \frac{1}{4}\sin(2\theta) + C$$

$$\int \sin^2 \theta \, d\theta = \frac{1}{2}\theta - \frac{1}{4}\sin(2\theta) + C$$

The theorem states that $0 \leq \beta - \alpha \leq 2\pi$. This ensures that region does not overlap itself, which would give a result that does not correspond directly to the area.

Example 9.5.3   Area of a polar region
Find the area of the circle defined by $r = \cos\theta$. (Recall this circle has radius $1/2$.)

SOLUTION   This is a direct application of Theorem 9.5.1. The circle is traced out on $[0, \pi]$, leading to the integral

$$\begin{aligned}
\text{Area} &= \frac{1}{2} \int_0^{\pi} \cos^2\theta \, d\theta \\
&= \frac{1}{2} \int_0^{\pi} \frac{1 + \cos(2\theta)}{2} \, d\theta \\
&= \frac{1}{4}\left(\theta + \frac{1}{2}\sin(2\theta)\right)\Big|_0^{\pi} \\
&= \frac{1}{4}\pi.
\end{aligned}$$

Of course, we already knew the area of a circle with radius $1/2$. We did this example to demonstrate that the area formula is correct.

Example 9.5.4   Area of a polar region
Find the area of the cardioid $r = 1 + \cos\theta$ bound between $\theta = \pi/6$ and $\theta = \pi/3$, as shown in Figure 9.5.4.

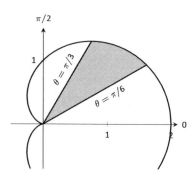

Figure 9.5.4: Finding the area of the shaded region of a cardioid in Example 9.5.4.

Notes:

SOLUTION   This is again a direct application of Theorem 9.5.1.

$$\begin{aligned}
\text{Area} &= \frac{1}{2}\int_{\pi/6}^{\pi/3}(1+\cos\theta)^2\,d\theta \\
&= \frac{1}{2}\int_{\pi/6}^{\pi/3}(1+2\cos\theta+\cos^2\theta)\,d\theta \\
&= \frac{1}{2}\left(\theta+2\sin\theta+\frac{1}{2}\theta+\frac{1}{4}\sin(2\theta)\right)\Bigg|_{\pi/6}^{\pi/3} \\
&= \frac{1}{8}\left(\pi+4\sqrt{3}-4\right)\approx 0.7587.
\end{aligned}$$

Area Between Curves

Our study of area in the context of rectangular functions led naturally to finding area bounded between curves. We consider the same in the context of polar functions.

Consider the shaded region shown in Figure 9.5.5. We can find the area of this region by computing the area bounded by $r_2 = f_2(\theta)$ and subtracting the area bounded by $r_1 = f_1(\theta)$ on $[\alpha, \beta]$. Thus

$$\text{Area} = \frac{1}{2}\int_\alpha^\beta r_2^2\,d\theta - \frac{1}{2}\int_\alpha^\beta r_1^2\,d\theta = \frac{1}{2}\int_\alpha^\beta \left(r_2^2 - r_1^2\right)\,d\theta.$$

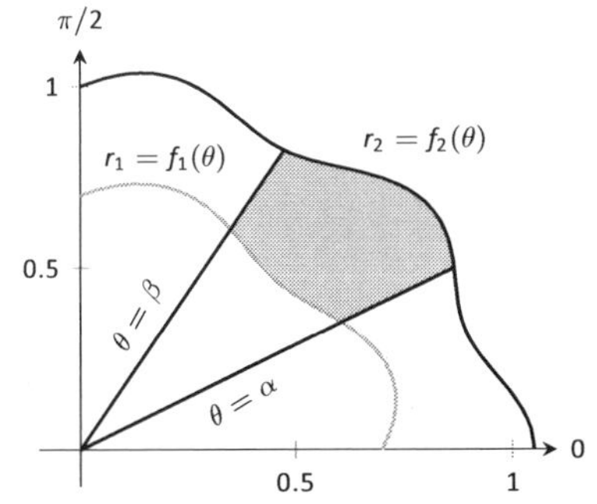

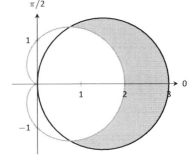

Figure 9.5.5: Illustrating area bound between two polar curves.

Key Idea 9.5.2   Area Between Polar Curves

The area A of the region bounded by $r_1 = f_1(\theta)$ and $r_2 = f_2(\theta)$, $\theta = \alpha$ and $\theta = \beta$ where $f_1(\theta) \leq f_2(\theta)$ on $[\alpha, \beta]$, is

$$A = \frac{1}{2}\int_\alpha^\beta \left(r_2^2 - r_1^2\right)\,d\theta.$$

Example 9.5.5   Area between polar curves
Find the area bounded between the curves $r = 1 + \cos\theta$ and $r = 3\cos\theta$, as shown in Figure 9.5.6.

SOLUTION   We need to find the points of intersection between these

Figure 9.5.6: Finding the area between polar curves in Example 9.5.5.

Notes:

two functions. Setting them equal to each other, we find:

$$1 + \cos\theta = 3\cos\theta$$
$$\cos\theta = 1/2$$
$$\theta = \pm\pi/3$$

Thus we integrate $\frac{1}{2}\left((3\cos\theta)^2 - (1+\cos\theta)^2\right)$ on $[-\pi/3, \pi/3]$.

$$\begin{aligned}
\text{Area} &= \frac{1}{2}\int_{-\pi/3}^{\pi/3}\left((3\cos\theta)^2 - (1+\cos\theta)^2\right)\,d\theta \\
&= \frac{1}{2}\int_{-\pi/3}^{\pi/3}\left(8\cos^2\theta - 2\cos\theta - 1\right)\,d\theta \\
&= \frac{1}{2}\left(2\sin(2\theta) - 2\sin\theta + 3\theta\right)\Big|_{-\pi/3}^{\pi/3} \\
&= \pi.
\end{aligned}$$

Amazingly enough, the area between these curves has a "nice" value.

Example 9.5.6   Area defined by polar curves
Find the area bounded between the polar curves $r = 1$ and $r = 2\cos(2\theta)$, as shown in Figure 9.5.7 (a).

    SOLUTION   We need to find the point of intersection between the two curves. Setting the two functions equal to each other, we have

$$2\cos(2\theta) = 1 \quad\Rightarrow\quad \cos(2\theta) = \frac{1}{2} \quad\Rightarrow\quad 2\theta = \pi/3 \quad\Rightarrow\quad \theta = \pi/6.$$

  In part (b) of the figure, we zoom in on the region and note that it is not really bounded *between* two polar curves, but rather *by* two polar curves, along with $\theta = 0$. The dashed line breaks the region into its component parts. Below the dashed line, the region is defined by $r = 1$, $\theta = 0$ and $\theta = \pi/6$. (Note: the dashed line lies on the line $\theta = \pi/6$.) Above the dashed line the region is bounded by $r = 2\cos(2\theta)$ and $\theta = \pi/6$. Since we have two separate regions, we find the area using two separate integrals.

    Call the area below the dashed line A_1 and the area above the dashed line A_2. They are determined by the following integrals:

$$A_1 = \frac{1}{2}\int_0^{\pi/6}(1)^2\,d\theta \qquad A_2 = \frac{1}{2}\int_{\pi/6}^{\pi/4}\left(2\cos(2\theta)\right)^2\,d\theta.$$

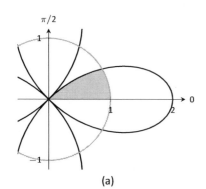

(a)

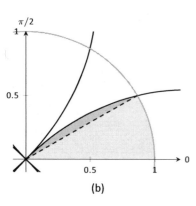

(b)

Figure 9.5.7: Graphing the region bounded by the functions in Example 9.5.6.

Notes:

(The upper bound of the integral computing A_2 is $\pi/4$ as $r = 2\cos(2\theta)$ is at the pole when $\theta = \pi/4$.)

We omit the integration details and let the reader verify that $A_1 = \pi/12$ and $A_2 = \pi/12 - \sqrt{3}/8$; the total area is $A = \pi/6 - \sqrt{3}/8$.

Arc Length

As we have already considered the arc length of curves defined by rectangular and parametric equations, we now consider it in the context of polar equations. Recall that the arc length L of the graph defined by the parametric equations $x = f(t)$, $y = g(t)$ on $[a, b]$ is

$$L = \int_a^b \sqrt{f'(t)^2 + g'(t)^2}\, dt = \int_a^b \sqrt{x'(t)^2 + y'(t)^2}\, dt. \qquad (9.1)$$

Now consider the polar function $r = f(\theta)$. We again use the identities $x = f(\theta)\cos\theta$ and $y = f(\theta)\sin\theta$ to create parametric equations based on the polar function. We compute $x'(\theta)$ and $y'(\theta)$ as done before when computing $\frac{dy}{dx}$, then apply Equation (9.1).

The expression $x'(\theta)^2 + y'(\theta)^2$ can be simplified a great deal; we leave this as an exercise and state that

$$x'(\theta)^2 + y'(\theta)^2 = f'(\theta)^2 + f(\theta)^2.$$

This leads us to the arc length formula.

Theorem 9.5.2   Arc Length of Polar Curves

Let $r = f(\theta)$ be a polar function with f' continuous on $[\alpha, \beta]$, on which the graph traces itself only once. The arc length L of the graph on $[\alpha, \beta]$ is

$$L = \int_\alpha^\beta \sqrt{f'(\theta)^2 + f(\theta)^2}\, d\theta = \int_\alpha^\beta \sqrt{(r')^2 + r^2}\, d\theta.$$

Example 9.5.7   Arc length of a limaçon
Find the arc length of the limaçon $r = 1 + 2\sin t$.

SOLUTION   With $r = 1 + 2\sin t$, we have $r' = 2\cos t$. The limaçon is traced out once on $[0, 2\pi]$, giving us our bounds of integration. Applying Theo-

---

Notes:

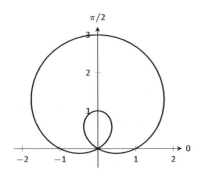

Figure 9.5.8: The limaçon in Example 9.5.7 whose arc length is measured.

rem 9.5.2, we have

$$L = \int_0^{2\pi} \sqrt{(2\cos\theta)^2 + (1 + 2\sin\theta)^2}\, d\theta$$

$$= \int_0^{2\pi} \sqrt{4\cos^2\theta + 4\sin^2\theta + 4\sin\theta + 1}\, d\theta$$

$$= \int_0^{2\pi} \sqrt{4\sin\theta + 5}\, d\theta$$

$$\approx 13.3649.$$

The final integral cannot be solved in terms of elementary functions, so we resorted to a numerical approximation. (Simpson's Rule, with $n = 4$, approximates the value with 13.0608. Using $n = 22$ gives the value above, which is accurate to 4 places after the decimal.)

Surface Area

The formula for arc length leads us to a formula for surface area. The following Theorem is based on Theorem 9.3.2.

Theorem 9.5.3   Surface Area of a Solid of Revolution

Consider the graph of the polar equation $r = f(\theta)$, where f' is continuous on $[\alpha, \beta]$, on which the graph does not cross itself.

1. The surface area of the solid formed by revolving the graph about the initial ray ($\theta = 0$) is:

$$\text{Surface Area} = 2\pi \int_\alpha^\beta f(\theta)\sin\theta\sqrt{f'(\theta)^2 + f(\theta)^2}\, d\theta.$$

2. The surface area of the solid formed by revolving the graph about the line $\theta = \pi/2$ is:

$$\text{Surface Area} = 2\pi \int_\alpha^\beta f(\theta)\cos\theta\sqrt{f'(\theta)^2 + f(\theta)^2}\, d\theta.$$

Notes:

Example 9.5.8   Surface area determined by a polar curve
Find the surface area formed by revolving one petal of the rose curve $r = \cos(2\theta)$ about its central axis (see Figure 9.5.9).

SOLUTION   We choose, as implied by the figure, to revolve the portion of the curve that lies on $[0, \pi/4]$ about the initial ray. Using Theorem 9.5.3 and the fact that $f'(\theta) = -2\sin(2\theta)$, we have

$$\text{Surface Area} = 2\pi \int_0^{\pi/4} \cos(2\theta)\sin(\theta)\sqrt{\left(-2\sin(2\theta)\right)^2 + \left(\cos(2\theta)\right)^2}\, d\theta$$
$$\approx 1.36707.$$

The integral is another that cannot be evaluated in terms of elementary functions. Simpson's Rule, with $n = 4$, approximates the value at 1.36751.

This chapter has been about curves in the plane. While there is great mathematics to be discovered in the two dimensions of a plane, we live in a three dimensional world and hence we should also look to do mathematics in 3D – that is, in *space*. The next chapter begins our exploration into space by introducing the topic of *vectors*, which are incredibly useful and powerful mathematical objects.

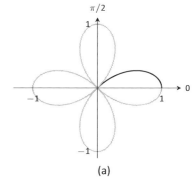
(a)

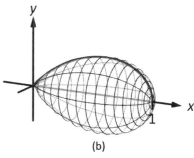
(b)

Figure 9.5.9: Finding the surface area of a rose–curve petal that is revolved around its central axis.

Notes:

Exercises 9.5

Terms and Concepts

1. Given polar equation $r = f(\theta)$, how can one create parametric equations of the same curve?

2. With rectangular coordinates, it is natural to approximate area with _____; with polar coordinates, it is natural to approximate area with _____.

Problems

In Exercises 3 – 10, find:

(a) $\dfrac{dy}{dx}$

(b) the equation of the tangent and normal lines to the curve at the indicated θ–value.

3. $r = 1; \quad \theta = \pi/4$

4. $r = \cos\theta; \quad \theta = \pi/4$

5. $r = 1 + \sin\theta; \quad \theta = \pi/6$

6. $r = 1 - 3\cos\theta; \quad \theta = 3\pi/4$

7. $r = \theta; \quad \theta = \pi/2$

8. $r = \cos(3\theta); \quad \theta = \pi/6$

9. $r = \sin(4\theta); \quad \theta = \pi/3$

10. $r = \dfrac{1}{\sin\theta - \cos\theta}; \quad \theta = \pi$

In Exercises 11 – 14, find the values of θ in the given interval where the graph of the polar function has horizontal and vertical tangent lines.

11. $r = 3; \quad [0, 2\pi]$

12. $r = 2\sin\theta; \quad [0, \pi]$

13. $r = \cos(2\theta); \quad [0, 2\pi]$

14. $r = 1 + \cos\theta; \quad [0, 2\pi]$

In Exercises 15 – 16, find the equation of the lines tangent to the graph at the pole.

15. $r = \sin\theta; \quad [0, \pi]$

16. $r = \sin(3\theta); \quad [0, \pi]$

In Exercises 17 – 28, find the area of the described region.

17. Enclosed by the circle: $r = 4\sin\theta$

18. Enclosed by the circle $r = 5$

19. Enclosed by one petal of $r = \sin(3\theta)$

20. Enclosed by one petal of the rose curve $r = \cos(n\,\theta)$, where n is a positive integer.

21. Enclosed by the cardioid $r = 1 - \sin\theta$

22. Enclosed by the inner loop of the limaçon $r = 1 + 2\cos\theta$

23. Enclosed by the outer loop of the limaçon $r = 1 + 2\cos\theta$ (including area enclosed by the inner loop)

24. Enclosed between the inner and outer loop of the limaçon $r = 1 + 2\cos\theta$

25. Enclosed by $r = 2\cos\theta$ and $r = 2\sin\theta$, as shown:

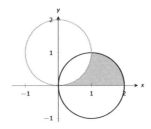

26. Enclosed by $r = \cos(3\theta)$ and $r = \sin(3\theta)$, as shown:

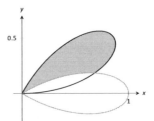

27. Enclosed by $r = \cos\theta$ and $r = \sin(2\theta)$, as shown:

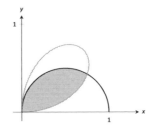

28. Enclosed by $r = \cos\theta$ and $r = 1 - \cos\theta$, as shown:

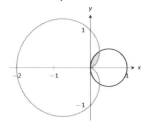

In Exercises 29 – 34, answer the questions involving arc length.

29. Use the arc length formula to compute the arc length of the circle $r = 2$.

30. Use the arc length formula to compute the arc length of the circle $r = 4\sin\theta$.

31. Use the arc length formula to compute the arc length of $r = \cos\theta + \sin\theta$.

32. Use the arc length formula to compute the arc length of the cardioid $r = 1 + \cos\theta$. (Hint: apply the formula, simplify, then use a Power–Reducing Formula to convert $1 + \cos\theta$ into a square.)

33. Approximate the arc length of one petal of the rose curve $r = \sin(3\theta)$ with Simpson's Rule and $n = 4$.

34. Let $x(\theta) = f(\theta)\cos\theta$ and $y(\theta) = f(\theta)\sin\theta$. Show, as suggested by the text, that

$$x'(\theta)^2 + y'(\theta)^2 = f'(\theta)^2 + f(\theta)^2.$$

In Exercises 35 – 40, answer the questions involving surface area.

35. Find the surface area of the sphere formed by revolving the circle $r = 2$ about the initial ray.

36. Find the surface area of the sphere formed by revolving the circle $r = 2\cos\theta$ about the initial ray.

37. Find the surface area of the solid formed by revolving the cardioid $r = 1 + \cos\theta$ about the initial ray.

38. Find the surface area of the solid formed by revolving the circle $r = 2\cos\theta$ about the line $\theta = \pi/2$.

39. Find the surface area of the solid formed by revolving the line $r = 3\sec\theta$, $-\pi/4 \le \theta \le \pi/4$, about the line $\theta = \pi/2$.

40. Find the surface area of the solid formed by revolving the line $r = 3\sec\theta$, $0 \le \theta \le \pi/4$, about the initial ray.

10: Vectors

This chapter introduces a new mathematical object, the **vector**. Defined in Section 10.2, we will see that vectors provide a powerful language for describing quantities that have magnitude and direction aspects. A simple example of such a quantity is force: when applying a force, one is generally interested in how much force is applied (i.e., the magnitude of the force) and the direction in which the force was applied. Vectors will play an important role in many of the subsequent chapters in this text.

This chapter begins with moving our mathematics out of the plane and into "space." That is, we begin to think mathematically not only in two dimensions, but in three. With this foundation, we can explore vectors both in the plane and in space.

10.1 Introduction to Cartesian Coordinates in Space

Up to this point in this text we have considered mathematics in a 2–dimensional world. We have plotted graphs on the x-y plane using rectangular and polar coordinates and found the area of regions in the plane. We have considered properties of *solid* objects, such as volume and surface area, but only by first defining a curve in the plane and then rotating it out of the plane.

While there is wonderful mathematics to explore in "2D," we live in a "3D" world and eventually we will want to apply mathematics involving this third dimension. In this section we introduce Cartesian coordinates in space and explore basic surfaces. This will lay a foundation for much of what we do in the remainder of the text.

Each point P in space can be represented with an ordered triple, $P = (a, b, c)$, where a, b and c represent the relative position of P along the x-, y- and z-axes, respectively. Each axis is perpendicular to the other two.

Visualizing points in space on paper can be problematic, as we are trying to represent a 3-dimensional concept on a 2–dimensional medium. We cannot draw three lines representing the three axes in which each line is perpendicular to the other two. Despite this issue, standard conventions exist for plotting shapes in space that we will discuss that are more than adequate.

One convention is that the axes must conform to the **right hand rule**. This rule states that when the index finger of the right hand is extended in the direction of the positive x-axis, and the middle finger (bent "inward" so it is perpendicular to the palm) points along the positive y-axis, then the extended thumb will point in the direction of the positive z-axis. (It may take some thought to

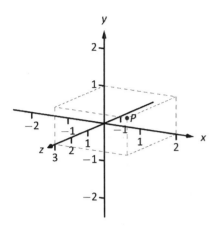

Figure 10.1.1: Plotting the point $P = (2, 1, 3)$ in space.

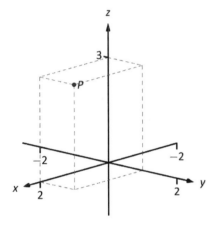

Figure 10.1.2: Plotting the point $P = (2, 1, 3)$ in space with a perspective used in this text.

verify this, but this system is inherently different from the one created by using the "left hand rule.")

As long as the coordinate axes are positioned so that they follow this rule, it does not matter how the axes are drawn on paper. There are two popular methods that we briefly discuss.

In Figure 10.1.1 we see the point $P = (2, 1, 3)$ plotted on a set of axes. The basic convention here is that the x-y plane is drawn in its standard way, with the z-axis down to the left. The perspective is that the paper represents the x-y plane and the positive z axis is coming up, off the page. This method is preferred by many engineers. Because it can be hard to tell where a single point lies in relation to all the axes, dashed lines have been added to let one see how far along each axis the point lies.

One can also consider the x-y plane as being a horizontal plane in, say, a room, where the positive z-axis is pointing up. When one steps back and looks at this room, one might draw the axes as shown in Figure 10.1.2. The same point P is drawn, again with dashed lines. This point of view is preferred by most mathematicians, and is the convention adopted by this text.

Just as the x- and y-axes divide the plane into four *quadrants*, the x-, y-, and z-coordinate planes divide space into eight *octants*. The octant in which x, y, and z are positive is called the **first octant**. We do not name the other seven octants in this text.

Measuring Distances

It is of critical importance to know how to measure distances between points in space. The formula for doing so is based on measuring distance in the plane, and is known (in both contexts) as the Euclidean measure of distance.

Definition 10.1.1   Distance In Space

Let $P = (x_1, y_1, z_1)$ and $Q = (x_2, y_2, z_2)$ be points in space. The distance D between P and Q is

$$D = \sqrt{(x_2 - x_1)^2 + (y_2 - y_1)^2 + (z_2 - z_1)^2}.$$

We refer to the line segment that connects points P and Q in space as $\overline{PQ}$, and refer to the length of this segment as $||\overline{PQ}||$. The above distance formula allows us to compute the length of this segment.

Example 10.1.1   Length of a line segment
Let $P = (1, 4, -1)$ and let $Q = (2, 1, 1)$. Draw the line segment $\overline{PQ}$ and find its length.

Notes:

SOLUTION The points P and Q are plotted in Figure 10.1.3; no special consideration need be made to draw the line segment connecting these two points; simply connect them with a straight line. One *cannot* actually measure this line on the page and deduce anything meaningful; its true length must be measured analytically. Applying Definition 10.1.1, we have

$$\|\overline{PQ}\| = \sqrt{(2-1)^2 + (1-4)^2 + (1-(-1))^2} = \sqrt{14} \approx 3.74.$$

Spheres

Just as a circle is the set of all points in the *plane* equidistant from a given point (its center), a sphere is the set of all points in *space* that are equidistant from a given point. Definition 10.1.1 allows us to write an equation of the sphere.

We start with a point $C = (a, b, c)$ which is to be the center of a sphere with radius r. If a point $P = (x, y, z)$ lies on the sphere, then P is r units from C; that is,

$$\|\overline{PC}\| = \sqrt{(x-a)^2 + (y-b)^2 + (z-c)^2} = r.$$

Squaring both sides, we get the standard equation of a sphere in space with center at $C = (a, b, c)$ with radius r, as given in the following Key Idea.

Key Idea 10.1.1   Standard Equation of a Sphere in Space

The standard equation of the sphere with radius r, centered at $C = (a, b, c)$, is

$$(x-a)^2 + (y-b)^2 + (z-c)^2 = r^2.$$

Example 10.1.2   Equation of a sphere
Find the center and radius of the sphere defined by $x^2 + 2x + y^2 - 4y + z^2 - 6z = 2$.

SOLUTION To determine the center and radius, we must put the equation in standard form. This requires us to complete the square (three times).

$$x^2 + 2x + y^2 - 4y + z^2 - 6z = 2$$
$$(x^2 + 2x + 1) + (y^2 - 4y + 4) + (z^2 - 6z + 9) - 14 = 2$$
$$(x+1)^2 + (y-2)^2 + (z-3)^2 = 16$$

The sphere is centered at $(-1, 2, 3)$ and has a radius of 4.

The equation of a sphere is an example of an implicit function defining a surface in space. In the case of a sphere, the variables x, y and z are all used. We

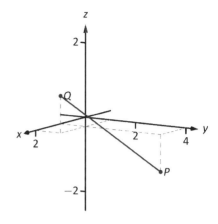

Figure 10.1.3: Plotting points P and Q in Example 10.1.1.

Notes:

now consider situations where surfaces are defined where one or two of these variables are absent.

Introduction to Planes in Space

The coordinate axes naturally define three planes (shown in Figure 10.1.4), the **coordinate planes**: the x-y plane, the y-z plane and the x-z plane. The x-y plane is characterized as the set of all points in space where the z-value is 0. This, in fact, gives us an equation that describes this plane: $z = 0$. Likewise, the x-z plane is all points where the y-value is 0, characterized by $y = 0$.

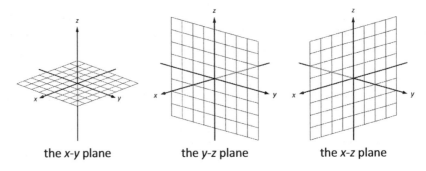

| the x-y plane | the y-z plane | the x-z plane |

Figure 10.1.4: The coordinate planes.

The equation $x = 2$ describes all points in space where the x-value is 2. This is a plane, parallel to the y-z coordinate plane, shown in Figure 10.1.5.

Example 10.1.3 Regions defined by planes

Sketch the region defined by the inequalities $-1 \leq y \leq 2$.

SOLUTION The region is all points between the planes $y = -1$ and $y = 2$. These planes are sketched in Figure 10.1.6, which are parallel to the x-z plane. Thus the region extends infinitely in the x and z directions, and is bounded by planes in the y direction.

Cylinders

The equation $x = 1$ obviously lacks the y and z variables, meaning it defines points where the y and z coordinates can take on any value. Now consider the equation $x^2 + y^2 = 1$ *in space*. In *the plane*, this equation describes a circle of radius 1, centered at the origin. In space, the z coordinate is not specified, meaning it can take on any value. In Figure 10.1.8 (a), we show part of the graph of the equation $x^2 + y^2 = 1$ by sketching 3 circles: the bottom one has a constant z-value of -1.5, the middle one has a z-value of 0 and the top circle has a z-value of 1. By plotting *all* possible z-values, we get the surface shown in Figure

Notes:

Figure 10.1.5: The plane $x = 2$.

Figure 10.1.6: Sketching the boundaries of a region in Example 10.1.3.

10.1.8(b). This surface looks like a "tube," or a "cylinder"; mathematicians call this surface a **cylinder** for an entirely different reason.

Definition 10.1.2   Cylinder

Let C be a curve in a plane and let L be a line not parallel to C. A **cylinder** is the set of all lines parallel to L that pass through C. The curve C is the **directrix** of the cylinder, and the lines are the **rulings**.

In this text, we consider curves C that lie in planes parallel to one of the coordinate planes, and lines L that are perpendicular to these planes, forming **right cylinders**. Thus the directrix can be defined using equations involving 2 variables, and the rulings will be parallel to the axis of the 3rd variable.

In the example preceding the definition, the curve $x^2 + y^2 = 1$ in the x-y plane is the directrix and the rulings are lines parallel to the z-axis. (Any circle shown in Figure 10.1.8 can be considered a directrix; we simply choose the one where $z = 0$.) Sample rulings can also be viewed in part (b) of the figure. More examples will help us understand this definition.

Example 10.1.4   Graphing cylinders

Graph the following cylinders.

1. $z = y^2$ 2. $x = \sin z$

SOLUTION

1. We can view the equation $z = y^2$ as a parabola in the y-z plane, as illustrated in Figure 10.1.7(a). As x does not appear in the equation, the rulings are lines through this parabola parallel to the x-axis, shown in (b). These rulings give an idea as to what the surface looks like, drawn in (c).

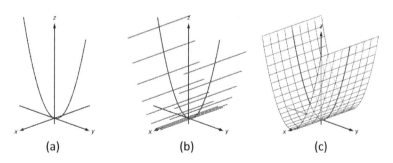

(a)   (b)   (c)

Figure 10.1.7: Sketching the cylinder defined by $z = y^2$.

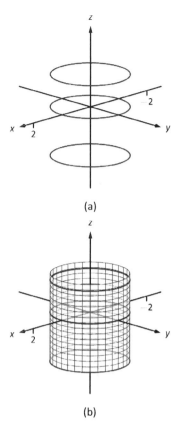

(a)

(b)

Figure 10.1.8: Sketching $x^2 + y^2 = 1$.

Notes:

2. We can view the equation $x = \sin z$ as a sine curve that exists in the x-z plane, as shown in Figure 10.1.9 (a). The rules are parallel to the y axis as the variable y does not appear in the equation $x = \sin z$; some of these are shown in part (b). The surface is shown in part (c) of the figure.

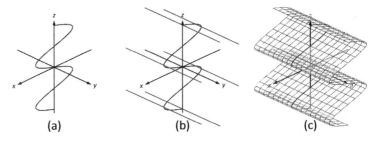

(a)   (b)   (c)

Figure 10.1.9: Sketching the cylinder defined by $x = \sin z$.

Surfaces of Revolution

One of the applications of integration we learned previously was to find the volume of solids of revolution – solids formed by revolving a curve about a horizontal or vertical axis. We now consider how to find the equation of the surface of such a solid.

Consider the surface formed by revolving $y = \sqrt{x}$ about the x-axis. Cross–sections of this surface parallel to the y-z plane are circles, as shown in Figure 10.1.10(a). Each circle has equation of the form $y^2 + z^2 = r^2$ for some radius r. The radius is a function of x; in fact, it is $r(x) = \sqrt{x}$. Thus the equation of the surface shown in Figure 10.1.10b is $y^2 + z^2 = (\sqrt{x})^2$.

We generalize the above principles to give the equations of surfaces formed by revolving curves about the coordinate axes.

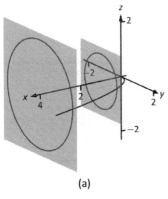

(a)

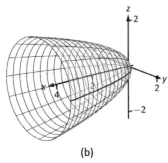

(b)

Figure 10.1.10: Introducing surfaces of revolution.

Key Idea 10.1.2   Surfaces of Revolution, Part 1

Let r be a radius function.

1. The equation of the surface formed by revolving $y = r(x)$ or $z = r(x)$ about the x-axis is $y^2 + z^2 = r(x)^2$.

2. The equation of the surface formed by revolving $x = r(y)$ or $z = r(y)$ about the y-axis is $x^2 + z^2 = r(y)^2$.

3. The equation of the surface formed by revolving $x = r(z)$ or $y = r(z)$ about the z-axis is $x^2 + y^2 = r(z)^2$.

Notes:

Example 10.1.5   Finding equation of a surface of revolution
Let $y = \sin z$ on $[0, \pi]$. Find the equation of the surface of revolution formed by revolving $y = \sin z$ about the z-axis.

SOLUTION   Using Key Idea 10.1.2, we find the surface has equation $x^2 + y^2 = \sin^2 z$. The curve is sketched in Figure 10.1.11(a) and the surface is drawn in Figure 10.1.11(b).

Note how the surface (and hence the resulting equation) is the same if we began with the curve $x = \sin z$, which is also drawn in Figure 10.1.11(a).

This particular method of creating surfaces of revolution is limited. For instance, in Example 7.3.4 of Section 7.3 we found the volume of the solid formed by revolving $y = \sin x$ about the y-axis. Our current method of forming surfaces can only rotate $y = \sin x$ about the x-axis. Trying to rewrite $y = \sin x$ as a function of y is not trivial, as simply writing $x = \sin^{-1} y$ only gives part of the region we desire.

What we desire is a way of writing the surface of revolution formed by rotating $y = f(x)$ about the y-axis. We start by first recognizing this surface is the same as revolving $z = f(x)$ about the z-axis. This will give us a more natural way of viewing the surface.

A value of x is a measurement of distance from the z-axis. At the distance r, we plot a z-height of $f(r)$. When rotating $f(x)$ about the z-axis, we want all points a distance of r from the z-axis in the x-y plane to have a z-height of $f(r)$. All such points satisfy the equation $r^2 = x^2 + y^2$; hence $r = \sqrt{x^2 + y^2}$. Replacing r with $\sqrt{x^2 + y^2}$ in $f(r)$ gives $z = f(\sqrt{x^2 + y^2})$. This is the equation of the surface.

Key Idea 10.1.3   Surfaces of Revolution, Part 2

Let $z = f(x)$, $x \geq 0$, be a curve in the x-z plane. The surface formed by revolving this curve about the z-axis has equation $z = f(\sqrt{x^2 + y^2})$.

Example 10.1.6   Finding equation of surface of revolution
Find the equation of the surface found by revolving $z = \sin x$ about the z-axis.

SOLUTION   Using Key Idea 10.1.3, the surface has equation $z = \sin\left(\sqrt{x^2 + y^2}\right)$. The curve and surface are graphed in Figure 10.1.12.

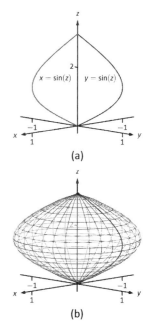

Figure 10.1.11: Revolving $y = \sin z$ about the z-axis in Example 10.1.5.

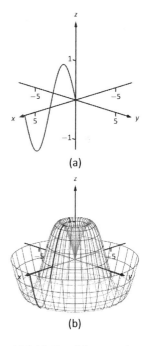

Figure 10.1.12: Revolving $z = \sin x$ about the z-axis in Example 10.1.6.

Notes:

Quadric Surfaces

Spheres, planes and cylinders are important surfaces to understand. We now consider one last type of surface, a **quadric surface**. The definition may look intimidating, but we will show how to analyze these surfaces in an illuminating way.

Definition 10.1.3   **Quadric Surface**

A **quadric surface** is the graph of the general second–degree equation in three variables:

$$Ax^2 + By^2 + Cz^2 + Dxy + Exz + Fyz + Gx + Hy + Iz + J = 0.$$

When the coefficients D, E or F are not zero, the basic shapes of the quadric surfaces are rotated in space. We will focus on quadric surfaces where these coefficients are 0; we will not consider rotations. There are six basic quadric surfaces: the elliptic paraboloid, elliptic cone, ellipsoid, hyperboloid of one sheet, hyperboloid of two sheets, and the hyperbolic paraboloid.

We study each shape by considering **traces**, that is, intersections of each surface with a plane parallel to a coordinate plane. For instance, consider the elliptic paraboloid $z = x^2/4 + y^2$, shown in Figure 10.1.13. If we intersect this shape with the plane $z = d$ (i.e., replace z with d), we have the equation:

$$d = \frac{x^2}{4} + y^2.$$

Divide both sides by d:

$$1 = \frac{x^2}{4d} + \frac{y^2}{d}.$$

This describes an ellipse – so cross sections parallel to the x-y coordinate plane are ellipses. This ellipse is drawn in the figure.

Now consider cross sections parallel to the x-z plane. For instance, letting $y = 0$ gives the equation $z = x^2/4$, clearly a parabola. Intersecting with the plane $x = 0$ gives a cross section defined by $z = y^2$, another parabola. These parabolas are also sketched in the figure.

Thus we see where the elliptic paraboloid gets its name: some cross sections are ellipses, and others are parabolas.

Such an analysis can be made with each of the quadric surfaces. We give a sample equation of each, provide a sketch with representative traces, and describe these traces.

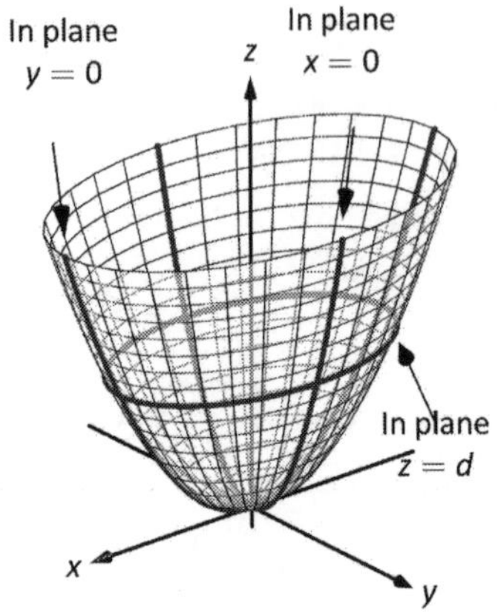

Figure 10.1.13: The elliptic paraboloid $z = x^2/4 + y^2$.

Notes:

Elliptic Paraboloid, $\quad z = \dfrac{x^2}{a^2} + \dfrac{y^2}{b^2}$

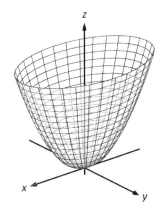

Plane	Trace
$x = d$	Parabola
$y = d$	Parabola
$z = d$	Ellipse

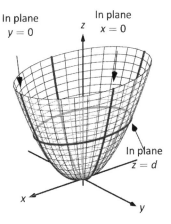

One variable in the equation of the elliptic paraboloid will be raised to the first power; above, this is the z variable. The paraboloid will "open" in the direction of this variable's axis. Thus $x = y^2/a^2 + z^2/b^2$ is an elliptic paraboloid that opens along the x-axis.

Multiplying the right hand side by (-1) defines an elliptic paraboloid that "opens" in the opposite direction.

Elliptic Cone, $\quad z^2 = \dfrac{x^2}{a^2} + \dfrac{y^2}{b^2}$

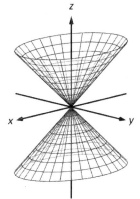

Plane	Trace
$x = 0$	Crossed Lines
$y = 0$	Crossed Lines
$x = d$	Hyperbola
$y = d$	Hyperbola
$z = d$	Ellipse

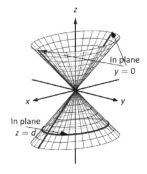
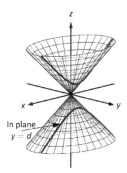

One can rewrite the equation as $z^2 - x^2/a^2 - y^2/b^2 = 0$. The one variable with a positive coefficient corresponds to the axis that the cones "open" along.

Ellipsoid, $\dfrac{x^2}{a^2} + \dfrac{y^2}{b^2} + \dfrac{z^2}{c^2} = 1$

Plane	Trace
$x = d$	Ellipse
$y = d$	Ellipse
$z = d$	Ellipse

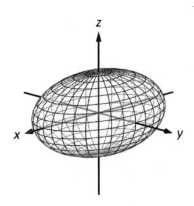

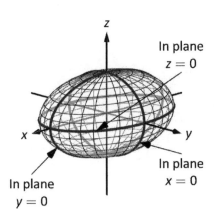

If $a = b = c \neq 0$, the ellipsoid is a sphere with radius a; compare to Key Idea 10.1.1.

Hyperboloid of One Sheet, $\dfrac{x^2}{a^2} + \dfrac{y^2}{b^2} - \dfrac{z^2}{c^2} = 1$

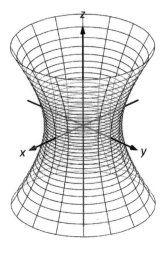

Plane	Trace
$x = d$	Hyperbola
$y = d$	Hyperbola
$z = d$	Ellipse

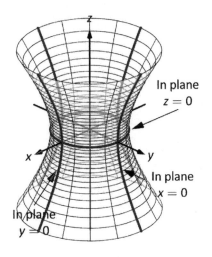

The one variable with a negative coefficient corresponds to the axis that the hyperboloid "opens" along.

Hyperboloid of Two Sheets, $\quad \dfrac{z^2}{c^2} - \dfrac{x^2}{a^2} - \dfrac{y^2}{b^2} = 1$

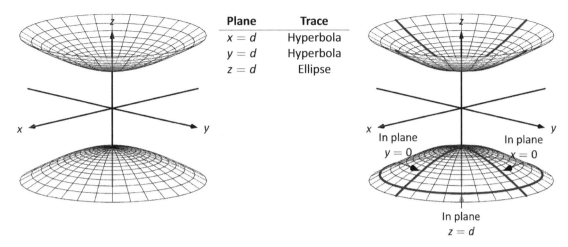

Plane	Trace
$x = d$	Hyperbola
$y = d$	Hyperbola
$z = d$	Ellipse

The one variable with a positive coefficient corresponds to the axis that the hyperboloid "opens" along. In the case illustrated, when $|d| < |c|$, there is no trace.

Hyperbolic Paraboloid, $\quad z = \dfrac{x^2}{a^2} - \dfrac{y^2}{b^2}$

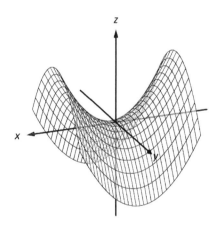

Plane	Trace
$x = d$	Parabola
$y = d$	Parabola
$z = d$	Hyperbola

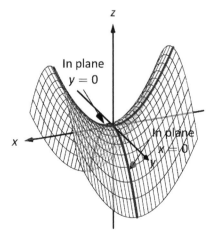

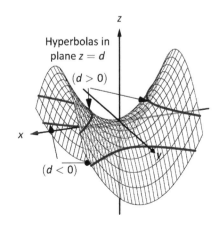

The parabolic traces will open along the axis of the one variable that is raised to the first power.

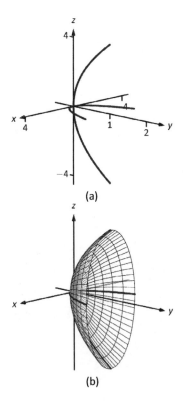

(a)

(b)

Figure 10.1.14: Sketching an elliptic paraboloid.

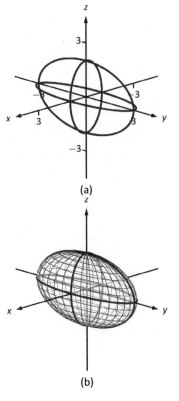

(a)

(b)

Figure 10.1.15: Sketching an ellipsoid.

Example 10.1.7   Sketching quadric surfaces
Sketch the quadric surface defined by the given equation.

1. $y = \dfrac{x^2}{4} + \dfrac{z^2}{16}$
2. $x^2 + \dfrac{y^2}{9} + \dfrac{z^2}{4} = 1.$
3. $z = y^2 - x^2.$

SOLUTION

1. $y = \dfrac{x^2}{4} + \dfrac{z^2}{16}$:

We first identify the quadric by pattern–matching with the equations given previously. Only two surfaces have equations where one variable is raised to the first power, the elliptic paraboloid and the hyperbolic paraboloid. In the latter case, the other variables have different signs, so we conclude that this describes a hyperbolic paraboloid. As the variable with the first power is y, we note the paraboloid opens along the y-axis.

To make a decent sketch by hand, we need only draw a few traces. In this case, the traces $x = 0$ and $z = 0$ form parabolas that outline the shape.

$x = 0$: The trace is the parabola $y = z^2/16$

$z = 0$: The trace is the parabola $y = x^2/4$.

Graphing each trace in the respective plane creates a sketch as shown in Figure 10.1.14(a). This is enough to give an idea of what the paraboloid looks like. The surface is filled in in (b).

2. $x^2 + \dfrac{y^2}{9} + \dfrac{z^2}{4} = 1$:

This is an ellipsoid. We can get a good idea of its shape by drawing the traces in the coordinate planes.

$x = 0$: The trace is the ellipse $\dfrac{y^2}{9} + \dfrac{z^2}{4} = 1$. The major axis is along the y–axis with length 6 (as $b = 3$, the length of the axis is 6); the minor axis is along the z-axis with length 4.

$y = 0$: The trace is the ellipse $x^2 + \dfrac{z^2}{4} = 1$. The major axis is along the z-axis, and the minor axis has length 2 along the x-axis.

$z = 0$: The trace is the ellipse $x^2 + \dfrac{y^2}{9} = 1$, with major axis along the y-axis.

Graphing each trace in the respective plane creates a sketch as shown in Figure 10.1.15(a). Filling in the surface gives Figure 10.1.15(b).

3. $z = y^2 - x^2$:

Notes:

This defines a hyperbolic paraboloid, very similar to the one shown in the gallery of quadric sections. Consider the traces in the $y-z$ and $x-z$ planes:

$x = 0$: The trace is $z = y^2$, a parabola opening up in the $y - z$ plane.

$y = 0$: The trace is $z = -x^2$, a parabola opening down in the $x - z$ plane.

Sketching these two parabolas gives a sketch like that in Figure 10.1.16(a), and filling in the surface gives a sketch like (b).

Example 10.1.8 Identifying quadric surfaces

Consider the quadric surface shown in Figure 10.1.17. Which of the following equations best fits this surface?

(a) $x^2 - y^2 - \dfrac{z^2}{9} = 0$ (c) $z^2 - x^2 - y^2 = 1$

(b) $x^2 - y^2 - z^2 = 1$ (d) $4x^2 - y^2 - \dfrac{z^2}{9} = 1$

SOLUTION The image clearly displays a hyperboloid of two sheets. The gallery informs us that the equation will have a form similar to $\frac{z^2}{c^2} - \frac{x^2}{a^2} - \frac{y^2}{b^2} = 1$.

We can immediately eliminate option (a), as the constant in that equation is not 1.

The hyperboloid "opens" along the x-axis, meaning x must be the only variable with a positive coefficient, eliminating (c).

The hyperboloid is wider in the z-direction than in the y-direction, so we need an equation where $c > b$. This eliminates (b), leaving us with (d). We should verify that the equation given in (d), $4x^2 - y^2 - \frac{z^2}{9} = 1$, fits.

We already established that this equation describes a hyperboloid of two sheets that opens in the x-direction and is wider in the z-direction than in the y. Now note the coefficient of the x-term. Rewriting $4x^2$ in standard form, we have: $4x^2 = \dfrac{x^2}{(1/2)^2}$. Thus when $y = 0$ and $z = 0$, x must be $1/2$; i.e., each hyperboloid "starts" at $x = 1/2$. This matches our figure.

We conclude that $4x^2 - y^2 - \dfrac{z^2}{9} = 1$ best fits the graph.

This section has introduced points in space and shown how equations can describe surfaces. The next sections explore *vectors*, an important mathematical object that we'll use to explore curves in space.

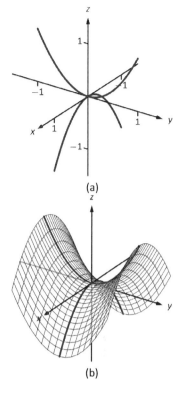

(a)

(b)

Figure 10.1.16: Sketching a hyperbolic paraboloid.

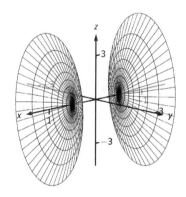

Figure 10.1.17: A possible equation of this quadric surface is found in Example 10.1.8.

Notes:

Exercises 10.1

Terms and Concepts

1. Axes drawn in space must conform to the _____ _____ rule.

2. In the plane, the equation $x = 2$ defines a _____; in space, $x = 2$ defines a _____.

3. In the plane, the equation $y = x^2$ defines a _____; in space, $y = x^2$ defines a _____.

4. Which quadric surface looks like a Pringles® chip?

5. Consider the hyperbola $x^2 - y^2 = 1$ in the plane. If this hyperbola is rotated about the x-axis, what quadric surface is formed?

6. Consider the hyperbola $x^2 - y^2 = 1$ in the plane. If this hyperbola is rotated about the y-axis, what quadric surface is formed?

Problems

7. The points $A = (1, 4, 2)$, $B = (2, 6, 3)$ and $C = (4, 3, 1)$ form a triangle in space. Find the distances between each pair of points and determine if the triangle is a right triangle.

8. The points $A = (1, 1, 3)$, $B = (3, 2, 7)$, $C = (2, 0, 8)$ and $D = (0, -1, 4)$ form a quadrilateral $ABCD$ in space. Is this a parallelogram?

9. Find the center and radius of the sphere defined by $x^2 - 8x + y^2 + 2y + z^2 + 8 = 0$.

10. Find the center and radius of the sphere defined by $x^2 + y^2 + z^2 + 4x - 2y - 4z + 4 = 0$.

In Exercises 11 – 14, describe the region in space defined by the inequalities.

11. $x^2 + y^2 + z^2 < 1$

12. $0 \leq x \leq 3$

13. $x \geq 0$, $y \geq 0$, $z \geq 0$

14. $y \geq 3$

In Exercises 15 – 18, sketch the cylinder in space.

15. $z = x^3$

16. $y = \cos z$

17. $\dfrac{x^2}{4} + \dfrac{y^2}{9} = 1$

18. $y = \dfrac{1}{x}$

In Exercises 19 – 22, give the equation of the surface of revolution described.

19. Revolve $z = \dfrac{1}{1 + y^2}$ about the y-axis.

20. Revolve $y = x^2$ about the x-axis.

21. Revolve $z = x^2$ about the z-axis.

22. Revolve $z = 1/x$ about the z-axis.

In Exercises 23 – 26, a quadric surface is sketched. Determine which of the given equations best fits the graph.

23.

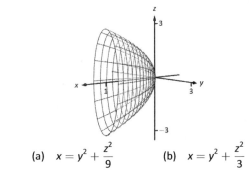

(a) $\quad x = y^2 + \dfrac{z^2}{9}$ (b) $\quad x = y^2 + \dfrac{z^2}{3}$

24.

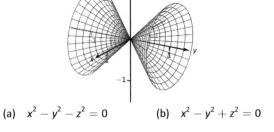

(a) $\quad x^2 - y^2 - z^2 = 0$ (b) $\quad x^2 - y^2 + z^2 = 0$

25.

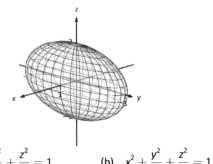

(a) $\quad x^2 + \dfrac{y^2}{3} + \dfrac{z^2}{2} = 1$ (b) $\quad x^2 + \dfrac{y^2}{9} + \dfrac{z^2}{4} = 1$

26.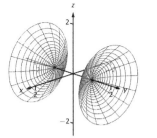

(a) $y^2 - x^2 - z^2 = 1$ (b) $y^2 + x^2 - z^2 = 1$

In Exercises 27 – 32, sketch the quadric surface.

27. $z - y^2 + x^2 = 0$

28. $z^2 = x^2 + \dfrac{y^2}{4}$

29. $x = -y^2 - z^2$

30. $16x^2 - 16y^2 - 16z^2 = 1$

31. $\dfrac{x^2}{9} - y^2 + \dfrac{z^2}{25} = 1$

32. $4x^2 + 2y^2 + z^2 = 4$

10.2 An Introduction to Vectors

Many quantities we think about daily can be described by a single number: temperature, speed, cost, weight and height. There are also many other concepts we encounter daily that cannot be described with just one number. For instance, a weather forecaster often describes wind with its speed and its direction ("... with winds from the southeast gusting up to 30 mph ..."). When applying a force, we are concerned with both the magnitude and direction of that force. In both of these examples, *direction* is important. Because of this, we study *vectors*, mathematical objects that convey both magnitude and direction information.

One "bare–bones" definition of a vector is based on what we wrote above: "a vector is a mathematical object with magnitude and direction parameters." This definition leaves much to be desired, as it gives no indication as to how such an object is to be used. Several other definitions exist; we choose here a definition rooted in a geometric visualization of vectors. It is very simplistic but readily permits further investigation.

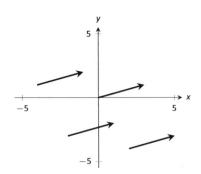

Figure 10.2.1: Drawing the same vector with different initial points.

Definition 10.2.1 Vector

A **vector** is a directed line segment.

Given points P and Q (either in the plane or in space), we denote with $\vec{PQ}$ the vector *from P to Q*. The point P is said to be the **initial point** of the vector, and the point Q is the **terminal point**.

The **magnitude**, **length** or **norm** of $\vec{PQ}$ is the length of the line segment $\overline{PQ}$: $||\,\vec{PQ}\,|| = ||\,\overline{PQ}\,||$.

Two vectors are **equal** if they have the same magnitude and direction.

Figure 10.2.1 shows multiple instances of the same vector. Each directed line segment has the same direction and length (magnitude), hence each is the same vector.

We use $\mathbb{R}^2$ (pronounced "r two") to represent all the vectors in the plane, and use $\mathbb{R}^3$ (pronounced "r three") to represent all the vectors in space.

Consider the vectors $\vec{PQ}$ and $\vec{RS}$ as shown in Figure 10.2.2. The vectors look to be equal; that is, they seem to have the same length and direction. Indeed, they are. Both vectors move 2 units to the right and 1 unit up from the initial point to reach the terminal point. One can analyze this movement to measure the

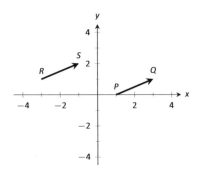

Figure 10.2.2: Illustrating how equal vectors have the same displacement.

Notes:

magnitude of the vector, and the movement itself gives direction information (one could also measure the slope of the line passing through P and Q or R and S). Since they have the same length and direction, these two vectors are equal.

This demonstrates that inherently all we care about is *displacement*; that is, how far in the x, y and possibly z directions the terminal point is from the initial point. Both the vectors $\vec{PQ}$ and $\vec{RS}$ in Figure 10.2.2 have an x-displacement of 2 and a y-displacement of 1. This suggests a standard way of describing vectors in the plane. A vector whose x-displacement is a and whose y-displacement is b will have terminal point (a, b) when the initial point is the origin, $(0, 0)$. This leads us to a definition of a standard and concise way of referring to vectors.

Definition 10.2.2   Component Form of a Vector

1. The **component form** of a vector $\vec{v}$ in $\mathbb{R}^2$, whose terminal point is (a, b) when its initial point is $(0, 0)$, is $\langle a, b \rangle$.

2. The **component form** of a vector $\vec{v}$ in $\mathbb{R}^3$, whose terminal point is (a, b, c) when its initial point is $(0, 0, 0)$, is $\langle a, b, c \rangle$.

The numbers a, b (and c, respectively) are the **components** of $\vec{v}$.

It follows from the definition that the component form of the vector $\vec{PQ}$, where $P = (x_1, y_1)$ and $Q = (x_2, y_2)$ is

$$\vec{PQ} = \langle x_2 - x_1, y_2 - y_1 \rangle\,;$$

in space, where $P = (x_1, y_1, z_1)$ and $Q = (x_2, y_2, z_2)$, the component form of $\vec{PQ}$ is

$$\vec{PQ} = \langle x_2 - x_1, y_2 - y_1, z_2 - z_1 \rangle\,.$$

We practice using this notation in the following example.

Example 10.2.1   Using component form notation for vectors

1. Sketch the vector $\vec{v} = \langle 2, -1 \rangle$ starting at $P = (3, 2)$ and find its magnitude.

2. Find the component form of the vector $\vec{w}$ whose initial point is $R = (-3, -2)$ and whose terminal point is $S = (-1, 2)$.

3. Sketch the vector $\vec{u} = \langle 2, -1, 3 \rangle$ starting at the point $Q = (1, 1, 1)$ and find its magnitude.

Notes:

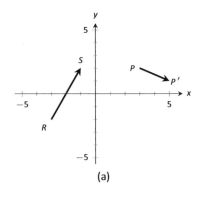

(a)

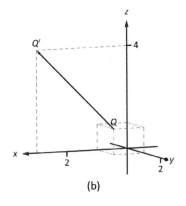

(b)

Figure 10.2.3: Graphing vectors in Example 10.2.1.

SOLUTION

1. Using P as the initial point, we move 2 units in the positive x-direction and -1 units in the positive y-direction to arrive at the terminal point $P' = (5, 1)$, as drawn in Figure 10.2.3(a).

   The magnitude of $\vec{v}$ is determined directly from the component form:

   $$\|\vec{v}\| = \sqrt{2^2 + (-1)^2} = \sqrt{5}.$$

2. Using the note following Definition 10.2.2, we have

   $$\vec{RS} = \langle -1 - (-3), 2 - (-2) \rangle = \langle 2, 4 \rangle.$$

   One can readily see from Figure 10.2.3(a) that the x- and y-displacement of $\vec{RS}$ is 2 and 4, respectively, as the component form suggests.

3. Using Q as the initial point, we move 2 units in the positive x-direction, -1 unit in the positive y-direction, and 3 units in the positive z-direction to arrive at the terminal point $Q' = (3, 0, 4)$, illustrated in Figure 10.2.3(b).

   The magnitude of $\vec{u}$ is:

   $$\|\vec{u}\| = \sqrt{2^2 + (-1)^2 + 3^2} = \sqrt{14}.$$

Now that we have defined vectors, and have created a nice notation by which to describe them, we start considering how vectors interact with each other. That is, we define an *algebra* on vectors.

Notes:

Definition 10.2.3   Vector Algebra

1. Let $\vec{u} = \langle u_1, u_2 \rangle$ and $\vec{v} = \langle v_1, v_2 \rangle$ be vectors in $\mathbb{R}^2$, and let c be a scalar.

    (a) The addition, or sum, of the vectors $\vec{u}$ and $\vec{v}$ is the vector
    $$\vec{u} + \vec{v} = \langle u_1 + v_1, u_2 + v_2 \rangle .$$

    (b) The scalar product of c and $\vec{v}$ is the vector
    $$c\vec{v} = c \langle v_1, v_2 \rangle = \langle cv_1, cv_2 \rangle .$$

2. Let $\vec{u} = \langle u_1, u_2, u_3 \rangle$ and $\vec{v} = \langle v_1, v_2, v_3 \rangle$ be vectors in $\mathbb{R}^3$, and let c be a scalar.

    (a) The addition, or sum, of the vectors $\vec{u}$ and $\vec{v}$ is the vector
    $$\vec{u} + \vec{v} = \langle u_1 + v_1, u_2 + v_2, u_3 + v_3 \rangle .$$

    (b) The scalar product of c and $\vec{v}$ is the vector
    $$c\vec{v} = c \langle v_1, v_2, v_3 \rangle = \langle cv_1, cv_2, cv_3 \rangle .$$

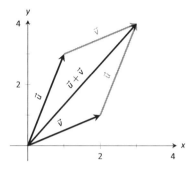

Figure 10.2.5: Illustrating how to add vectors using the Head to Tail Rule and Parallelogram Law.

In short, we say addition and scalar multiplication are computed "component–wise."

Example 10.2.2   Adding vectors

Sketch the vectors $\vec{u} = \langle 1, 3 \rangle$, $\vec{v} = \langle 2, 1 \rangle$ and $\vec{u} + \vec{v}$ all with initial point at the origin.

SOLUTION   We first compute $\vec{u} + \vec{v}$.

$$\begin{aligned} \vec{u} + \vec{v} &= \langle 1, 3 \rangle + \langle 2, 1 \rangle \\ &= \langle 3, 4 \rangle . \end{aligned}$$

These are all sketched in Figure 10.2.4.

As vectors convey magnitude and direction information, the sum of vectors also convey length and magnitude information. Adding $\vec{u} + \vec{v}$ suggests the following idea:

"Starting at an initial point, go out $\vec{u}$, then go out $\vec{v}$."

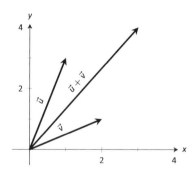

Figure 10.2.4: Graphing the sum of vectors in Example 10.2.2.

Notes:

This idea is sketched in Figure 10.2.5, where the initial point of $\vec{v}$ is the terminal point of $\vec{u}$. This is known as the "Head to Tail Rule" of adding vectors. Vector addition is very important. For instance, if the vectors $\vec{u}$ and $\vec{v}$ represent forces acting on a body, the sum $\vec{u} + \vec{v}$ gives the resulting force. Because of various physical applications of vector addition, the sum $\vec{u} + \vec{v}$ is often referred to as the **resultant vector**, or just the "resultant."

Analytically, it is easy to see that $\vec{u} + \vec{v} = \vec{v} + \vec{u}$. Figure 10.2.5 also gives a graphical representation of this, using gray vectors. Note that the vectors $\vec{u}$ and $\vec{v}$, when arranged as in the figure, form a parallelogram. Because of this, the Head to Tail Rule is also known as the Parallelogram Law: the vector $\vec{u} + \vec{v}$ is defined by forming the parallelogram defined by the vectors $\vec{u}$ and $\vec{v}$; the initial point of $\vec{u} + \vec{v}$ is the common initial point of parallelogram, and the terminal point of the sum is the common terminal point of the parallelogram.

While not illustrated here, the Head to Tail Rule and Parallelogram Law hold for vectors in $\mathbb{R}^3$ as well.

It follows from the properties of the real numbers and Definition 10.2.3 that

$$\vec{u} - \vec{v} = \vec{u} + (-1)\vec{v}.$$

The Parallelogram Law gives us a good way to visualize this subtraction. We demonstrate this in the following example.

Example 10.2.3   Vector Subtraction
Let $\vec{u} = \langle 3, 1 \rangle$ and $\vec{v} = \langle 1, 2 \rangle$. Compute and sketch $\vec{u} - \vec{v}$.

SOLUTION   The computation of $\vec{u} - \vec{v}$ is straightforward, and we show all steps below. Usually the formal step of multiplying by (-1) is omitted and we "just subtract."

$$\begin{aligned}\vec{u} - \vec{v} &= \vec{u} + (-1)\vec{v} \\ &= \langle 3, 1 \rangle + \langle -1, -2 \rangle \\ &= \langle 2, -1 \rangle.\end{aligned}$$

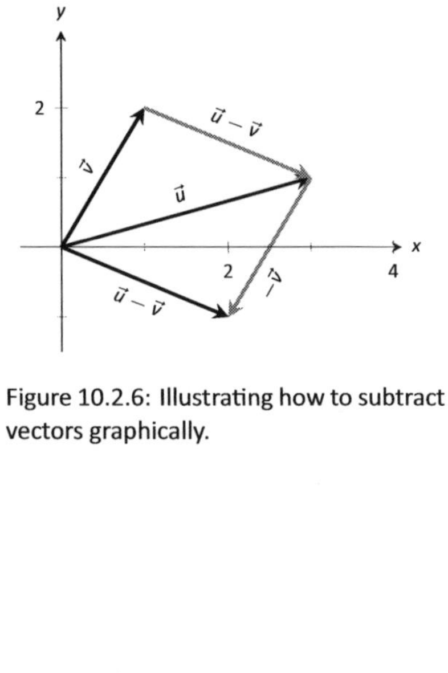

Figure 10.2.6 illustrates, using the Head to Tail Rule, how the subtraction can be viewed as the sum $\vec{u} + (-\vec{v})$. The figure also illustrates how $\vec{u} - \vec{v}$ can be obtained by looking only at the terminal points of $\vec{u}$ and $\vec{v}$ (when their initial points are the same).

Example 10.2.4   Scaling vectors

1. Sketch the vectors $\vec{v} = \langle 2, 1 \rangle$ and $2\vec{v}$ with initial point at the origin.

2. Compute the magnitudes of $\vec{v}$ and $2\vec{v}$.

Figure 10.2.6: Illustrating how to subtract vectors graphically.

Notes:

Solution

1. We compute $2\vec{v}$:

$$2\vec{v} = 2\langle 2, 1\rangle$$
$$= \langle 4, 2\rangle.$$

Both $\vec{v}$ and $2\vec{v}$ are sketched in Figure 10.2.7. Make note that $2\vec{v}$ does not start at the terminal point of $\vec{v}$; rather, its initial point is also the origin.

2. The figure suggests that $2\vec{v}$ is twice as long as $\vec{v}$. We compute their magnitudes to confirm this.

$$\|\vec{v}\| = \sqrt{2^2 + 1^2}$$
$$= \sqrt{5}.$$
$$\|2\vec{v}\| = \sqrt{4^2 + 2^2}$$
$$= \sqrt{20}$$
$$= \sqrt{4 \cdot 5} = 2\sqrt{5}.$$

As we suspected, $2\vec{v}$ is twice as long as $\vec{v}$.

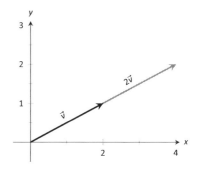

Figure 10.2.7: Graphing vectors $\vec{v}$ and $2\vec{v}$ in Example 10.2.4.

The **zero vector** is the vector whose initial point is also its terminal point. It is denoted by $\vec{0}$. Its component form, in $\mathbb{R}^2$, is $\langle 0, 0\rangle$; in $\mathbb{R}^3$, it is $\langle 0, 0, 0\rangle$. Usually the context makes is clear whether $\vec{0}$ is referring to a vector in the plane or in space.

Our examples have illustrated key principles in vector algebra: how to add and subtract vectors and how to multiply vectors by a scalar. The following theorem states formally the properties of these operations.

Notes:

Theorem 10.2.1   Properties of Vector Operations

The following are true for all scalars c and d, and for all vectors $\vec{u}$, $\vec{v}$ and $\vec{w}$, where $\vec{u}$, $\vec{v}$ and $\vec{w}$ are all in $\mathbb{R}^2$ or where $\vec{u}$, $\vec{v}$ and $\vec{w}$ are all in $\mathbb{R}^3$:

1. $\vec{u} + \vec{v} = \vec{v} + \vec{u}$ Commutative Property

2. $(\vec{u} + \vec{v}) + \vec{w} = \vec{u} + (\vec{v} + \vec{w})$ Associative Property

3. $\vec{v} + \vec{0} = \vec{v}$ Additive Identity

4. $(cd)\vec{v} = c(d\vec{v})$

5. $c(\vec{u} + \vec{v}) = c\vec{u} + c\vec{v}$ Distributive Property

6. $(c + d)\vec{v} = c\vec{v} + d\vec{v}$ Distributive Property

7. $0\vec{v} = \vec{0}$

8. $\|c\vec{v}\| = |c| \cdot \|\vec{v}\|$

9. $\|\vec{u}\| = 0$ if, and only if, $\vec{u} = \vec{0}$.

As stated before, each nonvector $\vec{v}$ conveys magnitude and direction information. We have a method of extracting the magnitude, which we write as $\|\vec{v}\|$. *Unit vectors* are a way of extracting just the direction information from a vector.

Definition 10.2.4   Unit Vector

A **unit vector** is a vector $\vec{v}$ with a magnitude of 1; that is,

$$\|\vec{v}\| = 1.$$

Consider this scenario: you are given a vector $\vec{v}$ and are told to create a vector of length 10 in the direction of $\vec{v}$. How does one do that? If we knew that $\vec{u}$ was the unit vector in the direction of $\vec{v}$, the answer would be easy: $10\vec{u}$. So how do we find $\vec{u}$?

Property 8 of Theorem 10.2.1 holds the key. If we divide $\vec{v}$ by its magnitude, it becomes a vector of length 1. Consider:

$$\left\| \frac{1}{\|\vec{v}\|}\vec{v} \right\| = \frac{1}{\|\vec{v}\|}\|\vec{v}\| \quad \text{(we can pull out } \frac{1}{\|\vec{v}\|} \text{ as it is a positive scalar)}$$
$$= 1.$$

Notes:

So the vector of length 10 in the direction of $\vec{v}$ is $10\dfrac{1}{\|\vec{v}\|}\vec{v}$. An example will make this more clear.

Example 10.2.5   Using Unit Vectors

Let $\vec{v} = \langle 3, 1\rangle$ and let $\vec{w} = \langle 1, 2, 2\rangle$.

1. Find the unit vector in the direction of $\vec{v}$.

2. Find the unit vector in the direction of $\vec{w}$.

3. Find the vector in the direction of $\vec{v}$ with magnitude 5.

SOLUTION

1. We find $\|\vec{v}\| = \sqrt{10}$. So the unit vector $\vec{u}$ in the direction of $\vec{v}$ is

$$\vec{u} = \frac{1}{\sqrt{10}}\vec{v} = \left\langle \frac{3}{\sqrt{10}}, \frac{1}{\sqrt{10}} \right\rangle.$$

2. We find $\|\vec{w}\| = 3$, so the unit vector $\vec{z}$ in the direction of $\vec{w}$ is

$$\vec{u} = \frac{1}{3}\vec{w} = \left\langle \frac{1}{3}, \frac{2}{3}, \frac{2}{3} \right\rangle.$$

3. To create a vector with magnitude 5 in the direction of $\vec{v}$, we multiply the unit vector $\vec{u}$ by 5. Thus $5\vec{u} = \langle 15/\sqrt{10}, 5/\sqrt{10}\rangle$ is the vector we seek. This is sketched in Figure 10.2.8.

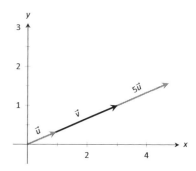

Figure 10.2.8: Graphing vectors in Example 10.2.5. All vectors shown have their initial point at the origin.

The basic formation of the unit vector $\vec{u}$ in the direction of a vector $\vec{v}$ leads to a interesting equation. It is:

$$\vec{v} = \|\vec{v}\|\frac{1}{\|\vec{v}\|}\vec{v}.$$

We rewrite the equation with parentheses to make a point:

$$\vec{v} = \underbrace{\|\vec{v}\|}_{\text{magnitude}} \cdot \underbrace{\left(\frac{1}{\|\vec{v}\|}\vec{v}\right)}_{\text{direction}}.$$

This equation illustrates the fact that a nonzero vector has both magnitude and direction, where we view a unit vector as supplying *only* direction information. Identifying unit vectors with direction allows us to define **parallel vectors**.

Notes:

Chapter 10 Vectors

Note: $\vec{0}$ is directionless; because $\|\vec{0}\| = 0$, there is no unit vector in the "direction" of $\vec{0}$.

Some texts define two vectors as being parallel if one is a scalar multiple of the other. By this definition, $\vec{0}$ is parallel to all vectors as $\vec{0} = 0\vec{v}$ for all $\vec{v}$.

We define what it means for two vectors to be perpendicular in Definition 10.3.2, which is written to exclude $\vec{0}$. It could be written to include $\vec{0}$; by such a definition, $\vec{0}$ is perpendicular to all vectors. While counter-intuitive, it is mathematically sound to allow $\vec{0}$ to be both parallel and perpendicular to all vectors.

We prefer the given definition of parallel as it is grounded in the fact that unit vectors provide direction information. One may adopt the convention that $\vec{0}$ is parallel to all vectors if they desire. (See also the marginal note on page 604.)

Definition 10.2.5   Parallel Vectors

1. Unit vectors $\vec{u}_1$ and $\vec{u}_2$ are **parallel** if $\vec{u}_1 = \pm\vec{u}_2$.

2. Nonzero vectors $\vec{v}_1$ and $\vec{v}_2$ are **parallel** if their respective unit vectors are parallel.

It is equivalent to say that vectors $\vec{v}_1$ and $\vec{v}_2$ are parallel if there is a scalar $c \neq 0$ such that $\vec{v}_1 = c\vec{v}_2$ (see marginal note).

If one graphed all unit vectors in $\mathbb{R}^2$ with the initial point at the origin, then the terminal points would all lie on the unit circle. Based on what we know from trigonometry, we can then say that the component form of all unit vectors in $\mathbb{R}^2$ is $\langle\cos\theta, \sin\theta\rangle$ for some angle θ.

A similar construction in $\mathbb{R}^3$ shows that the terminal points all lie on the unit sphere. These vectors also have a particular component form, but its derivation is not as straightforward as the one for unit vectors in $\mathbb{R}^2$. Important concepts about unit vectors are given in the following Key Idea.

Key Idea 10.2.1   Unit Vectors

1. The unit vector in the direction of a nonzero vector $\vec{v}$ is

$$\vec{u} = \frac{1}{\|\vec{v}\|}\vec{v}.$$

2. A vector $\vec{u}$ in $\mathbb{R}^2$ is a unit vector if, and only if, its component form is $\langle\cos\theta, \sin\theta\rangle$ for some angle θ.

3. A vector $\vec{u}$ in $\mathbb{R}^3$ is a unit vector if, and only if, its component form is $\langle\sin\theta\cos\varphi, \sin\theta\sin\varphi, \cos\theta\rangle$ for some angles θ and φ.

These formulas can come in handy in a variety of situations, especially the formula for unit vectors in the plane.

Example 10.2.6   Finding Component Forces
Consider a weight of 50lb hanging from two chains, as shown in Figure 10.2.9. One chain makes an angle of $30°$ with the vertical, and the other an angle of $45°$. Find the force applied to each chain.

SOLUTION   Knowing that gravity is pulling the 50lb weight straight down,

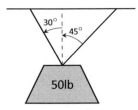

Figure 10.2.9: A diagram of a weight hanging from 2 chains in Example 10.2.6.

Notes:

582

we can create a vector $\vec{F}$ to represent this force.

$$\vec{F} = 50\,\langle 0, -1 \rangle = \langle 0, -50 \rangle.$$

We can view each chain as "pulling" the weight up, preventing it from falling. We can represent the force from each chain with a vector. Let $\vec{F}_1$ represent the force from the chain making an angle of $30°$ with the vertical, and let $\vec{F}_2$ represent the force form the other chain. Convert all angles to be measured from the horizontal (as shown in Figure 10.2.10), and apply Key Idea 10.2.1. As we do not yet know the magnitudes of these vectors, (that is the problem at hand), we use m_1 and m_2 to represent them.

$$\vec{F}_1 = m_1 \langle \cos 120°, \sin 120° \rangle$$

$$\vec{F}_2 = m_2 \langle \cos 45°, \sin 45° \rangle$$

As the weight is not moving, we know the sum of the forces is $\vec{0}$. This gives:

$$\vec{F} + \vec{F}_1 + \vec{F}_2 = \vec{0}$$

$$\langle 0, -50 \rangle + m_1 \langle \cos 120°, \sin 120° \rangle + m_2 \langle \cos 45°, \sin 45° \rangle = \vec{0}$$

The sum of the entries in the first component is 0, and the sum of the entries in the second component is also 0. This leads us to the following two equations:

$$m_1 \cos 120° + m_2 \cos 45° = 0$$
$$m_1 \sin 120° + m_2 \sin 45° = 50$$

This is a simple 2-equation, 2-unkown system of linear equations. We leave it to the reader to verify that the solution is

$$m_1 = 50(\sqrt{3} - 1) \approx 36.6; \qquad m_2 = \frac{50\sqrt{2}}{1 + \sqrt{3}} \approx 25.88.$$

It might seem odd that the sum of the forces applied to the chains is more than 50lb. We leave it to a physics class to discuss the full details, but offer this short explanation. Our equations were established so that the *vertical* components of each force sums to 50lb, thus supporting the weight. Since the chains are at an angle, they also pull against each other, creating an "additional" horizontal force while holding the weight in place.

Unit vectors were very important in the previous calculation; they allowed us to define a vector in the proper direction but with an unknown magnitude. Our computations were then computed component–wise. Because such calculations are often necessary, the *standard unit vectors* can be useful.

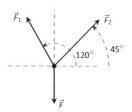

Figure 10.2.10: A diagram of the force vectors from Example 10.2.6.

Notes:

Definition 10.2.6   Standard Unit Vectors

1. In $\mathbb{R}^2$, the standard unit vectors are
$$\vec{i} = \langle 1, 0 \rangle \quad \text{and} \quad \vec{j} = \langle 0, 1 \rangle.$$

2. In $\mathbb{R}^3$, the standard unit vectors are
$$\vec{i} = \langle 1, 0, 0 \rangle \quad \text{and} \quad \vec{j} = \langle 0, 1, 0 \rangle \quad \text{and} \quad \vec{k} = \langle 0, 0, 1 \rangle.$$

Example 10.2.7   Using standard unit vectors

1. Rewrite $\vec{v} = \langle 2, -3 \rangle$ using the standard unit vectors.

2. Rewrite $\vec{w} = 4\vec{i} - 5\vec{j} + 2\vec{k}$ in component form.

SOLUTION

1.
$$\begin{aligned}\vec{v} &= \langle 2, -3 \rangle \\ &= \langle 2, 0 \rangle + \langle 0, -3 \rangle \\ &= 2\langle 1, 0 \rangle - 3\langle 0, 1 \rangle \\ &= 2\vec{i} - 3\vec{j}\end{aligned}$$

2.
$$\begin{aligned}\vec{w} &= 4\vec{i} - 5\vec{j} + 2\vec{k} \\ &= \langle 4, 0, 0 \rangle + \langle 0, -5, 0 \rangle + \langle 0, 0, 2 \rangle \\ &= \langle 4, -5, 2 \rangle\end{aligned}$$

These two examples demonstrate that converting between component form and the standard unit vectors is rather straightforward. Many mathematicians prefer component form, and it is the preferred notation in this text. Many engineers prefer using the standard unit vectors, and many engineering text use that notation.

Example 10.2.8   Finding Component Force

A weight of 25lb is suspended from a chain of length 2ft while a wind pushes the weight to the right with constant force of 5lb as shown in Figure 10.2.11. What angle will the chain make with the vertical as a result of the wind's pushing? How much higher will the weight be?

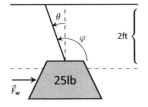

Figure 10.2.11: A figure of a weight being pushed by the wind in Example 10.2.8.

Notes:

SOLUTION   The force of the wind is represented by the vector $\vec{F}_w = 5\vec{i}$. The force of gravity on the weight is represented by $\vec{F}_g = -25\vec{j}$. The direction and magnitude of the vector representing the force on the chain are both unknown. We represent this force with

$$\vec{F}_c = m\,\langle \cos\varphi, \sin\varphi \rangle = m\cos\varphi\,\vec{i} + m\sin\varphi\,\vec{j}$$

for some magnitude m and some angle with the horizontal φ. (Note: θ is the angle the chain makes with the *vertical*; φ is the angle with the *horizontal*.)

As the weight is at equilibrium, the sum of the forces is $\vec{0}$:

$$\vec{F}_c + \vec{F}_w + \vec{F}_g = \vec{0}$$
$$m\cos\varphi\,\vec{i} + m\sin\varphi\,\vec{j} + 5\vec{i} - 25\vec{j} = \vec{0}$$

Thus the sum of the $\vec{i}$ and $\vec{j}$ components are 0, leading us to the following system of equations:

$$5 + m\cos\varphi = 0$$
$$-25 + m\sin\varphi = 0$$

$$(10.1)$$

This is enough to determine $\vec{F}_c$ already, as we know $m\cos\varphi = -5$ and $m\sin\varphi = 25$. Thus $F_c = \langle -5, 25 \rangle$. We can use this to find the magnitude m:

$$m = \sqrt{(-5)^2 + 25^2} = 5\sqrt{26} \approx 25.5\text{lb}.$$

We can then use either equality from Equation (10.1) to solve for φ. We choose the first equality as using arccosine will return an angle in the 2nd quadrant:

$$5 + 5\sqrt{26}\cos\varphi = 0 \quad \Rightarrow \quad \varphi = \cos^{-1}\left(\frac{-5}{5\sqrt{26}}\right) \approx 1.7682 \approx 101.31°.$$

Subtracting $90°$ from this angle gives us an angle of $11.31°$ with the vertical.

We can now use trigonometry to find out how high the weight is lifted. The diagram shows that a right triangle is formed with the 2ft chain as the hypotenuse with an interior angle of $11.31°$. The length of the adjacent side (in the diagram, the dashed vertical line) is $2\cos 11.31° \approx 1.96$ft. Thus the weight is lifted by about 0.04ft, almost 1/2in.

The algebra we have applied to vectors is already demonstrating itself to be very useful. There are two more fundamental operations we can perform with vectors, the *dot product* and the *cross product*. The next two sections explore each in turn.

Notes:

Exercises 10.2

Terms and Concepts

1. Name two different things that cannot be described with just one number, but rather need 2 or more numbers to fully describe them.

2. What is the difference between $(1, 2)$ and $\langle 1, 2 \rangle$?

3. What is a unit vector?

4. Unit vectors can be thought of as conveying what type of information?

5. What does it mean for two vectors to be parallel?

6. What effect does multiplying a vector by -2 have?

Problems

In Exercises 7 – 10, points P and Q are given. Write the vector $\vec{PQ}$ in component form and using the standard unit vectors.

7. $P = (2, -1)$, $Q = (3, 5)$

8. $P = (3, 2)$, $Q = (7, -2)$

9. $P = (0, 3, -1)$, $Q = (6, 2, 5)$

10. $P = (2, 1, 2)$, $Q = (4, 3, 2)$

11. Let $\vec{u} = \langle 1, -2 \rangle$ and $\vec{v} = \langle 1, 1 \rangle$.
    (a) Find $\vec{u} + \vec{v}$, $\vec{u} - \vec{v}$, $2\vec{u} - 3\vec{v}$.
    (b) Sketch the above vectors on the same axes, along with $\vec{u}$ and $\vec{v}$.
    (c) Find $\vec{x}$ where $\vec{u} + \vec{x} = 2\vec{v} - \vec{x}$.

12. Let $\vec{u} = \langle 1, 1, -1 \rangle$ and $\vec{v} = \langle 2, 1, 2 \rangle$.
    (a) Find $\vec{u} + \vec{v}$, $\vec{u} - \vec{v}$, $\pi\vec{u} - \sqrt{2}\vec{v}$.
    (b) Sketch the above vectors on the same axes, along with $\vec{u}$ and $\vec{v}$.
    (c) Find $\vec{x}$ where $\vec{u} + \vec{x} = \vec{v} + 2\vec{x}$.

In Exercises 13 – 16, sketch $\vec{u}$, $\vec{v}$, $\vec{u} + \vec{v}$ and $\vec{u} - \vec{v}$ on the same axes.

13.

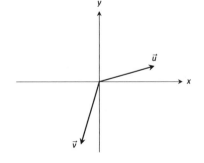

14.

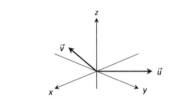

15.

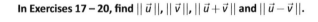

16.

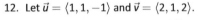

In Exercises 17 – 20, find $\|\vec{u}\|$, $\|\vec{v}\|$, $\|\vec{u} + \vec{v}\|$ and $\|\vec{u} - \vec{v}\|$.

17. $\vec{u} = \langle 2, 1 \rangle$, $\vec{v} = \langle 3, -2 \rangle$

18. $\vec{u} = \langle -3, 2, 2 \rangle$, $\vec{v} = \langle 1, -1, 1 \rangle$

19. $\vec{u} = \langle 1, 2 \rangle$, $\vec{v} = \langle -3, -6 \rangle$

20. $\vec{u} = \langle 2, -3, 6 \rangle$, $\vec{v} = \langle 10, -15, 30 \rangle$

21. Under what conditions is $\|\vec{u}\| + \|\vec{v}\| = \|\vec{u} + \vec{v}\|$?

In Exercises 22 – 25, find the unit vector $\vec{u}$ in the direction of $\vec{v}$.

22. $\vec{v} = \langle 3, 7 \rangle$

23. $\vec{v} = \langle 6, 8 \rangle$

24. $\vec{v} = \langle 1, -2, 2 \rangle$

25. $\vec{v} = \langle 2, -2, 2 \rangle$

26. Find the unit vector in the first quadrant of $\mathbb{R}^2$ that makes a 50° angle with the *x*-axis.

27. Find the unit vector in the second quadrant of $\mathbb{R}^2$ that makes a 30° angle with the *y*-axis.

28. Verify, from Key Idea 10.2.1, that

$$\vec{u} = \langle \sin\theta\cos\varphi, \sin\theta\sin\varphi, \cos\theta \rangle$$

is a unit vector for all angles θ and φ.

A weight of 100lb is suspended from two chains, making angles with the vertical of θ and φ as shown in the figure below.

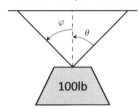

In Exercises 29 – 32, angles θ and φ are given. Find the magnitude of the force applied to each chain.

29. $\theta = 30°$, $\varphi = 30°$

30. $\theta = 60°$, $\varphi = 60°$

31. $\theta = 20°$, $\varphi = 15°$

32. $\theta = 0°$, $\varphi = 0°$

A weight of $p$lb is suspended from a chain of length ℓ while a constant force of $\vec{F}_w$ pushes the weight to the right, making an angle of θ with the vertical, as shown in the figure below.

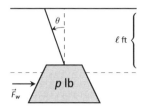

In Exercises 33 – 36, a force $\vec{F}_w$ and length ℓ are given. Find the angle θ and the height the weight is lifted as it moves to the right.

33. $\vec{F}_w = 1$lb, $\ell = 1$ft, $p = 1$lb

34. $\vec{F}_w = 1$lb, $\ell = 1$ft, $p = 10$lb

35. $\vec{F}_w = 1$lb, $\ell = 10$ft, $p = 1$lb

36. $\vec{F}_w = 10$lb, $\ell = 10$ft, $p = 1$lb

10.3   The Dot Product

The previous section introduced vectors and described how to add them together and how to multiply them by scalars. This section introduces *a* multiplication on vectors called the **dot product**.

Definition 10.3.1   **Dot Product**

1. Let $\vec{u} = \langle u_1, u_2 \rangle$ and $\vec{v} = \langle v_1, v_2 \rangle$ in $\mathbb{R}^2$. The **dot product** of $\vec{u}$ and $\vec{v}$, denoted $\vec{u} \cdot \vec{v}$, is

$$\vec{u} \cdot \vec{v} = u_1 v_1 + u_2 v_2.$$

2. Let $\vec{u} = \langle u_1, u_2, u_3 \rangle$ and $\vec{v} = \langle v_1, v_2, v_3 \rangle$ in $\mathbb{R}^3$. The **dot product** of $\vec{u}$ and $\vec{v}$, denoted $\vec{u} \cdot \vec{v}$, is

$$\vec{u} \cdot \vec{v} = u_1 v_1 + u_2 v_2 + u_3 v_3.$$

Note how this product of vectors returns a *scalar*, not another vector. We practice evaluating a dot product in the following example, then we will discuss why this product is useful.

Example 10.3.1   **Evaluating dot products**

1. Let $\vec{u} = \langle 1, 2 \rangle$, $\vec{v} = \langle 3, -1 \rangle$ in $\mathbb{R}^2$. Find $\vec{u} \cdot \vec{v}$.

2. Let $\vec{x} = \langle 2, -2, 5 \rangle$ and $\vec{y} = \langle -1, 0, 3 \rangle$ in $\mathbb{R}^3$. Find $\vec{x} \cdot \vec{y}$.

SOLUTION

1. Using Definition 10.3.1, we have

$$\vec{u} \cdot \vec{v} = 1(3) + 2(-1) = 1.$$

2. Using the definition, we have

$$\vec{x} \cdot \vec{y} = 2(-1) - 2(0) + 5(3) = 13.$$

The dot product, as shown by the preceding example, is very simple to evaluate. It is only the sum of products. While the definition gives no hint as to why

---

Notes:

we would care about this operation, there is an amazing connection between the dot product and angles formed by the vectors. Before stating this connection, we give a theorem stating some of the properties of the dot product.

Theorem 10.3.1   Properties of the Dot Product

Let $\vec{u}$, $\vec{v}$ and $\vec{w}$ be vectors in $\mathbb{R}^2$ or $\mathbb{R}^3$ and let c be a scalar.

1. $\vec{u} \cdot \vec{v} = \vec{v} \cdot \vec{u}$ Commutative Property

2. $\vec{u} \cdot (\vec{v} + \vec{w}) = \vec{u} \cdot \vec{v} + \vec{u} \cdot \vec{w}$ Distributive Property

3. $c(\vec{u} \cdot \vec{v}) = (c\vec{u}) \cdot \vec{v} = \vec{u} \cdot (c\vec{v})$

4. $\vec{0} \cdot \vec{v} = 0$

5. $\vec{v} \cdot \vec{v} = \| \vec{v} \|^2$

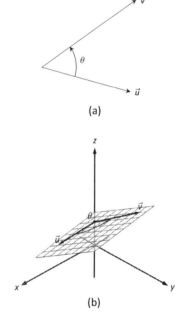

(a)

(b)

Figure 10.3.1: Illustrating the angle formed by two vectors with the same initial point.

The last statement of the theorem makes a handy connection between the magnitude of a vector and the dot product with itself. Our definition and theorem give properties of the dot product, but we are still likely wondering "What does the dot product *mean*?" It is helpful to understand that the dot product of a vector with itself is connected to its magnitude.

The next theorem extends this understanding by connecting the dot product to magnitudes and angles. Given vectors $\vec{u}$ and $\vec{v}$ in the plane, an angle θ is clearly formed when $\vec{u}$ and $\vec{v}$ are drawn with the same initial point as illustrated in Figure 10.3.1(a). (We always take θ to be the angle in $[0, \pi]$ as two angles are actually created.)

The same is also true of 2 vectors in space: given $\vec{u}$ and $\vec{v}$ in $\mathbb{R}^3$ with the same initial point, there is a plane that contains both $\vec{u}$ and $\vec{v}$. (When $\vec{u}$ and $\vec{v}$ are co-linear, there are infinitely many planes that contain both vectors.) In that plane, we can again find an angle θ between them (and again, $0 \leq \theta \leq \pi$). This is illustrated in Figure 10.3.1(b).

The following theorem connects this angle θ to the dot product of $\vec{u}$ and $\vec{v}$.

Notes:

> **Theorem 10.3.2   The Dot Product and Angles**
>
> Let $\vec{u}$ and $\vec{v}$ be nonzero vectors in $\mathbb{R}^2$ or $\mathbb{R}^3$. Then
>
> $$\vec{u} \cdot \vec{v} = \| \vec{u} \| \, \| \vec{v} \| \cos \theta,$$
>
> where θ, $0 \leq \theta \leq \pi$, is the angle between $\vec{u}$ and $\vec{v}$.

Using Theorem 10.3.1, we can rewrite this theorem as

$$\frac{\vec{u}}{\| \vec{u} \|} \cdot \frac{\vec{v}}{\| \vec{v} \|} = \cos \theta.$$

Note how on the left hand side of the equation, we are computing the dot product of two unit vectors. Recalling that unit vectors essentially only provide direction information, we can informally restate Theorem 10.3.2 as saying "The dot product of two directions gives the cosine of the angle between them."

When θ is an acute angle (i.e., $0 \leq \theta < \pi/2$), $\cos \theta$ is positive; when $\theta = \pi/2$, $\cos \theta = 0$; when θ is an obtuse angle ($\pi/2 < \theta \leq \pi$), $\cos \theta$ is negative. Thus the sign of the dot product gives a general indication of the angle between the vectors, illustrated in Figure 10.3.2.

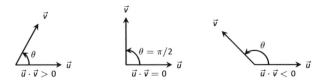

Figure 10.3.2: Illustrating the relationship between the angle between vectors and the sign of their dot product.

We *can* use Theorem 10.3.2 to compute the dot product, but generally this theorem is used to find the angle between known vectors (since the dot product is generally easy to compute). To this end, we rewrite the theorem's equation as

$$\cos \theta = \frac{\vec{u} \cdot \vec{v}}{\| \vec{u} \| \| \vec{v} \|} \quad \Leftrightarrow \quad \theta = \cos^{-1} \left(\frac{\vec{u} \cdot \vec{v}}{\| \vec{u} \| \| \vec{v} \|} \right).$$

We practice using this theorem in the following example.

Example 10.3.2   Using the dot product to find angles

Let $\vec{u} = \langle 3, 1 \rangle$, $\vec{v} = \langle -2, 6 \rangle$ and $\vec{w} = \langle -4, 3 \rangle$, as shown in Figure 10.3.3. Find the angles α, β and θ.

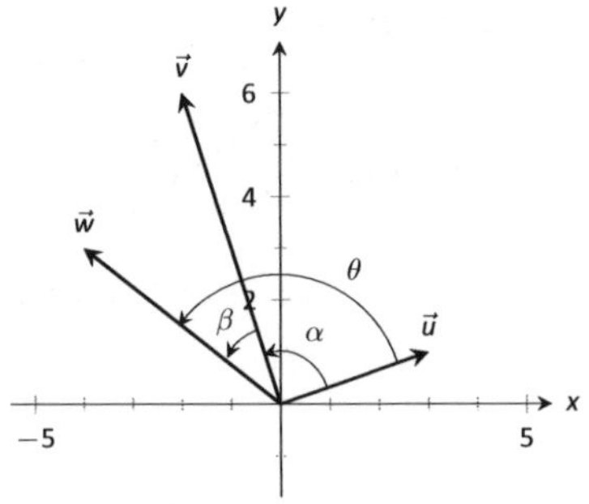

Figure 10.3.3: Vectors used in Example 10.3.2.

Notes:

SOLUTION We start by computing the magnitude of each vector.

$$\| \vec{u} \| = \sqrt{10}; \quad \| \vec{v} \| = 2\sqrt{10}; \quad \| \vec{w} \| = 5.$$

We now apply Theorem 10.3.2 to find the angles.

$$\alpha = \cos^{-1}\left(\frac{\vec{u} \cdot \vec{v}}{(\sqrt{10})(2\sqrt{10})} \right)$$
$$= \cos^{-1}(0) = \frac{\pi}{2} = 90°.$$

$$\beta = \cos^{-1}\left(\frac{\vec{v} \cdot \vec{w}}{(2\sqrt{10})(5)} \right)$$
$$= \cos^{-1}\left(\frac{26}{10\sqrt{10}} \right)$$
$$\approx 0.6055 \approx 34.7°.$$

$$\theta = \cos^{-1}\left(\frac{\vec{u} \cdot \vec{w}}{(\sqrt{10})(5)} \right)$$
$$= \cos^{-1}\left(\frac{-9}{5\sqrt{10}} \right)$$
$$\approx 2.1763 \approx 124.7°$$

We see from our computation that $\alpha + \beta = \theta$, as indicated by Figure 10.3.3. While we knew this should be the case, it is nice to see that this non-intuitive formula indeed returns the results we expected.

We do a similar example next in the context of vectors in space.

Example 10.3.3 Using the dot product to find angles
Let $\vec{u} = \langle 1, 1, 1 \rangle$, $\vec{v} = \langle -1, 3, -2 \rangle$ and $\vec{w} = \langle -5, 1, 4 \rangle$, as illustrated in Figure 10.3.4. Find the angle between each pair of vectors.

SOLUTION

1. Between $\vec{u}$ and $\vec{v}$:

$$\theta = \cos^{-1}\left(\frac{\vec{u} \cdot \vec{v}}{\| \vec{u} \| \| \vec{v} \|} \right)$$
$$= \cos^{-1}\left(\frac{0}{\sqrt{3}\sqrt{14}} \right)$$
$$= \frac{\pi}{2}.$$

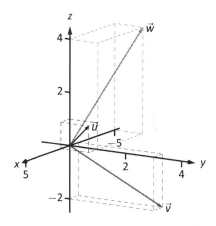

Figure 10.3.4: Vectors used in Example 10.3.3.

Notes:

2. Between $\vec{u}$ and $\vec{w}$:

$$\theta = \cos^{-1}\left(\frac{\vec{u}\cdot\vec{w}}{\|\vec{u}\|\|\vec{w}\|}\right)$$
$$= \cos^{-1}\left(\frac{0}{\sqrt{3}\sqrt{42}}\right)$$
$$= \frac{\pi}{2}.$$

3. Between $\vec{v}$ and $\vec{w}$:

$$\theta = \cos^{-1}\left(\frac{\vec{v}\cdot\vec{w}}{\|\vec{v}\|\|\vec{w}\|}\right)$$
$$= \cos^{-1}\left(\frac{0}{\sqrt{14}\sqrt{42}}\right)$$
$$= \frac{\pi}{2}.$$

While our work shows that each angle is $\pi/2$, i.e., $90°$, none of these angles looks to be a right angle in Figure 10.3.4. Such is the case when drawing three–dimensional objects on the page.

All three angles between these vectors was $\pi/2$, or $90°$. We know from geometry and everyday life that $90°$ angles are "nice" for a variety of reasons, so it should seem significant that these angles are all $\pi/2$. Notice the common feature in each calculation (and also the calculation of α in Example 10.3.2): the dot products of each pair of angles was 0. We use this as a basis for a definition of the term **orthogonal**, which is essentially synonymous to *perpendicular*.

Definition 10.3.2   Orthogonal

Nonzero vectors $\vec{u}$ and $\vec{v}$ are **orthogonal** if their dot product is 0.

Example 10.3.4   Finding orthogonal vectors

Let $\vec{u} = \langle 3, 5 \rangle$ and $\vec{v} = \langle 1, 2, 3 \rangle$.

1. Find two vectors in $\mathbb{R}^2$ that are orthogonal to $\vec{u}$.

2. Find two non–parallel vectors in $\mathbb{R}^3$ that are orthogonal to $\vec{v}$.

SOLUTION

Note: The term *perpendicular* originally referred to lines. As mathematics progressed, the concept of "being at right angles to" was applied to other objects, such as vectors and planes, and the term *orthogonal* was introduced. It is especially used when discussing objects that are hard, or impossible, to visualize: two vectors in 5-dimensional space are orthogonal if their dot product is 0. It is not wrong to say they are *perpendicular*, but common convention gives preference to the word *orthogonal*.

Notes:

1. Recall that a line perpendicular to a line with slope m has slope $-1/m$, the "opposite reciprocal slope." We can think of the slope of $\vec{u}$ as $5/3$, its "rise over run." A vector orthogonal to $\vec{u}$ will have slope $-3/5$. There are many such choices, though all parallel:

$$\langle -5, 3\rangle \quad \text{or} \quad \langle 5, -3\rangle \quad \text{or} \quad \langle -10, 6\rangle \quad \text{or} \quad \langle 15, -9\rangle, \text{etc.}$$

2. There are infinitely many directions in space orthogonal to any given direction, so there are an infinite number of non–parallel vectors orthogonal to $\vec{v}$. Since there are so many, we have great leeway in finding some.

One way is to arbitrarily pick values for the first two components, leaving the third unknown. For instance, let $\vec{v}_1 = \langle 2, 7, z\rangle$. If $\vec{v}_1$ is to be orthogonal to $\vec{v}$, then $\vec{v}_1 \cdot \vec{v} = 0$, so

$$2 + 14 + 3z = 0 \quad \Rightarrow z = \frac{-16}{3}.$$

So $\vec{v}_1 = \langle 2, 7, -16/3\rangle$ is orthogonal to $\vec{v}$. We can apply a similar technique by leaving the first or second component unknown.

Another method of finding a vector orthogonal to $\vec{v}$ mirrors what we did in part 1. Let $\vec{v}_2 = \langle -2, 1, 0\rangle$. Here we switched the first two components of $\vec{v}$, changing the sign of one of them (similar to the "opposite reciprocal" concept before). Letting the third component be 0 effectively ignores the third component of $\vec{v}$, and it is easy to see that

$$\vec{v}_2 \cdot \vec{v} = \langle -2, 1, 0\rangle \cdot \langle 1, 2, 3\rangle = 0.$$

Clearly $\vec{v}_1$ and $\vec{v}_2$ are not parallel.

An important construction is illustrated in Figure 10.3.5, where vectors $\vec{u}$ and $\vec{v}$ are sketched. In part (a), a dotted line is drawn from the tip of $\vec{u}$ to the line containing $\vec{v}$, where the dotted line is orthogonal to $\vec{v}$. In part (b), the dotted line is replaced with the vector $\vec{z}$ and $\vec{w}$ is formed, parallel to $\vec{v}$. It is clear by the diagram that $\vec{u} = \vec{w} + \vec{z}$. What is important about this construction is this: $\vec{u}$ is *decomposed* as the sum of two vectors, one of which is parallel to $\vec{v}$ and one that is perpendicular to $\vec{v}$. It is hard to overstate the importance of this construction (as we'll see in upcoming examples).

The vectors $\vec{w}$, $\vec{z}$ and $\vec{u}$ as shown in Figure 10.3.5 (b) form a right triangle, where the angle between $\vec{v}$ and $\vec{u}$ is labeled θ. We can find $\vec{w}$ in terms of $\vec{v}$ and $\vec{u}$.

Using trigonometry, we can state that

$$\|\vec{w}\| = \|\vec{u}\|\cos\theta. \tag{10.2}$$

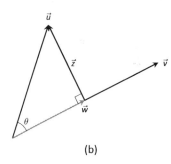

(a)

(b)

Figure 10.3.5: Developing the construction of the *orthogonal projection*.

Notes:

We also know that $\vec{w}$ is parallel to to $\vec{v}$; that is, the direction of $\vec{w}$ is the direction of $\vec{v}$, described by the unit vector $\vec{v}/\|\vec{v}\|$. The vector $\vec{w}$ is the vector in the direction $\vec{v}/\|\vec{v}\|$ with magnitude $\|\vec{u}\|\cos\theta$:

$$\vec{w} = \left(\|\vec{u}\|\cos\theta\right)\frac{1}{\|\vec{v}\|}\vec{v}.$$

Replace $\cos\theta$ using Theorem 10.3.2:

$$= \left(\|\vec{u}\|\frac{\vec{u}\cdot\vec{v}}{\|\vec{u}\|\|\vec{v}\|}\right)\frac{1}{\|\vec{v}\|}\vec{v}$$

$$= \frac{\vec{u}\cdot\vec{v}}{\|\vec{v}\|^2}\vec{v}.$$

Now apply Theorem 10.3.1.

$$= \frac{\vec{u}\cdot\vec{v}}{\vec{v}\cdot\vec{v}}\vec{v}.$$

Since this construction is so important, it is given a special name.

Definition 10.3.3   Orthogonal Projection

Let nonzero vectors $\vec{u}$ and $\vec{v}$ be given. The **orthogonal projection of $\vec{u}$ onto $\vec{v}$**, denoted $\text{proj}_{\vec{v}}\,\vec{u}$, is

$$\text{proj}_{\vec{v}}\,\vec{u} = \frac{\vec{u}\cdot\vec{v}}{\vec{v}\cdot\vec{v}}\vec{v}.$$

Example 10.3.5   Computing the orthogonal projection

1. Let $\vec{u} = \langle -2, 1\rangle$ and $\vec{v} = \langle 3, 1\rangle$. Find $\text{proj}_{\vec{v}}\,\vec{u}$, and sketch all three vectors with initial points at the origin.

2. Let $\vec{w} = \langle 2, 1, 3\rangle$ and $\vec{x} = \langle 1, 1, 1\rangle$. Find $\text{proj}_{\vec{x}}\,\vec{w}$, and sketch all three vectors with initial points at the origin.

SOLUTION

Notes:

1. Applying Definition 10.3.3, we have

$$\text{proj}_{\vec{v}}\,\vec{u} = \frac{\vec{u}\cdot\vec{v}}{\vec{v}\cdot\vec{v}}\vec{v}$$
$$= \frac{-5}{10}\,\langle 3, 1\rangle$$
$$= \left\langle -\frac{3}{2}, -\frac{1}{2}\right\rangle.$$

Vectors $\vec{u}$, $\vec{v}$ and $\text{proj}_{\vec{v}}\,\vec{u}$ are sketched in Figure 10.3.6(a). Note how the projection is parallel to $\vec{v}$; that is, it lies on the same line through the origin as $\vec{v}$, although it points in the opposite direction. That is because the angle between $\vec{u}$ and $\vec{v}$ is obtuse (i.e., greater than $90°$).

2. Apply the definition:

$$\text{proj}_{\vec{x}}\,\vec{w} = \frac{\vec{w}\cdot\vec{x}}{\vec{x}\cdot\vec{x}}\vec{x}$$
$$= \frac{6}{3}\,\langle 1, 1, 1\rangle$$
$$= \langle 2, 2, 2\rangle\,.$$

These vectors are sketched in Figure 10.3.6(b), and again in part (c) from a different perspective. Because of the nature of graphing these vectors, the sketch in part (b) makes it difficult to recognize that the drawn projection has the geometric properties it should. The graph shown in part (c) illustrates these properties better.

We can use the properties of the dot product found in Theorem 10.3.1 to rearrange the formula found in Definition 10.3.3:

$$\text{proj}_{\vec{v}}\,\vec{u} = \frac{\vec{u}\cdot\vec{v}}{\vec{v}\cdot\vec{v}}\vec{v}$$
$$= \frac{\vec{u}\cdot\vec{v}}{\|\vec{v}\|^2}\vec{v}$$
$$= \left(\vec{u}\cdot\frac{\vec{v}}{\|\vec{v}\|}\right)\frac{\vec{v}}{\|\vec{v}\|}.$$

The above formula shows that the orthogonal projection of $\vec{u}$ onto $\vec{v}$ is only concerned with the *direction* of $\vec{v}$, as both instances of $\vec{v}$ in the formula come in the form $\vec{v}/\|\vec{v}\|$, the unit vector in the direction of $\vec{v}$.

A special case of orthogonal projection occurs when $\vec{v}$ is a unit vector. In this situation, the formula for the orthogonal projection of a vector $\vec{u}$ onto $\vec{v}$ reduces to just $\text{proj}_{\vec{v}}\,\vec{u} = (\vec{u}\cdot\vec{v})\vec{v}$, as $\vec{v}\cdot\vec{v} = 1$.

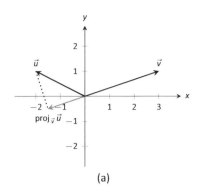

(a)

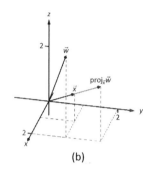

(b)

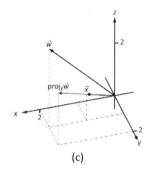

(c)

Figure 10.3.6: Graphing the vectors used in Example 10.3.5.

Notes:

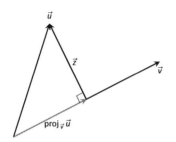

Figure 10.3.7: Illustrating the orthogonal projection.

This gives us a new understanding of the dot product. When $\vec{v}$ is a unit vector, essentially providing only direction information, the dot product of $\vec{u}$ and $\vec{v}$ gives "how much of $\vec{u}$ is in the direction of $\vec{v}$." This use of the dot product will be very useful in future sections.

Now consider Figure 10.3.7 where the concept of the orthogonal projection is again illustrated. It is clear that

$$\vec{u} = \text{proj}_{\vec{v}}\,\vec{u} + \vec{z}. \tag{10.3}$$

As we know what $\vec{u}$ and $\text{proj}_{\vec{v}}\,\vec{u}$ are, we can solve for $\vec{z}$ and state that

$$\vec{z} = \vec{u} - \text{proj}_{\vec{v}}\,\vec{u}.$$

This leads us to rewrite Equation (10.3) in a seemingly silly way:

$$\vec{u} = \text{proj}_{\vec{v}}\,\vec{u} + (\vec{u} - \text{proj}_{\vec{v}}\,\vec{u}).$$

This is not nonsense, as pointed out in the following Key Idea. (Notation note: the expression "$\parallel \vec{y}$" means "is parallel to $\vec{y}$." We can use this notation to state "$\vec{x} \parallel \vec{y}$" which means "$\vec{x}$ is parallel to $\vec{y}$." The expression "$\perp \vec{y}$" means "is orthogonal to $\vec{y}$," and is used similarly.)

Key Idea 10.3.1   Orthogonal Decomposition of Vectors

Let nonzero vectors $\vec{u}$ and $\vec{v}$ be given. Then $\vec{u}$ can be written as the sum of two vectors, one of which is parallel to $\vec{v}$, and one of which is orthogonal to $\vec{v}$:

$$\vec{u} = \underbrace{\text{proj}_{\vec{v}}\,\vec{u}}_{\parallel\,\vec{v}} + \underbrace{(\vec{u} - \text{proj}_{\vec{v}}\,\vec{u})}_{\perp\,\vec{v}}.$$

We illustrate the use of this equality in the following example.

Example 10.3.6   Orthogonal decomposition of vectors

1. Let $\vec{u} = \langle -2, 1\rangle$ and $\vec{v} = \langle 3, 1\rangle$ as in Example 10.3.5. Decompose $\vec{u}$ as the sum of a vector parallel to $\vec{v}$ and a vector orthogonal to $\vec{v}$.

2. Let $\vec{w} = \langle 2, 1, 3\rangle$ and $\vec{x} = \langle 1, 1, 1\rangle$ as in Example 10.3.5. Decompose $\vec{w}$ as the sum of a vector parallel to $\vec{x}$ and a vector orthogonal to $\vec{x}$.

SOLUTION

Notes:

1. In Example 10.3.5, we found that $\text{proj}_{\vec{v}}\,\vec{u} = \langle -1.5, -0.5\rangle$. Let

$$\vec{z} = \vec{u} - \text{proj}_{\vec{v}}\,\vec{u} = \langle -2, 1\rangle - \langle -1.5, -0.5\rangle = \langle -0.5, 1.5\rangle\,.$$

   Is $\vec{z}$ orthogonal to $\vec{v}$? (I.e, is $\vec{z} \perp \vec{v}$?) We check for orthogonality with the dot product:

$$\vec{z} \cdot \vec{v} = \langle -0.5, 1.5\rangle \cdot \langle 3, 1\rangle = 0.$$

   Since the dot product is 0, we know $\vec{z} \perp \vec{v}$. Thus:

$$\vec{u} = \text{proj}_{\vec{v}}\,\vec{u} + (\vec{u} - \text{proj}_{\vec{v}}\,\vec{u})$$
$$\langle -2, 1\rangle = \underbrace{\langle -1.5, -0.5\rangle}_{\|\,\vec{v}} + \underbrace{\langle -0.5, 1.5\rangle}_{\perp\,\vec{v}}\,.$$

2. We found in Example 10.3.5 that $\text{proj}_{\vec{x}}\,\vec{w} = \langle 2, 2, 2\rangle$. Applying the Key Idea, we have:

$$\vec{z} = \vec{w} - \text{proj}_{\vec{x}}\,\vec{w} = \langle 2, 1, 3\rangle - \langle 2, 2, 2\rangle = \langle 0, -1, 1\rangle\,.$$

   We check to see if $\vec{z} \perp \vec{x}$:

$$\vec{z} \cdot \vec{x} = \langle 0, -1, 1\rangle \cdot \langle 1, 1, 1\rangle = 0.$$

   Since the dot product is 0, we know the two vectors are orthogonal. We now write $\vec{w}$ as the sum of two vectors, one parallel and one orthogonal to $\vec{x}$:

$$\vec{w} = \text{proj}_{\vec{x}}\,\vec{w} + (\vec{w} - \text{proj}_{\vec{x}}\,\vec{w})$$
$$\langle 2, 1, 3\rangle = \underbrace{\langle 2, 2, 2\rangle}_{\|\,\vec{x}} + \underbrace{\langle 0, -1, 1\rangle}_{\perp\,\vec{x}}$$

We give an example of where this decomposition is useful.

Example 10.3.7   Orthogonally decomposing a force vector

Consider Figure 10.3.8(a), showing a box weighing 50lb on a ramp that rises 5ft over a span of 20ft. Find the components of force, and their magnitudes, acting on the box (as sketched in part (b) of the figure):

1. in the direction of the ramp, and

2. orthogonal to the ramp.

SOLUTION   As the ramp rises 5ft over a horizontal distance of 20ft, we can represent the direction of the ramp with the vector $\vec{r} = \langle 20, 5\rangle$. Gravity pulls down with a force of 50lb, which we represent with $\vec{g} = \langle 0, -50\rangle$.

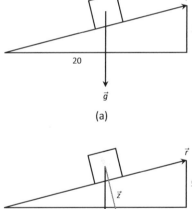

(a)

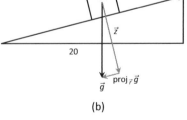

(b)

Figure 10.3.8: Sketching the ramp and box in Example 10.3.7. Note: *The vectors are not drawn to scale.*

Notes:

1. To find the force of gravity in the direction of the ramp, we compute $\text{proj}_{\vec{r}}\,\vec{g}$:

$$\begin{aligned}
\text{proj}_{\vec{r}}\,\vec{g} &= \frac{\vec{g}\cdot\vec{r}}{\vec{r}\cdot\vec{r}}\vec{r} \\
&= \frac{-250}{425}\langle 20,5\rangle \\
&= \left\langle -\frac{200}{17}, -\frac{50}{17}\right\rangle \approx \langle -11.76, -2.94\rangle .
\end{aligned}$$

The magnitude of $\text{proj}_{\vec{r}}\,\vec{g}$ is $\|\,\text{proj}_{\vec{r}}\,\vec{g}\,\| = 50/\sqrt{17} \approx 12.13\text{lb}$. Though the box weighs 50lb, a force of about 12lb is enough to keep the box from sliding down the ramp.

2. To find the component $\vec{z}$ of gravity orthogonal to the ramp, we use Key Idea 10.3.1.

$$\begin{aligned}
\vec{z} &= \vec{g} - \text{proj}_{\vec{r}}\,\vec{g} \\
&= \left\langle \frac{200}{17}, -\frac{800}{17}\right\rangle \approx \langle 11.76, -47.06\rangle .
\end{aligned}$$

The magnitude of this force is $\|\vec{z}\| \approx 48.51\text{lb}$. In physics and engineering, knowing this force is important when computing things like static frictional force. (For instance, we could easily compute if the static frictional force alone was enough to keep the box from sliding down the ramp.)

Application to Work

In physics, the application of a force F to move an object in a straight line a distance d produces *work*; the amount of work W is $W = Fd$, (where F is in the direction of travel). The orthogonal projection allows us to compute work when the force is not in the direction of travel.

Consider Figure 10.3.9, where a force $\vec{F}$ is being applied to an object moving in the direction of $\vec{d}$. (The distance the object travels is the magnitude of $\vec{d}$.) The

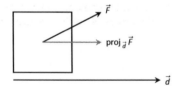

Figure 10.3.9: Finding work when the force and direction of travel are given as vectors.

work done is the amount of force in the direction of $\vec{d}$, $\| \operatorname{proj}_{\vec{d}} \vec{F} \|$, times $\| \vec{d} \|$:

$$\| \operatorname{proj}_{\vec{d}} \vec{F} \| \cdot \| \vec{d} \| = \left\| \frac{\vec{F} \cdot \vec{d}}{\vec{d} \cdot \vec{d}} \vec{d} \right\| \cdot \| \vec{d} \|$$

$$= \left| \frac{\vec{F} \cdot \vec{d}}{\| \vec{d} \|^2} \right| \cdot \| \vec{d} \| \cdot \| \vec{d} \|$$

$$= \frac{\left| \vec{F} \cdot \vec{d} \right|}{\| \vec{d} \|^2} \| \vec{d} \|^2$$

$$= \left| \vec{F} \cdot \vec{d} \right|.$$

The expression $\vec{F} \cdot \vec{d}$ will be positive if the angle between $\vec{F}$ and $\vec{d}$ is acute; when the angle is obtuse (hence $\vec{F} \cdot \vec{d}$ is negative), the force is causing motion in the opposite direction of $\vec{d}$, resulting in "negative work." We want to capture this sign, so we drop the absolute value and find that $W = \vec{F} \cdot \vec{d}$.

Definition 10.3.4   Work

Let $\vec{F}$ be a constant force that moves an object in a straight line from point P to point Q. Let $\vec{d} = \overrightarrow{PQ}$. The **work** W done by $\vec{F}$ along $\vec{d}$ is $W = \vec{F} \cdot \vec{d}$.

Example 10.3.8   Computing work

A man slides a box along a ramp that rises 3ft over a distance of 15ft by applying 50lb of force as shown in Figure 10.3.10. Compute the work done.

SOLUTION   The figure indicates that the force applied makes a 30° angle with the horizontal, so $\vec{F} = 50 \langle \cos 30°, \sin 30° \rangle \approx \langle 43.3, 25 \rangle$. The ramp is represented by $\vec{d} = \langle 15, 3 \rangle$. The work done is simply

$$\vec{F} \cdot \vec{d} = 50 \langle \cos 30°, \sin 30° \rangle \cdot \langle 15, 3 \rangle \approx 724.5 \text{ft–lb.}$$

Note how we did not actually compute the distance the object traveled, nor the magnitude of the force in the direction of travel; this is all inherently computed by the dot product!

The dot product is a powerful way of evaluating computations that depend on angles without actually using angles. The next section explores another "product" on vectors, the *cross product*. Once again, angles play an important role, though in a much different way.

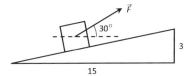

Figure 10.3.10: Computing work when sliding a box up a ramp in Example 10.3.8.

Notes:

Exercises 10.3

Terms and Concepts

1. The dot product of two vectors is a _____, not a vector.

2. How are the concepts of the dot product and vector magnitude related?

3. How can one quickly tell if the angle between two vectors is acute or obtuse?

4. Give a synonym for "orthogonal."

Problems

In Exercises 5 – 10, find the dot product of the given vectors.

5. $\vec{u} = \langle 2, -4 \rangle$, $\vec{v} = \langle 3, 7 \rangle$

6. $\vec{u} = \langle 5, 3 \rangle$, $\vec{v} = \langle 6, 1 \rangle$

7. $\vec{u} = \langle 1, -1, 2 \rangle$, $\vec{v} = \langle 2, 5, 3 \rangle$

8. $\vec{u} = \langle 3, 5, -1 \rangle$, $\vec{v} = \langle 4, -1, 7 \rangle$

9. $\vec{u} = \langle 1, 1 \rangle$, $\vec{v} = \langle 1, 2, 3 \rangle$

10. $\vec{u} = \langle 1, 2, 3 \rangle$, $\vec{v} = \langle 0, 0, 0 \rangle$

11. Create your own vectors $\vec{u}$, $\vec{v}$ and $\vec{w}$ in $\mathbb{R}^2$ and show that $\vec{u} \cdot (\vec{v} + \vec{w}) = \vec{u} \cdot \vec{v} + \vec{u} \cdot \vec{w}$.

12. Create your own vectors $\vec{u}$ and $\vec{v}$ in $\mathbb{R}^3$ and scalar c and show that $c(\vec{u} \cdot \vec{v}) = \vec{u} \cdot (c\vec{v})$.

In Exercises 13 – 16, find the measure of the angle between the two vectors in both radians and degrees.

13. $\vec{u} = \langle 1, 1 \rangle$, $\vec{v} = \langle 1, 2 \rangle$

14. $\vec{u} = \langle -2, 1 \rangle$, $\vec{v} = \langle 3, 5 \rangle$

15. $\vec{u} = \langle 8, 1, -4 \rangle$, $\vec{v} = \langle 2, 2, 0 \rangle$

16. $\vec{u} = \langle 1, 7, 2 \rangle$, $\vec{v} = \langle 4, -2, 5 \rangle$

In Exercises 17 – 20, a vector $\vec{v}$ is given. Give two vectors that are orthogonal to $\vec{v}$.

17. $\vec{v} = \langle 4, 7 \rangle$

18. $\vec{v} = \langle -3, 5 \rangle$

19. $\vec{v} = \langle 1, 1, 1 \rangle$

20. $\vec{v} = \langle 1, -2, 3 \rangle$

In Exercises 21 – 26, vectors $\vec{u}$ and $\vec{v}$ are given. Find $\text{proj}_{\vec{v}}\,\vec{u}$, the orthogonal projection of $\vec{u}$ onto $\vec{v}$, and sketch all three vectors with the same initial point.

21. $\vec{u} = \langle 1, 2 \rangle$, $\vec{v} = \langle -1, 3 \rangle$

22. $\vec{u} = \langle 5, 5 \rangle$, $\vec{v} = \langle 1, 3 \rangle$

23. $\vec{u} = \langle -3, 2 \rangle$, $\vec{v} = \langle 1, 1 \rangle$

24. $\vec{u} = \langle -3, 2 \rangle$, $\vec{v} = \langle 2, 3 \rangle$

25. $\vec{u} = \langle 1, 5, 1 \rangle$, $\vec{v} = \langle 1, 2, 3 \rangle$

26. $\vec{u} = \langle 3, -1, 2 \rangle$, $\vec{v} = \langle 2, 2, 1 \rangle$

In Exercises 27 – 32, vectors $\vec{u}$ and $\vec{v}$ are given. Write $\vec{u}$ as the sum of two vectors, one of which is parallel to $\vec{v}$ and one of which is perpendicular to $\vec{v}$. Note: these are the same pairs of vectors as found in Exercises 21 – 26.

27. $\vec{u} = \langle 1, 2 \rangle$, $\vec{v} = \langle -1, 3 \rangle$

28. $\vec{u} = \langle 5, 5 \rangle$, $\vec{v} = \langle 1, 3 \rangle$

29. $\vec{u} = \langle -3, 2 \rangle$, $\vec{v} = \langle 1, 1 \rangle$

30. $\vec{u} = \langle -3, 2 \rangle$, $\vec{v} = \langle 2, 3 \rangle$

31. $\vec{u} = \langle 1, 5, 1 \rangle$, $\vec{v} = \langle 1, 2, 3 \rangle$

32. $\vec{u} = \langle 3, -1, 2 \rangle$, $\vec{v} = \langle 2, 2, 1 \rangle$

33. A 10lb box sits on a ramp that rises 4ft over a distance of 20ft. How much force is required to keep the box from sliding down the ramp?

34. A 10lb box sits on a 15ft ramp that makes a 30° angle with the horizontal. How much force is required to keep the box from sliding down the ramp?

35. How much work is performed in moving a box horizontally 10ft with a force of 20lb applied at an angle of 45° to the horizontal?

36. How much work is performed in moving a box horizontally 10ft with a force of 20lb applied at an angle of 10° to the horizontal?

37. How much work is performed in moving a box up the length of a ramp that rises 2ft over a distance of 10ft, with a force of 50lb applied horizontally?

38. How much work is performed in moving a box up the length of a ramp that rises 2ft over a distance of 10ft, with a force of 50lb applied at an angle of 45° to the horizontal?

39. How much work is performed in moving a box up the length of a 10ft ramp that makes a 5° angle with the horizontal, with 50lb of force applied in the direction of the ramp?

10.4   The Cross Product

"Orthogonality" is immensely important. A quick scan of your current environment will undoubtedly reveal numerous surfaces and edges that are perpendicular to each other (including the edges of this page). The dot product provides a quick test for orthogonality: vectors $\vec{u}$ and $\vec{v}$ are perpendicular if, and only if, $\vec{u} \cdot \vec{v} = 0$.

Given two non–parallel, nonzero vectors $\vec{u}$ and $\vec{v}$ in space, it is very useful to find a vector $\vec{w}$ that is perpendicular to both $\vec{u}$ and $\vec{v}$. There is a operation, called the **cross product**, that creates such a vector. This section defines the cross product, then explores its properties and applications.

Definition 10.4.1   Cross Product

Let $\vec{u} = \langle u_1, u_2, u_3 \rangle$ and $\vec{v} = \langle v_1, v_2, v_3 \rangle$ be vectors in $\mathbb{R}^3$. The **cross product of $\vec{u}$ and $\vec{v}$**, denoted $\vec{u} \times \vec{v}$, is the vector

$$\vec{u} \times \vec{v} = \langle u_2 v_3 - u_3 v_2, -(u_1 v_3 - u_3 v_1), u_1 v_2 - u_2 v_1 \rangle .$$

This definition can be a bit cumbersome to remember. After an example we will give a convenient method for computing the cross product. For now, careful examination of the products and differences given in the definition should reveal a pattern that is not too difficult to remember. (For instance, in the first component only 2 and 3 appear as subscripts; in the second component, only 1 and 3 appear as subscripts. Further study reveals the order in which they appear.)

Let's practice using this definition by computing a cross product.

Example 10.4.1   Computing a cross product
Let $\vec{u} = \langle 2, -1, 4 \rangle$ and $\vec{v} = \langle 3, 2, 5 \rangle$. Find $\vec{u} \times \vec{v}$, and verify that it is orthogonal to both $\vec{u}$ and $\vec{v}$.

SOLUTION   Using Definition 10.4.1, we have

$$\vec{u} \times \vec{v} = \langle (-1)5 - (4)2, -((2)5 - (4)3), (2)2 - (-1)3 \rangle = \langle -13, 2, 7 \rangle .$$

(We encourage the reader to compute this product on their own, then verify their result.)

We test whether or not $\vec{u} \times \vec{v}$ is orthogonal to $\vec{u}$ and $\vec{v}$ using the dot product:

$$(\vec{u} \times \vec{v}) \cdot \vec{u} = \langle -13, 2, 7 \rangle \cdot \langle 2, -1, 4 \rangle = 0,$$

$$(\vec{u} \times \vec{v}) \cdot \vec{v} = \langle -13, 2, 7 \rangle \cdot \langle 3, 2, 5 \rangle = 0.$$

Since both dot products are zero, $\vec{u} \times \vec{v}$ is indeed orthogonal to both $\vec{u}$ and $\vec{v}$.

Notes:

A convenient method of computing the cross product starts with forming a particular 3×3 *matrix*, or rectangular array. The first row comprises the standard unit vectors $\vec{i}, \vec{j}$, and $\vec{k}$. The second and third rows are the vectors $\vec{u}$ and $\vec{v}$, respectively. Using $\vec{u}$ and $\vec{v}$ from Example 10.4.1, we begin with:

$$
\begin{array}{ccc}
\vec{i} & \vec{j} & \vec{k} \\
2 & -1 & 4 \\
3 & 2 & 5
\end{array}
$$

Now repeat the first two columns after the original three:

$$
\begin{array}{ccccc}
\vec{i} & \vec{j} & \vec{k} & \vec{i} & \vec{j} \\
2 & -1 & 4 & 2 & -1 \\
3 & 2 & 5 & 3 & 2
\end{array}
$$

This gives three full "upper left to lower right" diagonals, and three full "upper right to lower left" diagonals, as shown. Compute the products along each diagonal, then add the products on the right and subtract the products on the left:

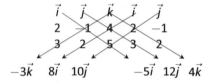

$-3\vec{k}$ $8\vec{i}$ $10\vec{j}$ $-5\vec{i}$ $12\vec{j}$ $4\vec{k}$

$$
\vec{u} \times \vec{v} = \left(-5\vec{i} + 12\vec{j} + 4\vec{k} \right) - \left(-3\vec{k} + 8\vec{i} + 10\vec{j} \right) = -13\vec{i} + 2\vec{j} + 7\vec{k} = \langle -13, 2, 7 \rangle .
$$

We practice using this method.

Example 10.4.2   Computing a cross product
Let $\vec{u} = \langle 1, 3, 6 \rangle$ and $\vec{v} = \langle -1, 2, 1 \rangle$. Compute both $\vec{u} \times \vec{v}$ and $\vec{v} \times \vec{u}$.

SOLUTION   To compute $\vec{u} \times \vec{v}$, we form the matrix as prescribed above, complete with repeated first columns:

$$
\begin{array}{ccccc}
\vec{i} & \vec{j} & \vec{k} & \vec{i} & \vec{j} \\
1 & 3 & 6 & 1 & 3 \\
-1 & 2 & 1 & -1 & 2
\end{array}
$$

We let the reader compute the products of the diagonals; we give the result:

$$
\vec{u} \times \vec{v} = \left(3\vec{i} - 6\vec{j} + 2\vec{k} \right) - \left(-3\vec{k} + 12\vec{i} + \vec{j} \right) = \langle -9, -7, 5 \rangle .
$$

Notes:

To compute $\vec{v} \times \vec{u}$, we switch the second and third rows of the above matrix, then multiply along diagonals and subtract:

$$
\begin{array}{ccccc}
\vec{i} & \vec{j} & \vec{k} & \vec{i} & \vec{j} \\
-1 & 2 & 1 & -1 & 2 \\
1 & 3 & 6 & 1 & 3
\end{array}
$$

Note how with the rows being switched, the products that once appeared on the right now appear on the left, and vice–versa. Thus the result is:

$$\vec{v} \times \vec{u} = \left(12\vec{i} + \vec{j} - 3\vec{k}\right) - \left(2\vec{k} + 3\vec{i} - 6\vec{j}\right) = \langle 9, 7, -5 \rangle,$$

which is the opposite of $\vec{u} \times \vec{v}$. We leave it to the reader to verify that each of these vectors is orthogonal to $\vec{u}$ and $\vec{v}$.

Properties of the Cross Product

It is not coincidence that $\vec{v} \times \vec{u} = -(\vec{u} \times \vec{v})$ in the preceding example; one can show using Definition 10.4.1 that this will always be the case. The following theorem states several useful properties of the cross product, each of which can be verified by referring to the definition.

Theorem 10.4.1   Properties of the Cross Product

Let $\vec{u}$, $\vec{v}$ and $\vec{w}$ be vectors in $\mathbb{R}^3$ and let c be a scalar. The following identities hold:

1. $\vec{u} \times \vec{v} = -(\vec{v} \times \vec{u})$ Anticommutative Property

2. (a) $(\vec{u} + \vec{v}) \times \vec{w} = \vec{u} \times \vec{w} + \vec{v} \times \vec{w}$ Distributive Properties
    (b) $\vec{u} \times (\vec{v} + \vec{w}) = \vec{u} \times \vec{v} + \vec{u} \times \vec{w}$

3. $c(\vec{u} \times \vec{v}) = (c\vec{u}) \times \vec{v} = \vec{u} \times (c\vec{v})$

4. (a) $(\vec{u} \times \vec{v}) \cdot \vec{u} = 0$ Orthogonality Properties
    (b) $(\vec{u} \times \vec{v}) \cdot \vec{v} = 0$

5. $\vec{u} \times \vec{u} = \vec{0}$

6. $\vec{u} \times \vec{0} = \vec{0}$

7. $\vec{u} \cdot (\vec{v} \times \vec{w}) = (\vec{u} \times \vec{v}) \cdot \vec{w}$ Triple Scalar Product

---

Notes:

We introduced the cross product as a way to find a vector orthogonal to two given vectors, but we did not give a proof that the construction given in Definition 10.4.1 satisfies this property. Theorem 10.4.1 asserts this property holds; we leave it as a problem in the Exercise section to verify this.

Property 5 from the theorem is also left to the reader to prove in the Exercise section, but it reveals something more interesting than "the cross product of a vector with itself is $\vec{0}$." Let $\vec{u}$ and $\vec{v}$ be parallel vectors; that is, let there be a scalar c such that $\vec{v} = c\vec{u}$. Consider their cross product:

$$
\begin{aligned}
\vec{u} \times \vec{v} &= \vec{u} \times (c\vec{u}) \\
&= c(\vec{u} \times \vec{u}) \qquad \text{(by Property 3 of Theorem 10.4.1)} \\
&= \vec{0}. \qquad\qquad \text{(by Property 5 of Theorem 10.4.1)}
\end{aligned}
$$

We have just shown that the cross product of parallel vectors is $\vec{0}$. This hints at something deeper. Theorem 10.3.2 related the angle between two vectors and their dot product; there is a similar relationship relating the cross product of two vectors and the angle between them, given by the following theorem.

Theorem 10.4.2   The Cross Product and Angles

Let $\vec{u}$ and $\vec{v}$ be nonzero vectors in $\mathbb{R}^3$. Then

$$\| \vec{u} \times \vec{v} \| = \| \vec{u} \| \, \| \vec{v} \| \sin\theta,$$

where θ, $0 \leq \theta \leq \pi$, is the angle between $\vec{u}$ and $\vec{v}$.

Note: We could rewrite Definition 10.3.2 and Theorem 10.4.2 to include $\vec{0}$, then define that $\vec{u}$ and $\vec{v}$ are parallel if $\vec{u} \times \vec{v} = \vec{0}$. Since $\vec{0} \cdot \vec{v} = 0$ and $\vec{0} \times \vec{v} = \vec{0}$, this would mean that $\vec{0}$ is both parallel *and* orthogonal to all vectors. Apparent paradoxes such as this are not uncommon in mathematics and can be very useful. (See also the marginal note on page 582.)

Note that this theorem makes a statement about the *magnitude* of the cross product. When the angle between $\vec{u}$ and $\vec{v}$ is 0 or π (i.e., the vectors are parallel), the magnitude of the cross product is 0. The only vector with a magnitude of 0 is $\vec{0}$ (see Property 9 of Theorem 10.2.1), hence the cross product of parallel vectors is $\vec{0}$.

We demonstrate the truth of this theorem in the following example.

Example 10.4.3   The cross product and angles
Let $\vec{u} = \langle 1, 3, 6 \rangle$ and $\vec{v} = \langle -1, 2, 1 \rangle$ as in Example 10.4.2. Verify Theorem 10.4.2 by finding θ, the angle between $\vec{u}$ and $\vec{v}$, and the magnitude of $\vec{u} \times \vec{v}$.

Notes:

SOLUTION   We use Theorem 10.3.2 to find the angle between $\vec{u}$ and $\vec{v}$.

$$\theta = \cos^{-1}\left(\frac{\vec{u}\cdot\vec{v}}{\|\vec{u}\|\,\|\vec{v}\|}\right)$$

$$= \cos^{-1}\left(\frac{11}{\sqrt{46}\sqrt{6}}\right)$$

$$\approx 0.8471 = 48.54°.$$

Our work in Example 10.4.2 showed that $\vec{u}\times\vec{v} = \langle -9, -7, 5\rangle$, hence $\|\vec{u}\times\vec{v}\| = \sqrt{155}$. Is $\|\vec{u}\times\vec{v}\| = \|\vec{u}\|\,\|\vec{v}\|\sin\theta$? Using numerical approximations, we find:

$$\|\vec{u}\times\vec{v}\| = \sqrt{155} \qquad\qquad \|\vec{u}\|\,\|\vec{v}\|\sin\theta = \sqrt{46}\sqrt{6}\sin 0.8471$$

$$\approx 12.45. \qquad\qquad\qquad\qquad \approx 12.45.$$

Numerically, they seem equal. Using a right triangle, one can show that

$$\sin\left(\cos^{-1}\left(\frac{11}{\sqrt{46}\sqrt{6}}\right)\right) = \frac{\sqrt{155}}{\sqrt{46}\sqrt{6}},$$

which allows us to verify the theorem exactly.

Right Hand Rule

The anticommutative property of the cross product demonstrates that $\vec{u}\times\vec{v}$ and $\vec{v}\times\vec{u}$ differ only by a sign – these vectors have the same magnitude but point in the opposite direction. When seeking a vector perpendicular to $\vec{u}$ and $\vec{v}$, we essentially have two directions to choose from, one in the direction of $\vec{u}\times\vec{v}$ and one in the direction of $\vec{v}\times\vec{u}$. Does it matter which we choose? How can we tell which one we will get without graphing, etc.?

Another wonderful property of the cross product, as defined, is that it follows the **right hand rule**. Given $\vec{u}$ and $\vec{v}$ in $\mathbb{R}^3$ with the same initial point, point the index finger of your right hand in the direction of $\vec{u}$ and let your middle finger point in the direction of $\vec{v}$ (much as we did when establishing the right hand rule for the 3-dimensional coordinate system). Your thumb will naturally extend in the direction of $\vec{u}\times\vec{v}$. One can "practice" this using Figure 10.4.1. If you switch, and point the index finder in the direction of $\vec{v}$ and the middle finger in the direction of $\vec{u}$, your thumb will now point in the opposite direction, allowing you to "visualize" the anticommutative property of the cross product.

Applications of the Cross Product

There are a number of ways in which the cross product is useful in mathematics, physics and other areas of science beyond "just" finding a vector perpendicular to two others. We highlight a few here.

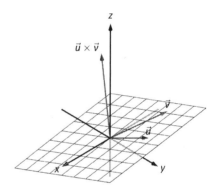

Figure 10.4.1: Illustrating the Right Hand Rule of the cross product.

Notes:

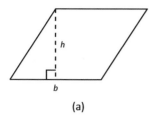

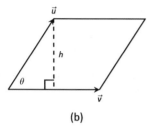

Figure 10.4.2: Using the cross product to find the area of a parallelogram.

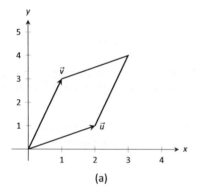

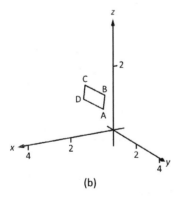

Figure 10.4.3: Sketching the parallelograms in Example 10.4.4.

Area of a Parallelogram

It is a standard geometry fact that the area of a parallelogram is $A = bh$, where b is the length of the base and h is the height of the parallelogram, as illustrated in Figure 10.4.2(a). As shown when defining the Parallelogram Law of vector addition, two vectors $\vec{u}$ and $\vec{v}$ define a parallelogram when drawn from the same initial point, as illustrated in Figure 10.4.2(b). Trigonometry tells us that $h = \|\vec{u}\|\sin\theta$, hence the area of the parallelogram is

$$A = \|\vec{u}\|\,\|\vec{v}\|\sin\theta = \|\vec{u}\times\vec{v}\|, \qquad (10.4)$$

where the second equality comes from Theorem 10.4.2. We illustrate using Equation (10.4) in the following example.

Example 10.4.4   Finding the area of a parallelogram

1. Find the area of the parallelogram defined by the vectors $\vec{u} = \langle 2,1\rangle$ and $\vec{v} = \langle 1,3\rangle$.

2. Verify that the points $A = (1,1,1)$, $B = (2,3,2)$, $C = (4,5,3)$ and $D = (3,3,2)$ are the vertices of a parallelogram. Find the area of the parallelogram.

SOLUTION

1. Figure 10.4.3(a) sketches the parallelogram defined by the vectors $\vec{u}$ and $\vec{v}$. We have a slight problem in that our vectors exist in $\mathbb{R}^2$, not $\mathbb{R}^3$, and the cross product is only defined on vectors in $\mathbb{R}^3$. We skirt this issue by viewing $\vec{u}$ and $\vec{v}$ as vectors in the $x-y$ plane of $\mathbb{R}^3$, and rewrite them as $\vec{u} = \langle 2,1,0\rangle$ and $\vec{v} = \langle 1,3,0\rangle$. We can now compute the cross product. It is easy to show that $\vec{u}\times\vec{v} = \langle 0,0,5\rangle$; therefore the area of the parallelogram is $A = \|\vec{u}\times\vec{v}\| = 5$.

2. To show that the quadrilateral $ABCD$ is a parallelogram (shown in Figure 10.4.3(b)), we need to show that the opposite sides are parallel. We can quickly show that $\overrightarrow{AB} = \overrightarrow{DC} = \langle 1,2,1\rangle$ and $\overrightarrow{BC} = \overrightarrow{AD} = \langle 2,2,1\rangle$. We find the area by computing the magnitude of the cross product of $\overrightarrow{AB}$ and $\overrightarrow{BC}$:

$$\overrightarrow{AB}\times\overrightarrow{BC} = \langle 0,1,-2\rangle \quad\Rightarrow\quad \|\overrightarrow{AB}\times\overrightarrow{BC}\| = \sqrt{5} \approx 2.236.$$

This application is perhaps more useful in finding the area of a triangle (in short, triangles are used more often than parallelograms). We illustrate this in the following example.

Notes:

Example 10.4.5   Area of a triangle

Find the area of the triangle with vertices $A = (1, 2)$, $B = (2, 3)$ and $C = (3, 1)$, as pictured in Figure 10.4.4.

SOLUTION   We found the area of this triangle in Example 7.1.4 to be 1.5 using integration. There we discussed the fact that finding the area of a triangle can be inconvenient using the "$\frac{1}{2}bh$" formula as one has to compute the height, which generally involves finding angles, etc. Using a cross product is much more direct.

We can choose any two sides of the triangle to use to form vectors; we choose $\overrightarrow{AB} = \langle 1, 1 \rangle$ and $\overrightarrow{AC} = \langle 2, -1 \rangle$. As in the previous example, we will rewrite these vectors with a third component of 0 so that we can apply the cross product. The area of the triangle is

$$\frac{1}{2}\| \overrightarrow{AB} \times \overrightarrow{AC} \| = \frac{1}{2}\| \langle 1, 1, 0 \rangle \times \langle 2, -1, 0 \rangle \| = \frac{1}{2}\| \langle 0, 0, -3 \rangle \| = \frac{3}{2}.$$

We arrive at the same answer as before with less work.

Volume of a Parallelepiped

The three dimensional analogue to the parallelogram is the **parallelepiped**. Each face is parallel to the opposite face, as illustrated in Figure 10.4.5. By crossing $\vec{v}$ and $\vec{w}$, one gets a vector whose magnitude is the area of the base. Dotting this vector with $\vec{u}$ computes the volume of parallelepiped! (Up to a sign; take the absolute value.)

Thus the volume of a parallelepiped defined by vectors $\vec{u}$, $\vec{v}$ and $\vec{w}$ is

$$V = |\vec{u} \cdot (\vec{v} \times \vec{w})|. \tag{10.5}$$

Note how this is the Triple Scalar Product, first seen in Theorem 10.4.1. Applying the identities given in the theorem shows that we can apply the Triple Scalar Product in any "order" we choose to find the volume. That is,

$$V = |\vec{u} \cdot (\vec{v} \times \vec{w})| = |\vec{u} \cdot (\vec{w} \times \vec{v})| = |(\vec{u} \times \vec{v}) \cdot \vec{w}|, \quad \text{etc.}$$

Example 10.4.6   Finding the volume of parallelepiped

Find the volume of the parallelepiped defined by the vectors $\vec{u} = \langle 1, 1, 0 \rangle$, $\vec{v} = \langle -1, 1, 0 \rangle$ and $\vec{w} = \langle 0, 1, 1 \rangle$.

SOLUTION   We apply Equation (10.5). We first find $\vec{v} \times \vec{w} = \langle 1, 1, -1 \rangle$. Then

$$|\vec{u} \cdot (\vec{v} \times \vec{w})| = |\langle 1, 1, 0 \rangle \cdot \langle 1, 1, -1 \rangle| = 2.$$

So the volume of the parallelepiped is 2 cubic units.

Notes:

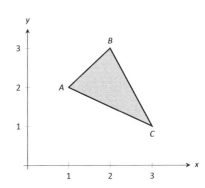

Figure 10.4.4: Finding the area of a triangle in Example 10.4.5.

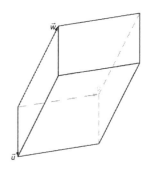

Figure 10.4.5: A parallelepiped is the three dimensional analogue to the parallelogram.

Note: The word "parallelepiped" is pronounced "parallel–eh–pipe–ed."

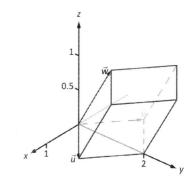

Figure 10.4.6: A parallelepiped in Example 10.4.6.

While this application of the Triple Scalar Product is interesting, it is not used all that often: parallelepipeds are not a common shape in physics and engineering. The last application of the cross product is very applicable in engineering.

Torque

Torque is a measure of the turning force applied to an object. A classic scenario involving torque is the application of a wrench to a bolt. When a force is applied to the wrench, the bolt turns. When we represent the force and wrench with vectors $\vec{F}$ and $\vec{\ell}$, we see that the bolt moves (because of the threads) in a direction orthogonal to $\vec{F}$ and $\vec{\ell}$. Torque is usually represented by the Greek letter τ, or tau, and has units of N·m, a Newton–meter, or ft·lb, a foot–pound.

While a full understanding of torque is beyond the purposes of this book, when a force $\vec{F}$ is applied to a lever arm $\vec{\ell}$, the resulting torque is

$$\vec{\tau} = \vec{\ell} \times \vec{F}. \tag{10.6}$$

Example 10.4.7   Computing torque
A lever of length 2ft makes an angle with the horizontal of $45°$. Find the resulting torque when a force of 10lb is applied to the end of the level where:

1. the force is perpendicular to the lever, and

2. the force makes an angle of $60°$ with the lever, as shown in Figure 10.4.7.

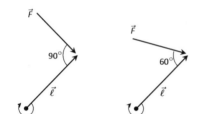

Figure 10.4.7: Showing a force being applied to a lever in Example 10.4.7.

SOLUTION

1. We start by determining vectors for the force and lever arm. Since the lever arm makes a $45°$ angle with the horizontal and is 2ft long, we can state that $\vec{\ell} = 2 \langle \cos 45°, \sin 45° \rangle = \langle \sqrt{2}, \sqrt{2} \rangle$.

    Since the force vector is perpendicular to the lever arm (as seen in the left hand side of Figure 10.4.7), we can conclude it is making an angle of $-45°$ with the horizontal. As it has a magnitude of 10lb, we can state $\vec{F} = 10 \langle \cos(-45°), \sin(-45°) \rangle = \langle 5\sqrt{2}, -5\sqrt{2} \rangle$.

    Using Equation (10.6) to find the torque requires a cross product. We again let the third component of each vector be 0 and compute the cross product:

    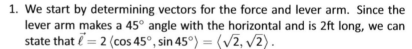

    $$\begin{aligned}
    \vec{\tau} &= \vec{\ell} \times \vec{F} \\
    &= \langle \sqrt{2}, \sqrt{2}, 0 \rangle \times \langle 5\sqrt{2}, -5\sqrt{2}, 0 \rangle \\
    &= \langle 0, 0, -20 \rangle
    \end{aligned}$$

Notes:

This clearly has a magnitude of 20 ft-lb.

We can view the force and lever arm vectors as lying "on the page"; our computation of $\vec{\tau}$ shows that the torque goes "into the page." This follows the Right Hand Rule of the cross product, and it also matches well with the example of the wrench turning the bolt. Turning a bolt clockwise moves it in.

2. Our lever arm can still be represented by $\vec{\ell} = \langle \sqrt{2}, \sqrt{2} \rangle$. As our force vector makes a $60°$ angle with $\vec{\ell}$, we can see (referencing the right hand side of the figure) that $\vec{F}$ makes a $-15°$ angle with the horizontal. Thus

$$\vec{F} = 10 \langle \cos -15°, \sin -15° \rangle = \left\langle \frac{5(1+\sqrt{3})}{\sqrt{2}}, \frac{5(-1+\sqrt{3})}{\sqrt{2}} \right\rangle$$
$$\approx \langle 9.659, -2.588 \rangle .$$

We again make the third component 0 and take the cross product to find the torque:

$$\vec{\tau} = \vec{\ell} \times \vec{F}$$
$$= \langle \sqrt{2}, \sqrt{2}, 0 \rangle \times \left\langle \frac{5(1+\sqrt{3})}{\sqrt{2}}, \frac{5(-1+\sqrt{3})}{\sqrt{2}}, 0 \right\rangle$$
$$= \langle 0, 0, -10\sqrt{3} \rangle$$
$$\approx \langle 0, 0, -17.321 \rangle .$$

As one might expect, when the force and lever arm vectors *are* orthogonal, the magnitude of force is greater than when the vectors *are not* orthogonal.

While the cross product has a variety of applications (as noted in this chapter), its fundamental use is finding a vector perpendicular to two others. Knowing a vector is orthogonal to two others is of incredible importance, as it allows us to find the equations of lines and planes in a variety of contexts. The importance of the cross product, in some sense, relies on the importance of lines and planes, which see widespread use throughout engineering, physics and mathematics. We study lines and planes in the next two sections.

Notes:

Exercises 10.4

Terms and Concepts

1. The cross product of two vectors is a _____, not a scalar.

2. One can visualize the direction of $\vec{u} \times \vec{v}$ using the _____ _____ _____.

3. Give a synonym for "orthogonal."

4. T/F: A fundamental principle of the cross product is that $\vec{u} \times \vec{v}$ is orthogonal to $\vec{u}$ and $\vec{v}$.

5. _____ is a measure of the turning force applied to an object.

6. T/F: If $\vec{u}$ and $\vec{v}$ are parallel, then $\vec{u} \times \vec{v} = \vec{0}$.

Problems

In Exercises 7 – 16, vectors $\vec{u}$ and $\vec{v}$ are given. Compute $\vec{u} \times \vec{v}$ and show this is orthogonal to both $\vec{u}$ and $\vec{v}$.

7. $\vec{u} = \langle 3, 2, -2 \rangle$, $\vec{v} = \langle 0, 1, 5 \rangle$

8. $\vec{u} = \langle 5, -4, 3 \rangle$, $\vec{v} = \langle 2, -5, 1 \rangle$

9. $\vec{u} = \langle 4, -5, -5 \rangle$, $\vec{v} = \langle 3, 3, 4 \rangle$

10. $\vec{u} = \langle -4, 7, -10 \rangle$, $\vec{v} = \langle 4, 4, 1 \rangle$

11. $\vec{u} = \langle 1, 0, 1 \rangle$, $\vec{v} = \langle 5, 0, 7 \rangle$

12. $\vec{u} = \langle 1, 5, -4 \rangle$, $\vec{v} = \langle -2, -10, 8 \rangle$

13. $\vec{u} = \langle a, b, 0 \rangle$, $\vec{v} = \langle c, d, 0 \rangle$

14. $\vec{u} = \vec{i}$, $\vec{v} = \vec{j}$

15. $\vec{u} = \vec{i}$, $\vec{v} = \vec{k}$

16. $\vec{u} = \vec{j}$, $\vec{v} = \vec{k}$

17. Pick any vectors $\vec{u}$, $\vec{v}$ and $\vec{w}$ in $\mathbb{R}^3$ and show that $\vec{u} \times (\vec{v} + \vec{w}) = \vec{u} \times \vec{v} + \vec{u} \times \vec{w}$.

18. Pick any vectors $\vec{u}$, $\vec{v}$ and $\vec{w}$ in $\mathbb{R}^3$ and show that $\vec{u} \cdot (\vec{v} \times \vec{w}) = (\vec{u} \times \vec{v}) \cdot \vec{w}$.

In Exercises 19 – 22, the magnitudes of vectors $\vec{u}$ and $\vec{v}$ in $\mathbb{R}^3$ are given, along with the angle θ between them. Use this information to find the magnitude of $\vec{u} \times \vec{v}$.

19. $\| \vec{u} \| = 2$, $\| \vec{v} \| = 5$, $\theta = 30°$

20. $\| \vec{u} \| = 3$, $\| \vec{v} \| = 7$, $\theta = \pi/2$

21. $\| \vec{u} \| = 3$, $\| \vec{v} \| = 4$, $\theta = \pi$

22. $\| \vec{u} \| = 2$, $\| \vec{v} \| = 5$, $\theta = 5\pi/6$

In Exercises 23 – 26, find the area of the parallelogram defined by the given vectors.

23. $\vec{u} = \langle 1, 1, 2 \rangle$, $\vec{v} = \langle 2, 0, 3 \rangle$

24. $\vec{u} = \langle -2, 1, 5 \rangle$, $\vec{v} = \langle -1, 3, 1 \rangle$

25. $\vec{u} = \langle 1, 2 \rangle$, $\vec{v} = \langle 2, 1 \rangle$

26. $\vec{u} = \langle 2, 0 \rangle$, $\vec{v} = \langle 0, 3 \rangle$

In Exercises 27 – 30, find the area of the triangle with the given vertices.

27. Vertices: $(0, 0, 0)$, $(1, 3, -1)$ and $(2, 1, 1)$.

28. Vertices: $(5, 2, -1)$, $(3, 6, 2)$ and $(1, 0, 4)$.

29. Vertices: $(1, 1)$, $(1, 3)$ and $(2, 2)$.

30. Vertices: $(3, 1)$, $(1, 2)$ and $(4, 3)$.

In Exercises 31 – 32, find the area of the quadrilateral with the given vertices. (Hint: break the quadrilateral into 2 triangles.)

31. Vertices: $(0, 0)$, $(1, 2)$, $(3, 0)$ and $(4, 3)$.

32. Vertices: $(0, 0, 0)$, $(2, 1, 1)$, $(-1, 2, -8)$ and $(1, -1, 5)$.

In Exercises 33 – 34, find the volume of the parallelepiped defined by the given vectors.

33. $\vec{u} = \langle 1, 1, 1 \rangle$, $\vec{v} = \langle 1, 2, 3 \rangle$, $\vec{w} = \langle 1, 0, 1 \rangle$

34. $\vec{u} = \langle -1, 2, 1 \rangle$, $\vec{v} = \langle 2, 2, 1 \rangle$, $\vec{w} = \langle 3, 1, 3 \rangle$

In Exercises 35 – 38, find a unit vector orthogonal to both $\vec{u}$ and $\vec{v}$.

35. $\vec{u} = \langle 1, 1, 1 \rangle$, $\vec{v} = \langle 2, 0, 1 \rangle$

36. $\vec{u} = \langle 1, -2, 1 \rangle$, $\vec{v} = \langle 3, 2, 1 \rangle$

37. $\vec{u} = \langle 5, 0, 2 \rangle$, $\vec{v} = \langle -3, 0, 7 \rangle$

38. $\vec{u} = \langle 1, -2, 1 \rangle$, $\vec{v} = \langle -2, 4, -2 \rangle$

39. A bicycle rider applies 150lb of force, straight down, onto a pedal that extends 7in horizontally from the crankshaft. Find the magnitude of the torque applied to the crankshaft.

40. A bicycle rider applies 150lb of force, straight down, onto a pedal that extends 7in from the crankshaft, making a 30° angle with the horizontal. Find the magnitude of the torque applied to the crankshaft.

41. To turn a stubborn bolt, 80lb of force is applied to a 10in wrench. What is the maximum amount of torque that can be applied to the bolt?

42. To turn a stubborn bolt, 80lb of force is applied to a 10in wrench in a confined space, where the direction of applied force makes a 10° angle with the wrench. How much torque is subsequently applied to the wrench?

43. Show, using the definition of the Cross Product, that $\vec{u} \cdot (\vec{u} \times \vec{v}) = 0$; that is, that $\vec{u}$ is orthogonal to the cross product of $\vec{u}$ and $\vec{v}$.

44. Show, using the definition of the Cross Product, that $\vec{u} \times \vec{u} = \vec{0}$.

10.5 Lines

To find the equation of a line in the *x-y* plane, we need two pieces of information: a point and the slope. The slope conveys *direction* information. As vertical lines have an undefined slope, the following statement is more accurate:

> To define a line, one needs a point on the line and the direction of the line.

This holds true for lines in space.

Let P be a point in space, let $\vec{p}$ be the vector with initial point at the origin and terminal point at P (i.e., $\vec{p}$ "points" to P), and let $\vec{d}$ be a vector. Consider the points on the line through P in the direction of $\vec{d}$.

Clearly one point on the line is P; we can say that the *vector* $\vec{p}$ lies at this point on the line. To find another point on the line, we can start at $\vec{p}$ and move in a direction parallel to $\vec{d}$. For instance, starting at $\vec{p}$ and traveling one length of $\vec{d}$ places one at another point on the line. Consider Figure 10.5.2 where certain points along the line are indicated.

The figure illustrates how every point on the line can be obtained by starting with $\vec{p}$ and moving a certain distance in the direction of $\vec{d}$. That is, we can define the line as a function of t:

$$\vec{\ell}(t) = \vec{p} + t\,\vec{d}. \tag{10.7}$$

In many ways, this is *not* a new concept. Compare Equation (10.7) to the familiar "$y = mx + b$" equation of a line:

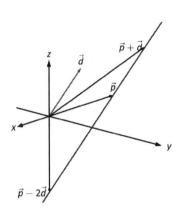

Figure 10.5.2: Defining a line in space.

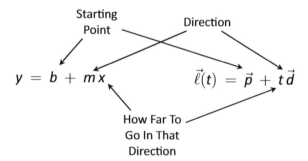

Figure 10.5.1: Understanding the vector equation of a line.

The equations exhibit the same structure: they give a starting point, define a direction, and state how far in that direction to travel.

Equation (10.7) is an example of a **vector–valued function**; the input of the function is a real number and the output is a vector. We will cover vector–valued functions extensively in the next chapter.

Notes:

There are other ways to represent a line. Let $\vec{p} = \langle x_0, y_0, z_0 \rangle$ and let $\vec{d} = \langle a, b, c \rangle$. Then the equation of the line through $\vec{p}$ in the direction of $\vec{d}$ is:

$$\vec{\ell}(t) = \vec{p} + t\vec{d}$$
$$= \langle x_0, y_0, z_0 \rangle + t \langle a, b, c \rangle$$
$$= \langle x_0 + at, y_0 + bt, z_0 + ct \rangle .$$

The last line states that the x values of the line are given by $x = x_0 + at$, the y values are given by $y = y_0 + bt$, and the z values are given by $z = z_0 + ct$. These three equations, taken together, are the **parametric equations of the line** through $\vec{p}$ in the direction of $\vec{d}$.

Finally, each of the equations for x, y and z above contain the variable t. We can solve for t in each equation:

$$x = x_0 + at \quad \Rightarrow \quad t = \frac{x - x_0}{a},$$
$$y = y_0 + bt \quad \Rightarrow \quad t = \frac{y - y_0}{b},$$
$$z = z_0 + ct \quad \Rightarrow \quad t = \frac{z - z_0}{c},$$

assuming $a, b, c \neq 0$. Since t is equal to each expression on the right, we can set these equal to each other, forming the **symmetric equations of the line** through $\vec{p}$ in the direction of $\vec{d}$:

$$\frac{x - x_0}{a} = \frac{y - y_0}{b} = \frac{z - z_0}{c}.$$

Each representation has its own advantages, depending on the context. We summarize these three forms in the following definition, then give examples of their use.

Notes:

Definition 10.5.1   Equations of Lines in Space

Consider the line in space that passes through $\vec{p} = \langle x_0, y_0, z_0 \rangle$ in the direction of $\vec{d} = \langle a, b, c \rangle$.

1. The **vector equation** of the line is

$$\vec{\ell}(t) = \vec{p} + t\vec{d}.$$

2. The **parametric equations** of the line are

$$x = x_0 + at, \quad y = y_0 + bt, \quad z = z_0 + ct.$$

3. The **symmetric equations** of the line are

$$\frac{x - x_0}{a} = \frac{y - y_0}{b} = \frac{z - z_0}{c}.$$

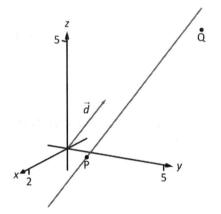

Figure 10.5.3: Graphing a line in Example 10.5.1.

Example 10.5.1   Finding the equation of a line

Give all three equations, as given in Definition 10.5.1, of the line through $P = (2, 3, 1)$ in the direction of $\vec{d} = \langle -1, 1, 2 \rangle$. Does the point $Q = (-1, 6, 6)$ lie on this line?

SOLUTION   We identify the point $P = (2, 3, 1)$ with the vector $\vec{p} = \langle 2, 3, 1 \rangle$. Following the definition, we have

- the vector equation of the line is $\vec{\ell}(t) = \langle 2, 3, 1 \rangle + t\langle -1, 1, 2 \rangle$;

- the parametric equations of the line are

$$x = 2 - t, \quad y = 3 + t, \quad z = 1 + 2t; \text{ and}$$

- the symmetric equations of the line are

$$\frac{x - 2}{-1} = \frac{y - 3}{1} = \frac{z - 1}{2}.$$

The first two equations of the line are useful when a t value is given: one can immediately find the corresponding point on the line. These forms are good when calculating with a computer; most software programs easily handle equations in these formats. (For instance, the graphics program that made Figure 10.5.3 can be given the input "(2-t,3+t,1+2*t)" for $-1 \leq t \leq 3$.).

Does the point $Q = (-1, 6, 6)$ lie on the line? The graph in Figure 10.5.3 makes it clear that it does not. We can answer this question without the graph

Notes:

using any of the three equation forms. Of the three, the symmetric equations are probably best suited for this task. Simply plug in the values of x, y and z and see if equality is maintained:

$$\frac{-1-2}{-1} \overset{?}{=} \frac{6-3}{1} \overset{?}{=} \frac{6-1}{2} \quad \Rightarrow \quad 3 = 3 \neq 2.5.$$

We see that Q does not lie on the line as it did not satisfy the symmetric equations.

Example 10.5.2   Finding the equation of a line through two points
Find the parametric equations of the line through the points $P = (2, -1, 2)$ and $Q = (1, 3, -1)$.

SOLUTION   Recall the statement made at the beginning of this section: to find the equation of a line, we need a point and a direction. We have *two* points; either one will suffice. The direction of the line can be found by the vector with initial point P and terminal point Q: $\overrightarrow{PQ} = \langle -1, 4, -3 \rangle$.

The parametric equations of the line ℓ through P in the direction of $\overrightarrow{PQ}$ are:

$$\ell: \quad x = 2 - t \quad y = -1 + 4t \quad z = 2 - 3t.$$

A graph of the points and line are given in Figure 10.5.4. Note how in the given parametrization of the line, $t = 0$ corresponds to the point P, and $t = 1$ corresponds to the point Q. This relates to the understanding of the vector equation of a line described in Figure 10.5.1. The parametric equations "start" at the point P, and t determines how far in the direction of $\overrightarrow{PQ}$ to travel. When $t = 0$, we travel 0 lengths of $\overrightarrow{PQ}$; when $t = 1$, we travel one length of $\overrightarrow{PQ}$, resulting in the point Q.

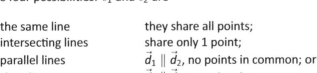

Figure 10.5.4: A graph of the line in Example 10.5.2.

Parallel, Intersecting and Skew Lines

In the plane, two *distinct* lines can either be parallel or they will intersect at exactly one point. In space, given equations of two lines, it can sometimes be difficult to tell whether the lines are distinct or not (i.e., the same line can be represented in different ways). Given lines $\vec{\ell}_1(t) = \vec{p}_1 + t\vec{d}_1$ and $\vec{\ell}_2(t) = \vec{p}_2 + t\vec{d}_2$, we have four possibilities: ℓ_1 and ℓ_2 are

the same line	they share all points;
intersecting lines	share only 1 point;
parallel lines	$\vec{d}_1 \parallel \vec{d}_2$, no points in common; or
skew lines	$\vec{d}_1 \nparallel \vec{d}_2$, no points in common.

Notes:

The next two examples investigate these possibilities.

Example 10.5.3   Comparing lines

Consider lines ℓ_1 and ℓ_2, given in parametric equation form:

$$\ell_1 : \begin{aligned} x &= 1 + 3t \\ y &= 2 - t \\ z &= t \end{aligned} \qquad \ell_2 : \begin{aligned} x &= -2 + 4s \\ y &= 3 + s \\ z &= 5 + 2s. \end{aligned}$$

Determine whether ℓ_1 and ℓ_2 are the same line, intersect, are parallel, or skew.

SOLUTION   We start by looking at the directions of each line. Line ℓ_1 has the direction given by $\vec{d}_1 = \langle 3, -1, 1 \rangle$ and line ℓ_2 has the direction given by $\vec{d}_2 = \langle 4, 1, 2 \rangle$. It should be clear that $\vec{d}_1$ and $\vec{d}_2$ are not parallel, hence ℓ_1 and ℓ_2 are not the same line, nor are they parallel. Figure 10.5.5 verifies this fact (where the points and directions indicated by the equations of each line are identified).

We next check to see if they intersect (if they do not, they are skew lines). To find if they intersect, we look for t and s values such that the respective x, y and z values are the same. That is, we want s and t such that:

$$\begin{aligned} 1 + 3t &= -2 + 4s \\ 2 - t &= 3 + s \\ t &= 5 + 2s. \end{aligned}$$

This is a relatively simple system of linear equations. Since the last equation is already solved for t, substitute that value of t into the equation above it:

$$2 - (5 + 2s) = 3 + s \quad \Rightarrow \quad s = -2, \; t = 1.$$

A key to remember is that we have *three* equations; we need to check if $s = -2$, $t = 1$ satisfies the first equation as well:

$$1 + 3(1) \neq -2 + 4(-2).$$

It does not. Therefore, we conclude that the lines ℓ_1 and ℓ_2 are skew.

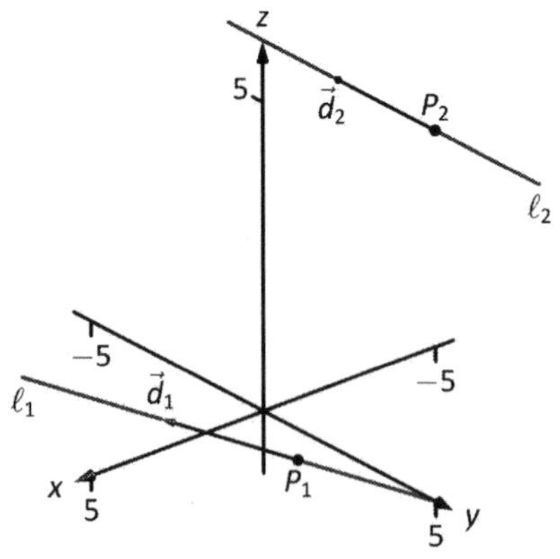

Figure 10.5.5: Sketching the lines from Example 10.5.3.

Example 10.5.4   Comparing lines

Consider lines ℓ_1 and ℓ_2, given in parametric equation form:

$$\ell_1 : \begin{aligned} x &= -0.7 + 1.6t \\ y &= 4.2 + 2.72t \\ z &= 2.3 - 3.36t \end{aligned} \qquad \ell_2 : \begin{aligned} x &= 2.8 - 2.9s \\ y &= 10.15 - 4.93s \\ z &= -5.05 + 6.09s. \end{aligned}$$

Determine whether ℓ_1 and ℓ_2 are the same line, intersect, are parallel, or skew.

Notes:

SOLUTION   It is obviously very difficult to simply look at these equations and discern anything. This is done intentionally. In the "real world," most equations that are used do not have nice, integer coefficients. Rather, there are lots of digits after the decimal and the equations can look "messy."

We again start by deciding whether or not each line has the same direction. The direction of ℓ_1 is given by $\vec{d}_1 = \langle 1.6, 2.72, -3.36 \rangle$ and the direction of ℓ_2 is given by $\vec{d}_2 = \langle -2.9, -4.93, 6.09 \rangle$. When it is not clear through observation whether two vectors are parallel or not, the standard way of determining this is by comparing their respective unit vectors. Using a calculator, we find:

$$\vec{u}_1 = \frac{\vec{d}_1}{\| \vec{d}_1 \|} = \langle 0.3471, 0.5901, -0.7289 \rangle$$

$$\vec{u}_2 = \frac{\vec{d}_2}{\| \vec{d}_2 \|} = \langle -0.3471, -0.5901, 0.7289 \rangle .$$

The two vectors seem to be parallel (at least, their components are equal to 4 decimal places). In most situations, it would suffice to conclude that the lines are at least parallel, if not the same. One way to be sure is to rewrite $\vec{d}_1$ and $\vec{d}_2$ in terms of fractions, not decimals. We have

$$\vec{d}_1 = \left\langle \frac{16}{10}, \frac{272}{100}, -\frac{336}{100} \right\rangle \qquad \vec{d}_2 = \left\langle -\frac{29}{10}, -\frac{493}{100}, \frac{609}{100} \right\rangle .$$

One can then find the magnitudes of each vector in terms of fractions, then compute the unit vectors likewise. After a lot of manual arithmetic (or after briefly using a computer algebra system), one finds that

$$\vec{u}_1 = \left\langle \sqrt{\frac{10}{83}}, \frac{17}{\sqrt{830}}, -\frac{21}{\sqrt{830}} \right\rangle \qquad \vec{u}_2 = \left\langle -\sqrt{\frac{10}{83}}, -\frac{17}{\sqrt{830}}, \frac{21}{\sqrt{830}} \right\rangle .$$

We can now say without equivocation that these lines are parallel.

Are they the same line? The parametric equations for a line describe one point that lies on the line, so we know that the point $P_1 = (-0.7, 4.2, 2.3)$ lies on ℓ_1. To determine if this point also lies on ℓ_2, plug in the x, y and z values of P_1 into the symmetric equations for ℓ_2:

$$\frac{(-0.7) - 2.8}{-2.9} \overset{?}{=} \frac{(4.2) - 10.15}{-4.93} \overset{?}{=} \frac{(2.3) - (-5.05)}{6.09} \quad \Rightarrow \quad 1.2069 = 1.2069 = 1.2069.$$

The point P_1 lies on both lines, so we conclude they are the same line, just parametrized differently. Figure 10.5.6 graphs this line along with the points and vectors described by the parametric equations. Note how $\vec{d}_1$ and $\vec{d}_2$ are parallel, though point in opposite directions (as indicated by their unit vectors above).

Notes:

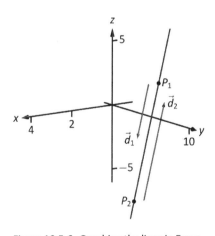

Figure 10.5.6: Graphing the lines in Example 10.5.4.

Distances

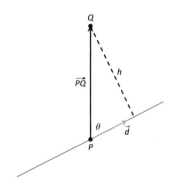

Figure 10.5.7: Establishing the distance from a point to a line.

Given a point Q and a line $\vec{\ell}(t) = \vec{p} + t\vec{d}$ in space, it is often useful to know the distance from the point to the line. (Here we use the standard definition of "distance," i.e., the length of the shortest line segment from the point to the line.) Identifying $\vec{p}$ with the point P, Figure 10.5.7 will help establish a general method of computing this distance h.

From trigonometry, we know $h = \| \overrightarrow{PQ} \| \sin \theta$. We have a similar identity involving the cross product: $\| \overrightarrow{PQ} \times \vec{d} \| = \| \overrightarrow{PQ} \| \| \vec{d} \| \sin \theta$. Divide both sides of this latter equation by $\| \vec{d} \|$ to obtain h:

$$ h = \frac{\| \overrightarrow{PQ} \times \vec{d} \|}{\| \vec{d} \|}. \tag{10.8} $$

It is also useful to determine the distance between lines, which we define as the length of the shortest line segment that connects the two lines (an argument from geometry shows that this line segments is perpendicular to both lines). Let lines $\vec{\ell}_1(t) = \vec{p}_1 + t\vec{d}_1$ and $\vec{\ell}_2(t) = \vec{p}_2 + t\vec{d}_2$ be given, as shown in Figure 10.5.8. To find the direction orthogonal to both $\vec{d}_1$ and $\vec{d}_2$, we take the cross product: $\vec{c} = \vec{d}_1 \times \vec{d}_2$. The magnitude of the orthogonal projection of $\overrightarrow{P_1P_2}$ onto $\vec{c}$ is the distance h we seek:

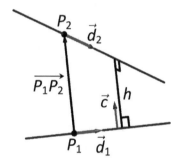

Figure 10.5.8: Establishing the distance between lines.

$$ \begin{aligned} h &= \| \operatorname{proj}_{\vec{c}} \overrightarrow{P_1P_2} \| \\ &= \left\| \frac{\overrightarrow{P_1P_2} \cdot \vec{c}}{\vec{c} \cdot \vec{c}} \vec{c} \right\| \\ &= \frac{|\overrightarrow{P_1P_2} \cdot \vec{c}|}{\| \vec{c} \|^2} \| \vec{c} \| \\ &= \frac{|\overrightarrow{P_1P_2} \cdot \vec{c}|}{\| \vec{c} \|}. \end{aligned} $$

A problem in the Exercise section is to show that this distance is 0 when the lines intersect. Note the use of the Triple Scalar Product: $\overrightarrow{P_1P_2} \cdot \vec{c} = \overrightarrow{P_1P_2} \cdot (\vec{d}_1 \times \vec{d}_2)$.

The following Key Idea restates these two distance formulas.

Notes:

> **Key Idea 10.5.1   Distances to Lines**
>
> 1. Let P be a point on a line ℓ that is parallel to $\vec{d}$. The distance h from a point Q to the line ℓ is:
>
> $$h = \frac{\|\overrightarrow{PQ} \times \vec{d}\|}{\|\vec{d}\|}.$$
>
> 2. Let P_1 be a point on line ℓ_1 that is parallel to $\vec{d_1}$, and let P_2 be a point on line ℓ_2 parallel to $\vec{d_2}$, and let $\vec{c} = \vec{d_1} \times \vec{d_2}$, where lines ℓ_1 and ℓ_2 are not parallel. The distance h between the two lines is:
>
> $$h = \frac{|\overrightarrow{P_1P_2} \cdot \vec{c}|}{\|\vec{c}\|}.$$

Example 10.5.5   Finding the distance from a point to a line

Find the distance from the point $Q = (1, 1, 3)$ to the line $\vec{\ell}(t) = \langle 1, -1, 1 \rangle + t\langle 2, 3, 1 \rangle$.

SOLUTION   The equation of the line gives us the point $P = (1, -1, 1)$ that lies on the line, hence $\overrightarrow{PQ} = \langle 0, 2, 2 \rangle$. The equation also gives $\vec{d} = \langle 2, 3, 1 \rangle$. Following Key Idea 10.5.1, we have the distance as

$$
\begin{aligned}
h &= \frac{\|\overrightarrow{PQ} \times \vec{d}\|}{\|\vec{d}\|} \\
&= \frac{\|\langle -4, 4, -4 \rangle\|}{\sqrt{14}} \\
&= \frac{4\sqrt{3}}{\sqrt{14}} \approx 1.852.
\end{aligned}
$$

The point Q is approximately 1.852 units from the line $\vec{\ell}(t)$.

Example 10.5.6   Finding the distance between lines

Find the distance between the lines

$$\ell_1 : \begin{array}{rcl} x &=& 1 + 3t \\ y &=& 2 - t \\ z &=& t \end{array} \qquad \ell_2 : \begin{array}{rcl} x &=& -2 + 4s \\ y &=& 3 + s \\ z &=& 5 + 2s. \end{array}$$

SOLUTION   These are the sames lines as given in Example 10.5.3, where

---

Notes:

we showed them to be skew. The equations allow us to identify the following points and vectors:

$$P_1 = (1, 2, 0) \quad P_2 = (-2, 3, 5) \quad \Rightarrow \quad \overrightarrow{P_1P_2} = \langle -3, 1, 5 \rangle .$$

$$\vec{d_1} = \langle 3, -1, 1 \rangle \quad \vec{d_2} = \langle 4, 1, 2 \rangle \quad \Rightarrow \quad \vec{c} = \vec{d_1} \times \vec{d_2} = \langle -3, -2, 7 \rangle .$$

From Key Idea 10.5.1 we have the distance h between the two lines is

$$h = \frac{|\overrightarrow{P_1P_2} \cdot \vec{c}|}{\|\vec{c}\|}$$
$$= \frac{42}{\sqrt{62}} \approx 5.334.$$

The lines are approximately 5.334 units apart.

One of the key points to understand from this section is this: to describe a line, we need a point and a direction. Whenever a problem is posed concerning a line, one needs to take whatever information is offered and glean point and direction information. Many questions can be asked (and *are* asked in the Exercise section) whose answer immediately follows from this understanding.

Lines are one of two fundamental objects of study in space. The other fundamental object is the *plane*, which we study in detail in the next section. Many complex three dimensional objects are studied by approximating their surfaces with lines and planes.

Notes:

Exercises 10.5

Terms and Concepts

1. To find an equation of a line, what two pieces of information are needed?

2. Two distinct lines in the plane can intersect or be _____.

3. Two distinct lines in space can intersect, be _____ or be _____.

4. Use your own words to describe what it means for two lines in space to be skew.

Problems

In Exercises 5 – 14, write the vector, parametric and symmetric equations of the lines described.

5. Passes through $P = (2, -4, 1)$, parallel to $\vec{d} = \langle 9, 2, 5 \rangle$.

6. Passes through $P = (6, 1, 7)$, parallel to $\vec{d} = \langle -3, 2, 5 \rangle$.

7. Passes through $P = (2, 1, 5)$ and $Q = (7, -2, 4)$.

8. Passes through $P = (1, -2, 3)$ and $Q = (5, 5, 5)$.

9. Passes through $P = (0, 1, 2)$ and orthogonal to both $\vec{d_1} = \langle 2, -1, 7 \rangle$ and $\vec{d_2} = \langle 7, 1, 3 \rangle$.

10. Passes through $P = (5, 1, 9)$ and orthogonal to both $\vec{d_1} = \langle 1, 0, 1 \rangle$ and $\vec{d_2} = \langle 2, 0, 3 \rangle$.

11. Passes through the point of intersection of $\vec{\ell_1}(t)$ and $\vec{\ell_2}(t)$ and orthogonal to both lines, where $\vec{\ell_1}(t) = \langle 2, 1, 1 \rangle + t \langle 5, 1, -2 \rangle$ and $\vec{\ell_2}(t) = \langle -2, -1, 2 \rangle + t \langle 3, 1, -1 \rangle$.

12. Passes through the point of intersection of $\ell_1(t)$ and $\ell_2(t)$ and orthogonal to both lines, where
$$\ell_1 = \begin{cases} x = t \\ y = -2 + 2t \\ z = 1 + t \end{cases} \quad \text{and} \quad \ell_2 = \begin{cases} x = 2 + t \\ y = 2 - t \\ z = 3 + 2t \end{cases}.$$

13. Passes through $P = (1, 1)$, parallel to $\vec{d} = \langle 2, 3 \rangle$.

14. Passes through $P = (-2, 5)$, parallel to $\vec{d} = \langle 0, 1 \rangle$.

In Exercises 15 – 22, determine if the described lines are the same line, parallel lines, intersecting or skew lines. If intersecting, give the point of intersection.

15. $\vec{\ell_1}(t) = \langle 1, 2, 1 \rangle + t \langle 2, -1, 1 \rangle$,
$\vec{\ell_2}(t) = \langle 3, 3, 3 \rangle + t \langle -4, 2, -2 \rangle$.

16. $\vec{\ell_1}(t) = \langle 2, 1, 1 \rangle + t \langle 5, 1, 3 \rangle$,
$\vec{\ell_2}(t) = \langle 14, 5, 9 \rangle + t \langle 1, 1, 1 \rangle$.

17. $\vec{\ell_1}(t) = \langle 3, 4, 1 \rangle + t \langle 2, -3, 4 \rangle$,
$\vec{\ell_2}(t) = \langle -3, 3, -3 \rangle + t \langle 3, -2, 4 \rangle$.

18. $\vec{\ell_1}(t) = \langle 1, 1, 1 \rangle + t \langle 3, 1, 3 \rangle$,
$\vec{\ell_2}(t) = \langle 7, 3, 7 \rangle + t \langle 6, 2, 6 \rangle$.

19. $\ell_1 = \begin{cases} x = 1 + 2t \\ y = 3 - 2t \\ z = t \end{cases}$ and $\ell_2 = \begin{cases} x = 3 - t \\ y = 3 + 5t \\ z = 2 + 7t \end{cases}$

20. $\ell_1 = \begin{cases} x = 1.1 + 0.6t \\ y = 3.77 + 0.9t \\ z = -2.3 + 1.5t \end{cases}$ and $\ell_2 = \begin{cases} x = 3.11 + 3.4t \\ y = 2 + 5.1t \\ z = 2.5 + 8.5t \end{cases}$

21. $\ell_1 = \begin{cases} x = 0.2 + 0.6t \\ y = 1.33 - 0.45t \\ z = -4.2 + 1.05t \end{cases}$ and $\ell_2 = \begin{cases} x = 0.86 + 9.2t \\ y = 0.835 - 6.9t \\ z = -3.045 + 16.1t \end{cases}$

22. $\ell_1 = \begin{cases} x = 0.1 + 1.1t \\ y = 2.9 - 1.5t \\ z = 3.2 + 1.6t \end{cases}$ and $\ell_2 = \begin{cases} x = 4 - 2.1t \\ y = 1.8 + 7.2t \\ z = 3.1 + 1.1t \end{cases}$

In Exercises 23 – 26, find the distance from the point to the line.

23. $Q = (1, 1, 1)$, $\quad \vec{\ell}(t) = \langle 2, 1, 3 \rangle + t \langle 2, 1, -2 \rangle$

24. $Q = (2, 5, 6)$, $\quad \vec{\ell}(t) = \langle -1, 1, 1 \rangle + t \langle 1, 0, 1 \rangle$

25. $Q = (0, 3)$, $\quad \vec{\ell}(t) = \langle 2, 0 \rangle + t \langle 1, 1 \rangle$

26. $Q = (1, 1)$, $\quad \vec{\ell}(t) = \langle 4, 5 \rangle + t \langle -4, 3 \rangle$

In Exercises 27 – 28, find the distance between the two lines.

27. $\vec{\ell_1}(t) = \langle 1, 2, 1 \rangle + t \langle 2, -1, 1 \rangle$,
$\vec{\ell_2}(t) = \langle 3, 3, 3 \rangle + t \langle 4, 2, -2 \rangle$.

28. $\vec{\ell_1}(t) = \langle 0, 0, 1 \rangle + t \langle 1, 0, 0 \rangle$,
$\vec{\ell_2}(t) = \langle 0, 0, 3 \rangle + t \langle 0, 1, 0 \rangle$.

Exercises 29 – 31 explore special cases of the distance formulas found in Key Idea 10.5.1.

29. Let Q be a point on the line $\vec{\ell}(t)$. Show why the distance formula correctly gives the distance from the point to the line as 0.

30. Let lines $\vec{\ell_1}(t)$ and $\vec{\ell_2}(t)$ be intersecting lines. Show why the distance formula correctly gives the distance between these lines as 0.

31. Let lines $\vec{\ell}_1(t)$ and $\vec{\ell}_2(t)$ be parallel.

    (a) Show why the distance formula for distance between lines cannot be used as stated to find the distance between the lines.

(b) Show why letting $\vec{c} = (\overrightarrow{P_1P_2} \times \vec{d}_2) \times \vec{d}_2$ allows one to use the formula.

(c) Show how one can use the formula for the distance between a point and a line to find the distance between parallel lines.

10.6 Planes

Any flat surface, such as a wall, table top or stiff piece of cardboard can be thought of as representing part of a plane. Consider a piece of cardboard with a point P marked on it. One can take a nail and stick it into the cardboard at P such that the nail is perpendicular to the cardboard; see Figure 10.6.1.

This nail provides a "handle" for the cardboard. Moving the cardboard around moves P to different locations in space. Tilting the nail (but keeping P fixed) tilts the cardboard. Both moving and tilting the cardboard defines a different plane in space. In fact, we can define a plane by: 1) the location of P in space, and 2) the direction of the nail.

The previous section showed that one can define a line given a point on the line and the direction of the line (usually given by a vector). One can make a similar statement about planes: we can define a plane in space given a point on the plane and the direction the plane "faces" (using the description above, the direction of the nail). Once again, the direction information will be supplied by a vector, called a **normal vector**, that is orthogonal to the plane.

What exactly does "orthogonal to the plane" mean? Choose any two points P and Q in the plane, and consider the vector $\vec{PQ}$. We say a vector $\vec{n}$ is orthogonal to the plane if $\vec{n}$ is perpendicular to $\vec{PQ}$ for all choices of P and Q; that is, if $\vec{n} \cdot \vec{PQ} = 0$ for all P and Q.

This gives us way of writing an equation describing the plane. Let $P = (x_0, y_0, z_0)$ be a point in the plane and let $\vec{n} = \langle a, b, c \rangle$ be a normal vector to the plane. A point $Q = (x, y, z)$ lies in the plane defined by P and $\vec{n}$ if, and only if, $\vec{PQ}$ is orthogonal to $\vec{n}$. Knowing $\vec{PQ} = \langle x - x_0, y - y_0, z - z_0 \rangle$, consider:

$$\vec{PQ} \cdot \vec{n} = 0$$
$$\langle x - x_0, y - y_0, z - z_0 \rangle \cdot \langle a, b, c \rangle = 0$$
$$a(x - x_0) + b(y - y_0) + c(z - z_0) = 0 \tag{10.9}$$

Equation (10.9) defines an *implicit* function describing the plane. More algebra produces:

$$ax + by + cz = ax_0 + by_0 + cz_0.$$

The right hand side is just a number, so we replace it with d:

$$ax + by + cz = d. \tag{10.10}$$

As long as $c \neq 0$, we can solve for z:

$$z = \frac{1}{c}(d - ax - by). \tag{10.11}$$

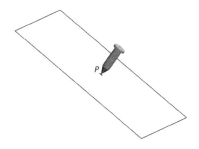

Figure 10.6.1: Illustrating defining a plane with a sheet of cardboard and a nail.

Notes:

Equation (10.11) is especially useful as many computer programs can graph functions in this form. Equations (10.9) and (10.10) have specific names, given next.

Definition 10.6.1 **Equations of a Plane in Standard and General Forms**

The plane passing through the point $P = (x_0, y_0, z_0)$ with normal vector $\vec{n} = \langle a, b, c \rangle$ can be described by an equation with **standard form**

$$a(x - x_0) + b(y - y_0) + c(z - z_0) = 0;$$

the equation's **general form** is

$$ax + by + cz = d.$$

A key to remember throughout this section is this: to find the equation of a plane, we need a point and a normal vector. We will give several examples of finding the equation of a plane, and in each one different types of information are given. In each case, we need to use the given information to find a point on the plane and a normal vector.

Example 10.6.1 **Finding the equation of a plane.**
Write the equation of the plane that passes through the points $P = (1, 1, 0)$, $Q = (1, 2, -1)$ and $R = (0, 1, 2)$ in standard form.

SOLUTION We need a vector $\vec{n}$ that is orthogonal to the plane. Since P, Q and R are in the plane, so are the vectors $\vec{PQ}$ and $\vec{PR}$; $\vec{PQ} \times \vec{PR}$ is orthogonal to $\vec{PQ}$ and $\vec{PR}$ and hence the plane itself.

It is straightforward to compute $\vec{n} = \vec{PQ} \times \vec{PR} = \langle 2, 1, 1 \rangle$. We can use any point we wish in the plane (any of P, Q or R will do) and we arbitrarily choose P. Following Definition 10.6.1, the equation of the plane in standard form is

$$2(x - 1) + (y - 1) + z = 0.$$

The plane is sketched in Figure 10.6.2.

We have just demonstrated the fact that any three non-collinear points define a plane. (This is why a three-legged stool does not "rock;" it's three feet always lie in a plane. A four-legged stool will rock unless all four feet lie in the same plane.)

Example 10.6.2 **Finding the equation of a plane.**
Verify that lines ℓ_1 and ℓ_2, whose parametric equations are given below, inter-

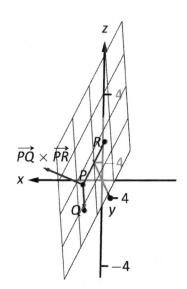

Figure 10.6.2: Sketching the plane in Example 10.6.1.

Notes:

sect, then give the equation of the plane that contains these two lines in general form.

$$\ell_1: \begin{array}{rcl} x &=& -5 + 2s \\ y &=& 1 + s \\ z &=& -4 + 2s \end{array} \qquad \ell_2: \begin{array}{rcl} x &=& 2 + 3t \\ y &=& 1 - 2t \\ z &=& 1 + t \end{array}$$

SOLUTION The lines clearly are not parallel. If they do not intersect, they are skew, meaning there is not a plane that contains them both. If they do intersect, there is such a plane.

To find their point of intersection, we set the x, y and z equations equal to each other and solve for s and t:

$$\begin{array}{rcl} -5 + 2s &=& 2 + 3t \\ 1 + s &=& 1 - 2t \\ -4 + 2s &=& 1 + t \end{array} \quad \Rightarrow \quad s = 2, \quad t = -1.$$

When $s = 2$ and $t = -1$, the lines intersect at the point $P = (-1, 3, 0)$.

Let $\vec{d_1} = \langle 2, 1, 2 \rangle$ and $\vec{d_2} = \langle 3, -2, 1 \rangle$ be the directions of lines ℓ_1 and ℓ_2, respectively. A normal vector to the plane containing these the two lines will also be orthogonal to $\vec{d_1}$ and $\vec{d_2}$. Thus we find a normal vector $\vec{n}$ by computing $\vec{n} = \vec{d_1} \times \vec{d_2} = \langle 5, 4 - 7 \rangle$.

We can pick any point in the plane with which to write our equation; each line gives us infinite choices of points. We choose P, the point of intersection. We follow Definition 10.6.1 to write the plane's equation in general form:

$$5(x + 1) + 4(y - 3) - 7z = 0$$
$$5x + 5 + 4y - 12 - 7z = 0$$
$$5x + 4y - 7z = 7.$$

The plane's equation in general form is $5x + 4y - 7z = 7$; it is sketched in Figure 10.6.3.

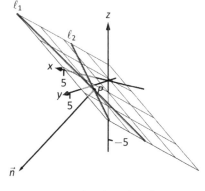

Figure 10.6.3: Sketching the plane in Example 10.6.2.

Example 10.6.3 Finding the equation of a plane

Give the equation, in standard form, of the plane that passes through the point $P = (-1, 0, 1)$ and is orthogonal to the line with vector equation $\vec{\ell}(t) = \langle -1, 0, 1 \rangle + t \langle 1, 2, 2 \rangle$.

SOLUTION As the plane is to be orthogonal to the line, the plane must be orthogonal to the direction of the line given by $\vec{d} = \langle 1, 2, 2 \rangle$. We use this as our normal vector. Thus the plane's equation, in standard form, is

$$(x + 1) + 2y + 2(z - 1) = 0.$$

The line and plane are sketched in Figure 10.6.4.

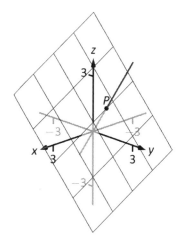

Figure 10.6.4: The line and plane in Example 10.6.3.

Notes:

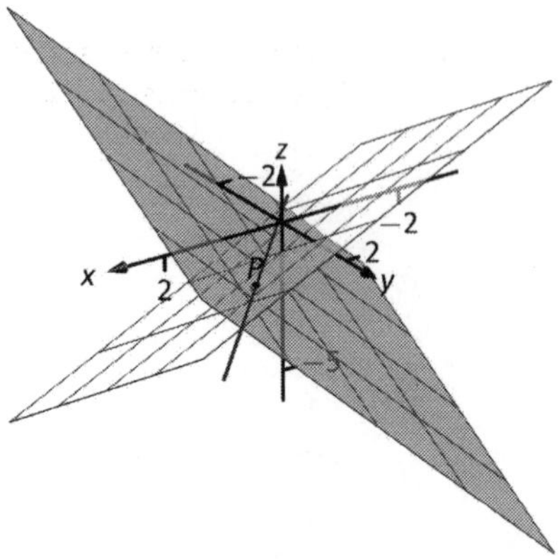

Figure 10.6.5: Graphing the planes and their line of intersection in Example 10.6.4.

Example 10.6.4   Finding the intersection of two planes

Give the parametric equations of the line that is the intersection of the planes p_1 and p_2, where:

$$p_1 : x - (y - 2) + (z - 1) = 0$$
$$p_2 : -2(x - 2) + (y + 1) + (z - 3) = 0$$

SOLUTION   To find an equation of a line, we need a point on the line and the direction of the line.

We can find a point on the line by solving each equation of the planes for z:

$$p_1 : z = -x + y - 1$$
$$p_2 : z = 2x - y - 2$$

We can now set these two equations equal to each other (i.e., we are finding values of x and y where the planes have the same z value):

$$-x + y - 1 = 2x - y - 2$$
$$2y = 3x - 1$$
$$y = \frac{1}{2}(3x - 1)$$

We can choose any value for x; we choose $x = 1$. This determines that $y = 1$. We can now use the equations of either plane to find z: when $x = 1$ and $y = 1$, $z = -1$ on both planes. We have found a point P on the line: $P = (1, 1, -1)$.

We now need the direction of the line. Since the line lies in each plane, its direction is orthogonal to a normal vector for each plane. Considering the equations for p_1 and p_2, we can quickly determine their normal vectors. For p_1, $\vec{n}_1 = \langle 1, -1, 1 \rangle$ and for p_2, $\vec{n}_2 = \langle -2, 1, 1 \rangle$. A direction orthogonal to both of these directions is their cross product: $\vec{d} = \vec{n}_1 \times \vec{n}_2 = \langle -2, -3, -1 \rangle$.

The parametric equations of the line through $P = (1, 1, -1)$ in the direction of $d = \langle -2, -3, -1 \rangle$ is:

$$\ell : \quad x = -2t + 1 \quad y = -3t + 1 \quad z = -t - 1.$$

The planes and line are graphed in Figure 10.6.5.

Example 10.6.5   Finding the intersection of a plane and a line

Find the point of intersection, if any, of the line $\ell(t) = \langle 3, -3, -1 \rangle + t \langle -1, 2, 1 \rangle$ and the plane with equation in general form $2x + y + z = 4$.

SOLUTION   The equation of the plane shows that the vector $\vec{n} = \langle 2, 1, 1 \rangle$ is a normal vector to the plane, and the equation of the line shows that the line

Notes:

moves parallel to $\vec{d} = \langle -1, 2, 1 \rangle$. Since these are not orthogonal, we know there is a point of intersection. (If there were orthogonal, it would mean that the plane and line were parallel to each other, either never intersecting or the line was in the plane itself.)

To find the point of intersection, we need to find a t value such that $\ell(t)$ satisfies the equation of the plane. Rewriting the equation of the line with parametric equations will help:

$$\ell(t) = \begin{cases} x = 3 - t \\ y = -3 + 2t \\ z = -1 + t \end{cases}.$$

Replacing x, y and z in the equation of the plane with the expressions containing t found in the equation of the line allows us to determine a t value that indicates the point of intersection:

$$2x + y + z = 4$$
$$2(3 - t) + (-3 + 2t) + (-1 + t) = 4$$
$$t = 2.$$

When $t = 2$, the point on the line satisfies the equation of the plane; that point is $\ell(2) = \langle 1, 1, 1 \rangle$. Thus the point $(1, 1, 1)$ is the point of intersection between the plane and the line, illustrated in Figure 10.6.6.

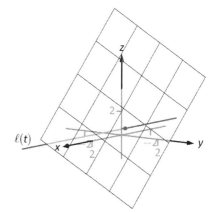

Figure 10.6.6: Illustrating the intersection of a line and a plane in Example 10.6.5.

Distances

Just as it was useful to find distances between points and lines in the previous section, it is also often necessary to find the distance from a point to a plane.

Consider Figure 10.6.7, where a plane with normal vector $\vec{n}$ is sketched containing a point P and a point Q, not on the plane, is given. We measure the distance from Q to the plane by measuring the length of the projection of $\overrightarrow{PQ}$ onto $\vec{n}$. That is, we want:

$$\left\| \text{proj}_{\vec{n}} \overrightarrow{PQ} \right\| = \left\| \frac{\vec{n} \cdot \overrightarrow{PQ}}{\| \vec{n} \|^2} \vec{n} \right\| = \frac{|\vec{n} \cdot \overrightarrow{PQ}|}{\| \vec{n} \|} \tag{10.12}$$

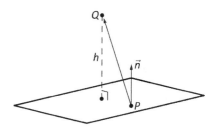

Figure 10.6.7: Illustrating finding the distance from a point to a plane.

Equation (10.12) is important as it does more than just give the distance between a point and a plane. We will see how it allows us to find several other distances as well: the distance between parallel planes and the distance from a line and a plane. Because Equation (10.12) is important, we restate it as a Key Idea.

Notes:

Key Idea 10.6.1   Distance from a Point to a Plane

Let a plane with normal vector $\vec{n}$ be given, and let Q be a point. The distance h from Q to the plane is

$$h = \frac{|\vec{n} \cdot \vec{PQ}|}{\|\vec{n}\|},$$

where P is any point in the plane.

Example 10.6.6   Distance between a point and a plane

Find the distance between the point $Q = (2, 1, 4)$ and the plane with equation $2x - 5y + 6z = 9$.

SOLUTION   Using the equation of the plane, we find the normal vector $\vec{n} = \langle 2, -5, 6 \rangle$. To find a point on the plane, we can let x and y be anything we choose, then let z be whatever satisfies the equation. Letting x and y be 0 seems simple; this makes $z = 1.5$. Thus we let $P = \langle 0, 0, 1.5 \rangle$, and $\vec{PQ} = \langle 2, 1, 2.5 \rangle$.

The distance h from Q to the plane is given by Key Idea 10.6.1:

$$
\begin{aligned}
h &= \frac{|\vec{n} \cdot \vec{PQ}|}{\|\vec{n}\|} \\
&= \frac{|\langle 2, -5, 6 \rangle \cdot \langle 2, 1, 2.5 \rangle|}{\|\langle 2, -5, 6 \rangle\|} \\
&= \frac{|14|}{\sqrt{65}} \\
&\approx 1.74.
\end{aligned}
$$

We can use Key Idea 10.6.1 to find other distances. Given two parallel planes, we can find the distance between these planes by letting P be a point on one plane and Q a point on the other. If ℓ is a line parallel to a plane, we can use the Key Idea to find the distance between them as well: again, let P be a point in the plane and let Q be any point on the line. (One can also use Key Idea 10.5.1.) The Exercise section contains problems of these types.

These past two sections have not explored lines and planes in space as an exercise of mathematical curiosity. However, there are many, many applications of these fundamental concepts. Complex shapes can be modeled (or, *approximated*) using planes. For instance, part of the exterior of an aircraft may have a complex, yet smooth, shape, and engineers will want to know how air flows across this piece as well as how heat might build up due to air friction. Many equations that help determine air flow and heat dissipation are difficult to apply to arbitrary surfaces, but simple to apply to planes. By approximating a surface with millions of small planes one can more readily model the needed behavior.

Notes:

Exercises 10.6

Terms and Concepts

1. In order to find the equation of a plane, what two pieces of information must one have?

2. What is the relationship between a plane and one of its normal vectors?

Problems

In Exercises 3 – 6, give any two points in the given plane.

3. $2x - 4y + 7z = 2$

4. $3(x + 2) + 5(y - 9) - 4z = 0$

5. $x = 2$

6. $4(y + 2) - (z - 6) = 0$

In Exercises 7 – 20, give the equation of the described plane in standard and general forms.

7. Passes through $(2, 3, 4)$ and has normal vector $\vec{n} = \langle 3, -1, 7 \rangle$.

8. Passes through $(1, 3, 5)$ and has normal vector $\vec{n} = \langle 0, 2, 4 \rangle$.

9. Passes through the points $(1, 2, 3)$, $(3, -1, 4)$ and $(1, 0, 1)$.

10. Passes through the points $(5, 3, 8)$, $(6, 4, 9)$ and $(3, 3, 3)$.

11. Contains the intersecting lines
$\vec{\ell_1}(t) = \langle 2, 1, 2 \rangle + t \langle 1, 2, 3 \rangle$ and
$\vec{\ell_2}(t) = \langle 2, 1, 2 \rangle + t \langle 2, 5, 4 \rangle$.

12. Contains the intersecting lines
$\vec{\ell_1}(t) = \langle 5, 0, 3 \rangle + t \langle -1, 1, 1 \rangle$ and
$\vec{\ell_2}(t) = \langle 1, 4, 7 \rangle + t \langle 3, 0, -3 \rangle$.

13. Contains the parallel lines
$\vec{\ell_1}(t) = \langle 1, 1, 1 \rangle + t \langle 1, 2, 3 \rangle$ and
$\vec{\ell_2}(t) = \langle 1, 1, 2 \rangle + t \langle 1, 2, 3 \rangle$.

14. Contains the parallel lines
$\vec{\ell_1}(t) = \langle 1, 1, 1 \rangle + t \langle 4, 1, 3 \rangle$ and
$\vec{\ell_2}(t) = \langle 4, 4, 4 \rangle + t \langle 4, 1, 3 \rangle$.

15. Contains the point $(2, -6, 1)$ and the line
$$\ell(t) = \begin{cases} x = 2 + 5t \\ y = 2 + 2t \\ z = -1 + 2t \end{cases}$$

16. Contains the point $(5, 7, 3)$ and the line
$$\ell(t) = \begin{cases} x = t \\ y = t \\ z = t \end{cases}$$

17. Contains the point $(5, 7, 3)$ and is orthogonal to the line
$\vec{\ell}(t) = \langle 4, 5, 6 \rangle + t \langle 1, 1, 1 \rangle$.

18. Contains the point $(4, 1, 1)$ and is orthogonal to the line
$$\ell(t) = \begin{cases} x = 4 + 4t \\ y = 1 + 1t \\ z = 1 + 1t \end{cases}$$

19. Contains the point $(-4, 7, 2)$ and is parallel to the plane
$3(x - 2) + 8(y + 1) - 10z = 0$.

20. Contains the point $(1, 2, 3)$ and is parallel to the plane
$x = 5$.

In Exercises 21 – 22, give the equation of the line that is the intersection of the given planes.

21. $p1 : 3(x - 2) + (y - 1) + 4z = 0$, and
$p2 : 2(x - 1) - 2(y + 3) + 6(z - 1) = 0$.

22. $p1 : 5(x - 5) + 2(y + 2) + 4(z - 1) = 0$, and
$p2 : 3x - 4(y - 1) + 2(z - 1) = 0$.

In Exercises 23 – 26, find the point of intersection between the line and the plane.

23. line: $\langle 5, 1, -1 \rangle + t \langle 2, 2, 1 \rangle$,
plane: $5x - y - z = -3$

24. line: $\langle 4, 1, 0 \rangle + t \langle 1, 0, -1 \rangle$,
plane: $3x + y - 2z = 8$

25. line: $\langle 1, 2, 3 \rangle + t \langle 3, 5, -1 \rangle$,
plane: $3x - 2y - z = 4$

26. line: $\langle 1, 2, 3 \rangle + t \langle 3, 5, -1 \rangle$,
plane: $3x - 2y - z = -4$

In Exercises 27 – 30, find the given distances.

27. The distance from the point $(1, 2, 3)$ to the plane
$3(x - 1) + (y - 2) + 5(z - 2) = 0$.

28. The distance from the point $(2, 6, 2)$ to the plane
$2(x - 1) - y + 4(z + 1) = 0$.

29. The distance between the parallel planes
$x + y + z = 0$ and
$(x - 2) + (y - 3) + (z + 4) = 0$

30. The distance between the parallel planes
    $2(x - 1) + 2(y + 1) + (z - 2) = 0$ and
    $2(x - 3) + 2(y - 1) + (z - 3) = 0$

31. Show why if the point Q lies in a plane, then the distance formula correctly gives the distance from the point to the plane as 0.

32. How is Exercise 30 in Section 10.5 easier to answer once we have an understanding of planes?

11: Vector Valued Functions

In the previous chapter, we learned about vectors and were introduced to the power of vectors within mathematics. In this chapter, we'll build on this foundation to define functions whose input is a real number and whose output is a vector. We'll see how to graph these functions and apply calculus techniques to analyze their behavior. Most importantly, we'll see *why* we are interested in doing this: we'll see beautiful applications to the study of moving objects.

11.1   Vector–Valued Functions

We are very familiar with **real valued functions**, that is, functions whose output is a real number. This section introduces **vector–valued functions** – functions whose output is a vector.

Definition 11.1.1   Vector–Valued Functions

A **vector–valued function** is a function of the form

$$\vec{r}(t) = \langle f(t), g(t) \rangle \quad \text{or} \quad \vec{r}(t) = \langle f(t), g(t), h(t) \rangle,$$

where f, g and h are real valued functions.

The **domain** of $\vec{r}$ is the set of all values of t for which $\vec{r}(t)$ is defined. The **range** of $\vec{r}$ is the set of all possible output vectors $\vec{r}(t)$.

Evaluating and Graphing Vector–Valued Functions

Evaluating a vector–valued function at a specific value of t is straightforward; simply evaluate each component function at that value of t. For instance, if $\vec{r}(t) = \langle t^2, t^2 + t - 1 \rangle$, then $\vec{r}(-2) = \langle 4, 1 \rangle$. We can sketch this vector, as is done in Figure 11.1.1(a). Plotting lots of vectors is cumbersome, though, so generally we do not sketch the whole vector but just the terminal point. The **graph** of a vector–valued function is the set of all terminal points of $\vec{r}(t)$, where the initial point of each vector is always the origin. In Figure 11.1.1(b) we sketch the graph of $\vec{r}$; we can indicate individual points on the graph with their respective vector, as shown.

Vector–valued functions are closely related to parametric equations of graphs. While in both methods we plot points $\big(x(t), y(t)\big)$ or $\big(x(t), y(t), z(t)\big)$ to produce a graph, in the context of vector–valued functions each such point represents a vector. The implications of this will be more fully realized in the next section as we apply calculus ideas to these functions.

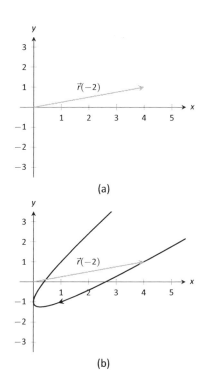

(a)

(b)

Figure 11.1.1: Sketching the graph of a vector–valued function.

t	$t^3 - t$	$\dfrac{1}{t^2+1}$
-2	-6	$1/5$
-1	0	$1/2$
0	0	1
1	0	$1/2$
2	6	$1/5$

(a)

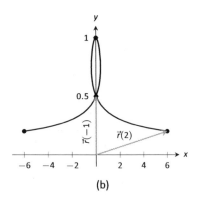

(b)

Figure 11.1.2:   Sketching the vector–valued function of Example 11.1.1.

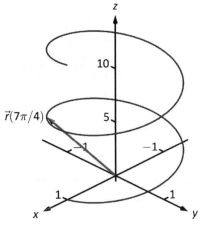

Figure 11.1.3: The graph of $\vec{r}(t)$ in Example 11.1.2.

Example 11.1.1   Graphing vector–valued functions

Graph $\vec{r}(t) = \left\langle t^3 - t, \dfrac{1}{t^2+1} \right\rangle$, for $-2 \leq t \leq 2$. Sketch $\vec{r}(-1)$ and $\vec{r}(2)$.

SOLUTION   We start by making a table of t, x and y values as shown in Figure 11.1.2(a). Plotting these points gives an indication of what the graph looks like. In Figure 11.1.2(b), we indicate these points and sketch the full graph. We also highlight $\vec{r}(-1)$ and $\vec{r}(2)$ on the graph.

Example 11.1.2   Graphing vector–valued functions.

Graph $\vec{r}(t) = \langle \cos t, \sin t, t \rangle$ for $0 \leq t \leq 4\pi$.

SOLUTION   We can again plot points, but careful consideration of this function is very revealing. Momentarily ignoring the third component, we see the x and y components trace out a circle of radius 1 centered at the origin. Noticing that the z component is t, we see that as the graph winds around the z-axis, it is also increasing at a constant rate in the positive z direction, forming a spiral. This is graphed in Figure 11.1.3. In the graph $\vec{r}(7\pi/4) \approx (0.707, -0.707, 5.498)$ is highlighted to help us understand the graph.

Algebra of Vector–Valued Functions

Definition 11.1.2   Operations on Vector–Valued Functions

Let $\vec{r}_1(t) = \langle f_1(t), g_1(t) \rangle$ and $\vec{r}_2(t) = \langle f_2(t), g_2(t) \rangle$ be vector–valued functions in $\mathbb{R}^2$ and let c be a scalar. Then:

1. $\vec{r}_1(t) \pm \vec{r}_2(t) = \langle f_1(t) \pm f_2(t), g_1(t) \pm g_2(t) \rangle$.

2. $c\vec{r}_1(t) = \langle cf_1(t), cg_1(t) \rangle$.

A similar definition holds for vector–valued functions in $\mathbb{R}^3$.

This definition states that we add, subtract and scale vector-valued functions component–wise. Combining vector–valued functions in this way can be very useful (as well as create interesting graphs).

Example 11.1.3   Adding and scaling vector–valued functions.

Let $\vec{r}_1(t) = \langle 0.2t, 0.3t \rangle$, $\vec{r}_2(t) = \langle \cos t, \sin t \rangle$ and $\vec{r}(t) = \vec{r}_1(t) + \vec{r}_2(t)$. Graph $\vec{r}_1(t)$, $\vec{r}_2(t)$, $\vec{r}(t)$ and $5\vec{r}(t)$ on $-10 \leq t \leq 10$.

Notes:

SOLUTION We can graph $\vec{r}_1$ and $\vec{r}_2$ easily by plotting points (or just using technology). Let's think about each for a moment to better understand how vector–valued functions work.

We can rewrite $\vec{r}_1(t) = \langle 0.2t, 0.3t \rangle$ as $\vec{r}_1(t) = t\langle 0.2, 0.3 \rangle$. That is, the function $\vec{r}_1$ scales the vector $\langle 0.2, 0.3 \rangle$ by t. This scaling of a vector produces a line in the direction of $\langle 0.2, 0.3 \rangle$.

We are familiar with $\vec{r}_2(t) = \langle \cos t, \sin t \rangle$; it traces out a circle, centered at the origin, of radius 1. Figure 11.1.4(a) graphs $\vec{r}_1(t)$ and $\vec{r}_2(t)$.

Adding $\vec{r}_1(t)$ to $\vec{r}_2(t)$ produces $\vec{r}(t) = \langle \cos t + 0.2t, \sin t + 0.3t \rangle$, graphed in Figure 11.1.4(b). The linear movement of the line combines with the circle to create loops that move in the direction of $\langle 0.2, 0.3 \rangle$. (We encourage the reader to experiment by changing $\vec{r}_1(t)$ to $\langle 2t, 3t \rangle$, etc., and observe the effects on the loops.)

Multiplying $\vec{r}(t)$ by 5 scales the function by 5, producing $5\vec{r}(t) = \langle 5\cos t + 1, 5\sin t + 1.5 \rangle$, which is graphed in Figure 11.1.4(c) along with $\vec{r}(t)$. The new function is "5 times bigger" than $\vec{r}(t)$. Note how the graph of $5\vec{r}(t)$ in (c) looks identical to the graph of $\vec{r}(t)$ in (b). This is due to the fact that the x and y bounds of the plot in (c) are exactly 5 times larger than the bounds in (b).

Example 11.1.4 Adding and scaling vector–valued functions.

A **cycloid** is a graph traced by a point p on a rolling circle, as shown in Figure 11.1.5. Find an equation describing the cycloid, where the circle has radius 1.

Figure 11.1.5: Tracing a cycloid.

SOLUTION This problem is not very difficult if we approach it in a clever way. We start by letting $\vec{p}(t)$ describe the position of the point p on the circle, where the circle is centered at the origin and only rotates clockwise (i.e., it does not roll). This is relatively simple given our previous experiences with parametric equations; $\vec{p}(t) = \langle \cos t, -\sin t \rangle$.

We now want the circle to roll. We represent this by letting $\vec{c}(t)$ represent the location of the center of the circle. It should be clear that the y component of $\vec{c}(t)$ should be 1; the center of the circle is always going to be 1 if it rolls on a horizontal surface.

The x component of $\vec{c}(t)$ is a linear function of t: $f(t) = mt$ for some scalar m. When $t = 0$, $f(t) = 0$ (the circle starts centered on the y-axis). When $t = 2\pi$, the circle has made one complete revolution, traveling a distance equal to its

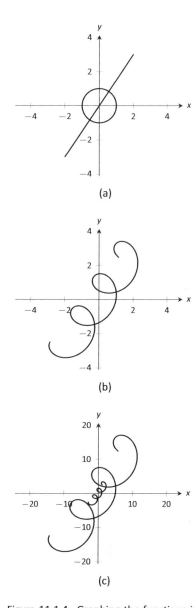

(a)

(b)

(c)

Figure 11.1.4: Graphing the functions in Example 11.1.3.

Notes:

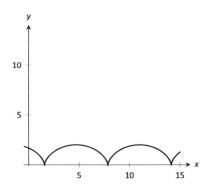

Figure 11.1.6: The cycloid in Example 11.1.4.

circumference, which is also 2π. This gives us a point on our line $f(t) = mt$, the point $(2\pi, 2\pi)$. It should be clear that $m = 1$ and $f(t) = t$. So $\vec{c}(t) = \langle t, 1 \rangle$.

We now combine $\vec{p}$ and $\vec{c}$ together to form the equation of the cycloid: $\vec{r}(t) = \vec{p}(t) + \vec{c}(t) = \langle \cos t + t, -\sin t + 1 \rangle$, which is graphed in Figure 11.1.6.

Displacement

A vector–valued function $\vec{r}(t)$ is often used to describe the position of a moving object at time t. At $t = t_0$, the object is at $\vec{r}(t_0)$; at $t = t_1$, the object is at $\vec{r}(t_1)$. Knowing the locations $\vec{r}(t_0)$ and $\vec{r}(t_1)$ give no indication of the path taken between them, but often we only care about the difference of the locations, $\vec{r}(t_1) - \vec{r}(t_0)$, the **displacement**.

Definition 11.1.3   Displacement

Let $\vec{r}(t)$ be a vector–valued function and let $t_0 < t_1$ be values in the domain. The **displacement** $\vec{d}$ of $\vec{r}$, from $t = t_0$ to $t = t_1$, is

$$\vec{d} = \vec{r}(t_1) - \vec{r}(t_0).$$

When the displacement vector is drawn with initial point at $\vec{r}(t_0)$, its terminal point is $\vec{r}(t_1)$. We think of it as the vector which points from a starting position to an ending position.

Example 11.1.5   Finding and graphing displacement vectors
Let $\vec{r}(t) = \left\langle \cos(\frac{\pi}{2}t), \sin(\frac{\pi}{2}t) \right\rangle$. Graph $\vec{r}(t)$ on $-1 \le t \le 1$, and find the displacement of $\vec{r}(t)$ on this interval.

SOLUTION   The function $\vec{r}(t)$ traces out the unit circle, though at a different rate than the "usual" $\langle \cos t, \sin t \rangle$ parametrization. At $t_0 = -1$, we have $\vec{r}(t_0) = \langle 0, -1 \rangle$; at $t_1 = 1$, we have $\vec{r}(t_1) = \langle 0, 1 \rangle$. The displacement of $\vec{r}(t)$ on $[-1, 1]$ is thus $\vec{d} = \langle 0, 1 \rangle - \langle 0, -1 \rangle = \langle 0, 2 \rangle$.

A graph of $\vec{r}(t)$ on $[-1, 1]$ is given in Figure 11.1.7, along with the displacement vector $\vec{d}$ on this interval.

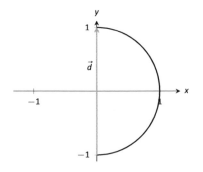

Figure 11.1.7: Graphing the displacement of a position function in Example 11.1.5.

Measuring displacement makes us contemplate related, yet very different, concepts. Considering the semi–circular path the object in Example 11.1.5 took, we can quickly verify that the object ended up a distance of 2 units from its initial location. That is, we can compute $|| \vec{d} || = 2$. However, measuring *distance from the starting point* is different from measuring *distance traveled*. Being a semi–

Notes:

circle, we can measure the distance traveled by this object as $\pi \approx 3.14$ units. Knowing *distance from the starting point* allows us to compute **average rate of change.**

Definition 11.1.4   Average Rate of Change

Let $\vec{r}(t)$ be a vector–valued function, where each of its component functions is continuous on its domain, and let $t_0 < t_1$. The **average rate of change** of $\vec{r}(t)$ on $[t_0, t_1]$ is

$$\text{average rate of change} = \frac{\vec{r}(t_1) - \vec{r}(t_0)}{t_1 - t_0}.$$

Example 11.1.6   Average rate of change

Let $\vec{r}(t) = \left\langle \cos(\frac{\pi}{2}t), \sin(\frac{\pi}{2}t) \right\rangle$ as in Example 11.1.5. Find the average rate of change of $\vec{r}(t)$ on $[-1, 1]$ and on $[-1, 5]$.

SOLUTION   We computed in Example 11.1.5 that the displacement of $\vec{r}(t)$ on $[-1, 1]$ was $\vec{d} = \langle 0, 2 \rangle$. Thus the average rate of change of $\vec{r}(t)$ on $[-1, 1]$ is:

$$\frac{\vec{r}(1) - \vec{r}(-1)}{1 - (-1)} = \frac{\langle 0, 2 \rangle}{2} = \langle 0, 1 \rangle.$$

We interpret this as follows: the object followed a semi–circular path, meaning it moved towards the right then moved back to the left, while climbing slowly, then quickly, then slowly again. *On average*, however, it progressed straight up at a constant rate of $\langle 0, 1 \rangle$ per unit of time.

We can quickly see that the displacement on $[-1, 5]$ is the same as on $[-1, 1]$, so $\vec{d} = \langle 0, 2 \rangle$. The average rate of change is different, though:

$$\frac{\vec{r}(5) - \vec{r}(-1)}{5 - (-1)} = \frac{\langle 0, 2 \rangle}{6} = \langle 0, 1/3 \rangle.$$

As it took "3 times as long" to arrive at the same place, this average rate of change on $[-1, 5]$ is $1/3$ the average rate of change on $[-1, 1]$.

We considered average rates of change in Sections 1.1 and 2.1 as we studied limits and derivatives. The same is true here; in the following section we apply calculus concepts to vector–valued functions as we find limits, derivatives, and integrals. Understanding the average rate of change will give us an understanding of the derivative; displacement gives us one application of integration.

Notes:

Exercises 11.1

Terms and Concepts

1. Vector–valued functions are closely related to _____ _____ of graphs.

2. When sketching vector–valued functions, technically one isn't graphing points, but rather _____.

3. It can be useful to think of _____ as a vector that points from a starting position to an ending position.

4. In the context of vector–valued functions, average rate of change is _____ divided by time.

Problems

In Exercises 5 – 12, sketch the vector–valued function on the given interval.

5. $\vec{r}(t) = \langle t^2, t^2 - 1 \rangle$, for $-2 \le t \le 2$.

6. $\vec{r}(t) = \langle t^2, t^3 \rangle$, for $-2 \le t \le 2$.

7. $\vec{r}(t) = \langle 1/t, 1/t^2 \rangle$, for $-2 \le t \le 2$.

8. $\vec{r}(t) = \langle \frac{1}{10}t^2, \sin t \rangle$, for $-2\pi \le t \le 2\pi$.

9. $\vec{r}(t) = \langle \frac{1}{10}t^2, \sin t \rangle$, for $-2\pi \le t \le 2\pi$.

10. $\vec{r}(t) = \langle 3\sin(\pi t), 2\cos(\pi t) \rangle$, on $[0, 2]$.

11. $\vec{r}(t) = \langle 3\cos t, 2\sin(2t) \rangle$, on $[0, 2\pi]$.

12. $\vec{r}(t) = \langle 2\sec t, \tan t \rangle$, on $[-\pi, \pi]$.

In Exercises 13 – 16, sketch the vector–valued function on the given interval in $\mathbb{R}^3$. Technology may be useful in creating the sketch.

13. $\vec{r}(t) = \langle 2\cos t, t, 2\sin t \rangle$, on $[0, 2\pi]$.

14. $\vec{r}(t) = \langle 3\cos t, \sin t, t/\pi \rangle$ on $[0, 2\pi]$.

15. $\vec{r}(t) = \langle \cos t, \sin t, \sin t \rangle$ on $[0, 2\pi]$.

16. $\vec{r}(t) = \langle \cos t, \sin t, \sin(2t) \rangle$ on $[0, 2\pi]$.

In Exercises 17 – 20, find $\| \vec{r}(t) \|$.

17. $\vec{r}(t) = \langle t, t^2 \rangle$.

18. $\vec{r}(t) = \langle 5\cos t, 3\sin t \rangle$.

19. $\vec{r}(t) = \langle 2\cos t, 2\sin t, t \rangle$.

20. $\vec{r}(t) = \langle \cos t, t, t^2 \rangle$.

In Exercises 21 – 30, create a vector–valued function whose graph matches the given description.

21. A circle of radius 2, centered at $(1, 2)$, traced counter–clockwise once on $[0, 2\pi]$.

22. A circle of radius 3, centered at $(5, 5)$, traced clockwise once on $[0, 2\pi]$.

23. An ellipse, centered at $(0, 0)$ with vertical major axis of length 10 and minor axis of length 3, traced once counter–clockwise on $[0, 2\pi]$.

24. An ellipse, centered at $(3, -2)$ with horizontal major axis of length 6 and minor axis of length 4, traced once clockwise on $[0, 2\pi]$.

25. A line through $(2, 3)$ with a slope of 5.

26. A line through $(1, 5)$ with a slope of $-1/2$.

27. The line through points $(1, 2, 3)$ and $(4, 5, 6)$, where $\vec{r}(0) = \langle 1, 2, 3 \rangle$ and $\vec{r}(1) = \langle 4, 5, 6 \rangle$.

28. The line through points $(1, 2)$ and $(4, 4)$, where $\vec{r}(0) = \langle 1, 2 \rangle$ and $\vec{r}(1) = \langle 4, 4 \rangle$.

29. A vertically oriented helix with radius of 2 that starts at $(2, 0, 0)$ and ends at $(2, 0, 4\pi)$ after 1 revolution on $[0, 2\pi]$.

30. A vertically oriented helix with radius of 3 that starts at $(3, 0, 0)$ and ends at $(3, 0, 3)$ after 2 revolutions on $[0, 1]$.

In Exercises 31 – 34, find the average rate of change of $\vec{r}(t)$ on the given interval.

31. $\vec{r}(t) = \langle t, t^2 \rangle$ on $[-2, 2]$.

32. $\vec{r}(t) = \langle t, t + \sin t \rangle$ on $[0, 2\pi]$.

33. $\vec{r}(t) = \langle 3\cos t, 2\sin t, t \rangle$ on $[0, 2\pi]$.

34. $\vec{r}(t) = \langle t, t^2, t^3 \rangle$ on $[-1, 3]$.

11.2 Calculus and Vector–Valued Functions

The previous section introduced us to a new mathematical object, the vector–valued function. We now apply calculus concepts to these functions. We start with the limit, then work our way through derivatives to integrals.

Limits of Vector–Valued Functions

The initial definition of the limit of a vector–valued function is a bit intimidating, as was the definition of the limit in Definition 1.2.1. The theorem following the definition shows that in practice, taking limits of vector–valued functions is no more difficult than taking limits of real–valued functions.

Definition 11.2.1   Limits of Vector–Valued Functions

Let I be an open interval containing c, and let $\vec{r}(t)$ be a vector–valued function defined on I, except possibly at c. The **limit of $\vec{r}(t)$, as t approaches c, is $\vec{L}$**, expressed as

$$\lim_{t \to c} \vec{r}(t) = \vec{L},$$

means that given any $\varepsilon > 0$, there exists a $\delta > 0$ such that for all $t \neq c$, if $|t - c| < \delta$, we have $\|\vec{r}(t) - \vec{L}\| < \varepsilon$.

Note: we can define one-sided limits in a manner very similar to Definition 11.2.1.

Note how the measurement of distance between real numbers is the absolute value of their difference; the measure of distance between vectors is the vector norm, or magnitude, of their difference.

Theorem 11.2.1 states that we can compute limits of vector–valued functions component–wise.

Theorem 11.2.1   Limits of Vector–Valued Functions

1. Let $\vec{r}(t) = \langle f(t), g(t) \rangle$ be a vector–valued function in $\mathbb{R}^2$ defined on an open interval I containing c, except possibly at c. Then

$$\lim_{t \to c} \vec{r}(t) = \left\langle \lim_{t \to c} f(t),\ \lim_{t \to c} g(t) \right\rangle.$$

2. Let $\vec{r}(t) = \langle f(t), g(t), h(t) \rangle$ be a vector–valued function in $\mathbb{R}^3$ defined on an open interval I containing c, except possibly at c. Then

$$\lim_{t \to c} \vec{r}(t) = \left\langle \lim_{t \to c} f(t),\ \lim_{t \to c} g(t),\ \lim_{t \to c} h(t) \right\rangle$$

Notes:

Example 11.2.1   Finding limits of vector–valued functions

Let $\vec{r}(t) = \left\langle \dfrac{\sin t}{t},\ t^2 - 3t + 3,\ \cos t \right\rangle$. Find $\lim\limits_{t \to 0} \vec{r}(t)$.

SOLUTION   We apply the theorem and compute limits component–wise.

$$\lim_{t \to 0} \vec{r}(t) = \left\langle \lim_{t \to 0} \frac{\sin t}{t},\ \lim_{t \to 0} t^2 - 3t + 3,\ \lim_{t \to 0} \cos t \right\rangle$$

$$= \langle 1, 3, 1 \rangle.$$

Continuity

Note: Using one-sided limits, we can also define continuity on closed intervals as done before.

Definition 11.2.2   Continuity of Vector–Valued Functions

Let $\vec{r}(t)$ be a vector–valued function defined on an open interval I containing c.

1. $\vec{r}(t)$ is **continuous at** c if $\lim\limits_{t \to c} \vec{r}(t) = \vec{r}(c)$.

2. If $\vec{r}(t)$ is continuous at all c in I, then $\vec{r}(t)$ is **continuous on** I.

We again have a theorem that lets us evaluate continuity component–wise.

Theorem 11.2.2   Continuity of Vector–Valued Functions

Let $\vec{r}(t)$ be a vector–valued function defined on an open interval I containing c. Then $\vec{r}(t)$ is continuous at c if, and only if, each of its component functions is continuous at c.

Example 11.2.2   Evaluating continuity of vector–valued functions

Let $\vec{r}(t) = \left\langle \dfrac{\sin t}{t},\ t^2 - 3t + 3,\ \cos t \right\rangle$. Determine whether $\vec{r}$ is continuous at $t = 0$ and $t = 1$.

SOLUTION   While the second and third components of $\vec{r}(t)$ are defined at $t = 0$, the first component, $(\sin t)/t$, is not. Since the first component is not even defined at $t = 0$, $\vec{r}(t)$ is not defined at $t = 0$, and hence it is not continuous at $t = 0$.

At $t = 1$ each of the component functions is continuous. Therefore $\vec{r}(t)$ is continuous at $t = 1$.

Notes:

Derivatives

Consider a vector–valued function $\vec{r}$ defined on an open interval I containing t_0 and t_1. We can compute the displacement of $\vec{r}$ on $[t_0, t_1]$, as shown in Figure 11.2.1(a). Recall that dividing the displacement vector by $t_1 - t_0$ gives the average rate of change on $[t_0, t_1]$, as shown in (b).

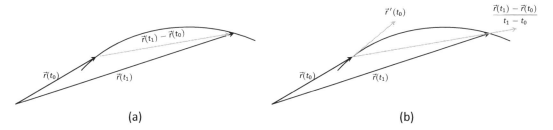

Figure 11.2.1: Illustrating displacement, leading to an understanding of the derivative of vector–valued functions.

The **derivative** of a vector–valued function is a measure of the *instantaneous* rate of change, measured by taking the limit as the length of $[t_0, t_1]$ goes to 0. Instead of thinking of an interval as $[t_0, t_1]$, we think of it as $[c, c + h]$ for some value of h (hence the interval has length h). The *average* rate of change is

$$\frac{\vec{r}(c + h) - \vec{r}(c)}{h}$$

for any value of $h \neq 0$. We take the limit as $h \to 0$ to measure the instantaneous rate of change; this is the derivative of $\vec{r}$.

Definition 11.2.3   Derivative of a Vector–Valued Function

Let $\vec{r}(t)$ be continuous on an open interval I containing c.

1. The **derivative of $\vec{r}$ at $t = c$** is

$$\vec{r}\,'(c) = \lim_{h \to 0} \frac{\vec{r}(c + h) - \vec{r}(c)}{h}.$$

2. The **derivative of $\vec{r}$** is

$$\vec{r}\,'(t) = \lim_{h \to 0} \frac{\vec{r}(t + h) - \vec{r}(t)}{h}.$$

Alternate notations for the derivative of $\vec{r}$ include:

$$\vec{r}\,'(t) = \frac{d}{dt}\big(\vec{r}(t)\big) = \frac{d\vec{r}}{dt}.$$

Notes:

Note: again, using one-sided limits, we can define differentiability on closed intervals. We'll make use of this a few times in this chapter.

If a vector–valued function has a derivative for all c in an open interval I, we say that $\vec{r}(t)$ is **differentiable** on I.

Once again we might view this definition as intimidating, but recall that we can evaluate limits component–wise. The following theorem verifies that this means we can compute derivatives component–wise as well, making the task not too difficult.

Theorem 11.2.3   Derivatives of Vector–Valued Functions

1. Let $\vec{r}(t) = \langle f(t), g(t) \rangle$. Then

$$\vec{r}\,'(t) = \langle f'(t), g'(t) \rangle.$$

2. Let $\vec{r}(t) = \langle f(t), g(t), h(t) \rangle$. Then

$$\vec{r}\,'(t) = \langle f'(t), g'(t), h'(t) \rangle.$$

Example 11.2.3   Derivatives of vector–valued functions
Let $\vec{r}(t) = \langle t^2, t \rangle$.

1. Sketch $\vec{r}(t)$ and $\vec{r}\,'(t)$ on the same axes.

2. Compute $\vec{r}\,'(1)$ and sketch this vector with its initial point at the origin and at $\vec{r}(1)$.

SOLUTION

1. Theorem 11.2.3 allows us to compute derivatives component–wise, so

$$\vec{r}\,'(t) = \langle 2t, 1 \rangle.$$

$\vec{r}(t)$ and $\vec{r}\,'(t)$ are graphed together in Figure 11.2.2(a). Note how plotting the two of these together, in this way, is not very illuminating. When dealing with real–valued functions, plotting $f(x)$ with $f'(x)$ gave us useful information as we were able to compare f and f' at the same x-values. When dealing with vector–valued functions, it is hard to tell which points on the graph of $\vec{r}\,'$ correspond to which points on the graph of $\vec{r}$.

2. We easily compute $\vec{r}\,'(1) = \langle 2, 1 \rangle$, which is drawn in Figure 11.2.2 with its initial point at the origin, as well as at $\vec{r}(1) = \langle 1, 1 \rangle$. These are sketched in Figure 11.2.2(b).

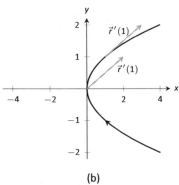

(a)

(b)

Figure 11.2.2: Graphing the derivative of a vector–valued function in Example 11.2.3.

Notes:

Example 11.2.4 **Derivatives of vector–valued functions**
Let $\vec{r}(t) = \langle \cos t, \sin t, t \rangle$. Compute $\vec{r}\,'(t)$ and $\vec{r}\,'(\pi/2)$. Sketch $\vec{r}\,'(\pi/2)$ with its initial point at the origin and at $\vec{r}(\pi/2)$.

SOLUTION We compute $\vec{r}\,'$ as $\vec{r}\,'(t) = \langle -\sin t, \cos t, 1 \rangle$. At $t = \pi/2$, we have $\vec{r}\,'(\pi/2) = \langle -1, 0, 1 \rangle$. Figure 11.2.3 shows two graphs of $\vec{r}(t)$, from different perspectives, with $\vec{r}\,'(\pi/2)$ plotted with its initial point at the origin and at $\vec{r}(\pi/2)$.

In Examples 11.2.3 and 11.2.4, sketching a particular derivative with its initial point at the origin did not seem to reveal anything significant. However, when we sketched the vector with its initial point on the corresponding point on the graph, we did see something significant: the vector appeared to be *tangent* to the graph. We have not yet defined what "tangent" means in terms of curves in space; in fact, we use the derivative to define this term.

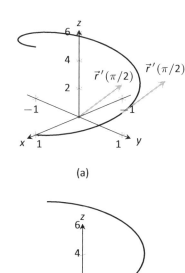

(a)

(b)

Figure 11.2.3: Viewing a vector–valued function, and its derivative at one point, from two different perspectives.

Definition 11.2.4 **Tangent Vector, Tangent Line**

Let $\vec{r}(t)$ be a differentiable vector–valued function on an open interval I containing c, where $\vec{r}\,'(c) \neq \vec{0}$.

1. A vector $\vec{v}$ is **tangent to the graph of** $\vec{r}(t)$ **at** $t = c$ if $\vec{v}$ is parallel to $\vec{r}\,'(c)$.

2. The **tangent line** to the graph of $\vec{r}(t)$ at $t = c$ is the line through $\vec{r}(c)$ with direction parallel to $\vec{r}\,'(c)$. An equation of the tangent line is
$$\vec{\ell}(t) = \vec{r}(c) + t\,\vec{r}\,'(c).$$

Example 11.2.5 **Finding tangent lines to curves in space**
Let $\vec{r}(t) = \langle t, t^2, t^3 \rangle$ on $[-1.5, 1.5]$. Find the vector equation of the line tangent to the graph of $\vec{r}$ at $t = -1$.

SOLUTION To find the equation of a line, we need a point on the line and the line's direction. The point is given by $\vec{r}(-1) = \langle -1, 1, -1 \rangle$. (To be clear, $\langle -1, 1, -1 \rangle$ is a *vector*, not a point, but we use the point "pointed to" by this vector.)

The direction comes from $\vec{r}\,'(-1)$. We compute, component–wise, $\vec{r}\,'(t) = \langle 1, 2t, 3t^2 \rangle$. Thus $\vec{r}\,'(-1) = \langle 1, -2, 3 \rangle$.

The vector equation of the line is $\ell(t) = \langle -1, 1, -1 \rangle + t\,\langle 1, -2, 3 \rangle$. This line and $\vec{r}(t)$ are sketched, from two perspectives, in Figure 11.2.4 (a) and (b).

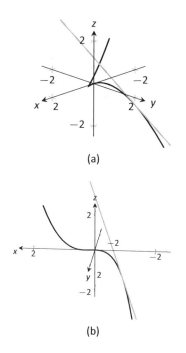

(a)

(b)

Figure 11.2.4: Graphing a curve in space with its tangent line.

Notes:

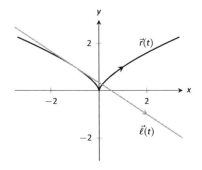

Figure 11.2.5: Graphing $\vec{r}(t)$ and its tangent line in Example 11.2.6.

Example 11.2.6   Finding tangent lines to curves
Find the equations of the lines tangent to $\vec{r}(t) = \left\langle t^3, t^2 \right\rangle$ at $t = -1$ and $t = 0$.

SOLUTION   We find that $\vec{r}\,'(t) = \left\langle 3t^2, 2t \right\rangle$. At $t = -1$, we have

$$\vec{r}(-1) = \langle -1, 1 \rangle \quad \text{and} \quad \vec{r}\,'(-1) = \langle 3, -2 \rangle,$$

so the equation of the line tangent to the graph of $\vec{r}(t)$ at $t = -1$ is

$$\ell(t) = \langle -1, 1 \rangle + t\,\langle 3, -2 \rangle.$$

This line is graphed with $\vec{r}(t)$ in Figure 11.2.5.

At $t = 0$, we have $\vec{r}\,'(0) = \langle 0, 0 \rangle = \vec{0}$! This implies that the tangent line "has no direction." We cannot apply Definition 11.2.4, hence cannot find the equation of the tangent line.

We were unable to compute the equation of the tangent line to $\vec{r}(t) = \left\langle t^3, t^2 \right\rangle$ at $t = 0$ because $\vec{r}\,'(0) = \vec{0}$. The graph in Figure 11.2.5 shows that there is a cusp at this point. This leads us to another definition of **smooth**, previously defined by Definition 9.2.2 in Section 9.2.

Definition 11.2.5   Smooth Vector–Valued Functions

Let $\vec{r}(t)$ be a differentiable vector–valued function on an open interval I where $\vec{r}\,'(t)$ is continuous on I. $\vec{r}(t)$ is **smooth** on I if $\vec{r}\,'(t) \neq \vec{0}$ on I.

Having established derivatives of vector–valued functions, we now explore the relationships between the derivative and other vector operations. The following theorem states how the derivative interacts with vector addition and the various vector products.

Notes:

Theorem 11.2.4   Properties of Derivatives of Vector–Valued Functions

Let $\vec{r}$ and $\vec{s}$ be differentiable vector–valued functions, let f be a differentiable real–valued function, and let c be a real number.

1. $\dfrac{d}{dt}\left(\vec{r}(t) \pm \vec{s}(t)\right) = \vec{r}\,'(t) \pm \vec{s}\,'(t)$

2. $\dfrac{d}{dt}\left(c\vec{r}(t)\right) = c\vec{r}\,'(t)$

3. $\dfrac{d}{dt}\left(f(t)\vec{r}(t)\right) = f'(t)\vec{r}(t) + f(t)\vec{r}\,'(t)$ **Product Rule**

4. $\dfrac{d}{dt}\left(\vec{r}(t) \cdot \vec{s}(t)\right) = \vec{r}\,'(t) \cdot \vec{s}(t) + \vec{r}(t) \cdot \vec{s}\,'(t)$ **Product Rule**

5. $\dfrac{d}{dt}\left(\vec{r}(t) \times \vec{s}(t)\right) = \vec{r}\,'(t) \times \vec{s}(t) + \vec{r}(t) \times \vec{s}\,'(t)$ **Product Rule**

6. $\dfrac{d}{dt}\left(\vec{r}(f(t))\right) = \vec{r}\,'(f(t))f'(t)$ **Chain Rule**

Example 11.2.7   Using derivative properties of vector–valued functions
Let $\vec{r}(t) = \langle t, t^2 - 1 \rangle$ and let $\vec{u}(t)$ be the unit vector that points in the direction of $\vec{r}(t)$.

1. Graph $\vec{r}(t)$ and $\vec{u}(t)$ on the same axes, on $[-2, 2]$.

2. Find $\vec{u}\,'(t)$ and sketch $\vec{u}\,'(-2)$, $\vec{u}\,'(-1)$ and $\vec{u}\,'(0)$. Sketch each with initial point the corresponding point on the graph of $\vec{u}$.

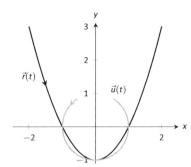

Figure 11.2.6: Graphing $\vec{r}(t)$ and $\vec{u}(t)$ in Example 11.2.7.

Solution

1. To form the unit vector that points in the direction of $\vec{r}$, we need to divide $\vec{r}(t)$ by its magnitude.

$$\|\,\vec{r}(t)\,\| = \sqrt{t^2 + (t^2 - 1)^2} \quad \Rightarrow \quad \vec{u}(t) = \frac{1}{\sqrt{t^2 + (t^2 - 1)^2}} \langle t, t^2 - 1 \rangle.$$

$\vec{r}(t)$ and $\vec{u}(t)$ are graphed in Figure 11.2.6. Note how the graph of $\vec{u}(t)$ forms part of a circle; this must be the case, as the length of $\vec{u}(t)$ is 1 for all t.

Notes:

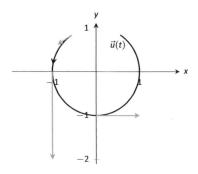

Figure 11.2.7: Graphing some of the derivatives of $\vec{u}(t)$ in Example 11.2.7.

2. To compute $\vec{u}'(t)$, we use Theorem 11.2.4, writing

$$\vec{u}(t) = f(t)\vec{r}(t), \quad \text{where} \quad f(t) = \frac{1}{\sqrt{t^2 + (t^2 - 1)^2}} = \left(t^2 + (t^2 - 1)^2\right)^{-1/2}.$$

(We *could* write

$$\vec{u}(t) = \left\langle \frac{t}{\sqrt{t^2 + (t^2 - 1)^2}}, \frac{t^2 - 1}{\sqrt{t^2 + (t^2 - 1)^2}} \right\rangle$$

and then take the derivative. It is a matter of preference; this latter method requires two applications of the Quotient Rule where our method uses the Product and Chain Rules.)

We find $f'(t)$ using the Chain Rule:

$$f'(t) = -\frac{1}{2}\left(t^2 + (t^2 - 1)^2\right)^{-3/2}\left(2t + 2(t^2 - 1)(2t)\right)$$

$$= -\frac{2t(2t^2 - 1)}{2\left(\sqrt{t^2 + (t^2 - 1)^2}\right)^3}$$

We now find $\vec{u}'(t)$ using part 3 of Theorem 11.2.4:

$$\vec{u}'(t) = f'(t)\vec{u}(t) + f(t)\vec{u}'(t)$$

$$= -\frac{2t(2t^2 - 1)}{2\left(\sqrt{t^2 + (t^2 - 1)^2}\right)^3}\langle t, t^2 - 1\rangle + \frac{1}{\sqrt{t^2 + (t^2 - 1)^2}}\langle 1, 2t\rangle.$$

This is admittedly very "messy;" such is usually the case when we deal with unit vectors. We can use this formula to compute $\vec{u}'(-2)$, $\vec{u}'(-1)$ and $\vec{u}'(0)$:

$$\vec{u}'(-2) = \left\langle -\frac{15}{13\sqrt{13}}, -\frac{10}{13\sqrt{13}} \right\rangle \approx \langle -0.320, -0.213\rangle$$

$$\vec{u}'(-1) = \langle 0, -2\rangle$$

$$\vec{u}'(0) = \langle 1, 0\rangle$$

Each of these is sketched in Figure 11.2.7. Note how the length of the vector gives an indication of how quickly the circle is being traced at that point. When $t = -2$, the circle is being drawn relatively slow; when $t = -1$, the circle is being traced much more quickly.

It is a basic geometric fact that a line tangent to a circle at a point P is perpendicular to the line passing through the center of the circle and P. This is

Notes:

illustrated in Figure 11.2.7; each tangent vector is perpendicular to the line that passes through its initial point and the center of the circle. Since the center of the circle is the origin, we can state this another way: $\vec{u}\,'(t)$ is orthogonal to $\vec{u}(t)$.

Recall that the dot product serves as a test for orthogonality: if $\vec{u} \cdot \vec{v} = 0$, then $\vec{u}$ is orthogonal to $\vec{v}$. Thus in the above example, $\vec{u}(t) \cdot \vec{u}\,'(t) = 0$.

This is true of any vector–valued function that has a constant length, that is, that traces out part of a circle. It has important implications later on, so we state it as a theorem (and leave its formal proof as an Exercise.)

Theorem 11.2.5   Vector–Valued Functions of Constant Length

Let $\vec{r}(t)$ be a vector–valued function of constant length that is differentiable on an open interval I. That is, $\| \vec{r}(t) \| = c$ for all t in I (equivalently, $\vec{r}(t) \cdot \vec{r}(t) = c^2$ for all t in I). Then $\vec{r}(t) \cdot \vec{r}\,'(t) = 0$ for all t in I.

Integration

Before formally defining integrals of vector–valued functions, consider the following equation that our calculus experience tells us *should* be true:

$$\int_a^b \vec{r}\,'(t)\, dt = \vec{r}(b) - \vec{r}(a).$$

That is, the integral of a rate of change function should give total change. In the context of vector–valued functions, this total change is displacement. The above equation *is* true; we now develop the theory to show why.

We can define antiderivatives and the indefinite integral of vector–valued functions in the same manner we defined indefinite integrals in Definition 5.1.1. However, we cannot define the definite integral of a vector–valued function as we did in Definition 5.2.1. That definition was based on the signed area between a function $y = f(x)$ and the x-axis. An area–based definition will not be useful in the context of vector–valued functions. Instead, we define the definite integral of a vector–valued function in a manner similar to that of Theorem 5.3.2, utilizing Riemann sums.

Notes:

Definition 11.2.6   **Antiderivatives, Indefinite and Definite Integrals of Vector–Valued Functions**

Let $\vec{r}(t)$ be a continuous vector–valued function on $[a, b]$. An **antiderivative** of $\vec{r}(t)$ is a function $\vec{R}(t)$ such that $\vec{R}'(t) = \vec{r}(t)$.

The set of all antiderivatives of $\vec{r}(t)$ is the **indefinite integral of $\vec{r}(t)$**, denoted by

$$\int \vec{r}(t)\, dt.$$

The definite integral of $\vec{r}(t)$ on $[a, b]$ is

$$\int_a^b \vec{r}(t)\, dt = \lim_{||\Delta t|| \to 0} \sum_{i=1}^{n} \vec{r}(c_i)\Delta t_i,$$

where Δt_i is the length of the ith subinterval of a partition of $[a, b]$, $||\Delta t||$ is the length of the largest subinterval in the partition, and c_i is any value in the ith subinterval of the partition.

It is probably difficult to infer meaning from the definition of the definite integral. The important thing to realize from the definition is that it is built upon limits, which we can evaluate component–wise.

The following theorem simplifies the computation of definite integrals; the rest of this section and the following section will give meaning and application to these integrals.

Theorem 11.2.6   **Indefinite and Definite Integrals of Vector–Valued Functions**

Let $\vec{r}(t) = \langle f(t), g(t) \rangle$ be a vector–valued function in $\mathbb{R}^2$ that are continuous on $[a, b]$.

1. $\displaystyle \int \vec{r}(t)\, dt = \left\langle \int f(t)\, dt, \int g(t)\, dt \right\rangle$

2. $\displaystyle \int_a^b \vec{r}(t)\, dt = \left\langle \int_a^b f(t)\, dt, \int_a^b g(t)\, dt \right\rangle$

A similar statement holds for vector–valued functions in $\mathbb{R}^3$.

Notes:

Example 11.2.8   **Evaluating a definite integral of a vector–valued function**

Let $\vec{r}(t) = \langle e^{2t}, \sin t \rangle$. Evaluate $\displaystyle\int_0^1 \vec{r}(t)\, dt$.

> **SOLUTION**   We follow Theorem 11.2.6.

$$\int_0^1 \vec{r}(t)\, dt = \int_0^1 \langle e^{2t}, \sin t \rangle\, dt$$

$$= \left\langle \int_0^1 e^{2t}\, dt\,,\, \int_0^1 \sin t\, dt \right\rangle$$

$$= \left\langle \left.\frac{1}{2} e^{2t}\right|_0^1\,,\, \left.-\cos t\right|_0^1 \right\rangle$$

$$= \left\langle \frac{1}{2}(e^2 - 1)\,,\, -\cos(1) + 1 \right\rangle$$

$$\approx \langle 3.19, 0.460 \rangle\,.$$

Example 11.2.9   **Solving an initial value problem**

Let $\vec{r}\,''(t) = \langle 2, \cos t, 12t \rangle$. Find $\vec{r}(t)$, where $\vec{r}(0) = \langle -7, -1, 2 \rangle$ and $\vec{r}\,'(0) = \langle 5, 3, 0 \rangle$.

> **SOLUTION**   Knowing $\vec{r}\,''(t) = \langle 2, \cos t, 12t \rangle$, we find $\vec{r}\,'(t)$ by evaluating the indefinite integral.

$$\int \vec{r}\,''(t)\, dt = \left\langle \int 2\, dt\,,\, \int \cos t\, dt\,,\, \int 12t\, dt \right\rangle$$

$$= \langle 2t + C_1, \sin t + C_2, 6t^2 + C_3 \rangle$$

$$= \langle 2t, \sin t, 6t^2 \rangle + \langle C_1, C_2, C_3 \rangle$$

$$= \langle 2t, \sin t, 6t^2 \rangle + \vec{C}.$$

Note how each indefinite integral creates its own constant which we collect as one constant vector $\vec{C}$. Knowing $\vec{r}\,'(0) = \langle 5, 3, 0 \rangle$ allows us to solve for $\vec{C}$:

$$\vec{r}\,'(t) = \langle 2t, \sin t, 6t^2 \rangle + \vec{C}$$

$$\vec{r}\,'(0) = \langle 0, 0, 0 \rangle + \vec{C}$$

$$\langle 5, 3, 0 \rangle = \vec{C}.$$

So $\vec{r}\,'(t) = \langle 2t, \sin t, 6t^2 \rangle + \langle 5, 3, 0 \rangle = \langle 2t + 5, \sin t + 3, 6t^2 \rangle$. To find $\vec{r}(t)$, we integrate once more.

$$\int \vec{r}\,'(t)\, dt = \left\langle \int 2t + 5\, dt\,,\, \int \sin t + 3\, dt\,,\, \int 6t^2\, dt \right\rangle$$

$$= \langle t^2 + 5t, -\cos t + 3t, 2t^3 \rangle + \vec{C}.$$

Notes:

With $\vec{r}(0) = \langle -7, -1, 2 \rangle$, we solve for $\vec{C}$:

$$\vec{r}(t) = \langle t^2 + 5t, -\cos t + 3t, 2t^3 \rangle + \vec{C}$$
$$\vec{r}(0) = \langle 0, -1, 0 \rangle + \vec{C}$$
$$\langle -7, -1, 2 \rangle = \langle 0, -1, 0 \rangle + \vec{C}$$
$$\langle -7, 0, 2 \rangle = \vec{C}.$$

So $\vec{r}(t) = \langle t^2 + 5t, -\cos t + 3t, 2t^3 \rangle + \langle -7, 0, 2 \rangle = \langle t^2 + 5t - 7, -\cos t + 3t, 2t^3 + 2 \rangle$.

What does the integration of a vector–valued function *mean*? There are many applications, but none as direct as "the area under the curve" that we used in understanding the integral of a real–valued function.

A key understanding for us comes from considering the integral of a derivative:

$$\int_a^b \vec{r}\,'(t)\,dt = \vec{r}(t)\Big|_a^b = \vec{r}(b) - \vec{r}(a).$$

Integrating a *rate of change* function gives *displacement*.

Noting that vector–valued functions are closely related to parametric equations, we can describe the arc length of the graph of a vector–valued function as an integral. Given parametric equations $x = f(t), y = g(t)$, the arc length on $[a, b]$ of the graph is

$$\text{Arc Length} = \int_a^b \sqrt{f'(t)^2 + g'(t)^2}\,dt,$$

as stated in Theorem 9.3.1. If $\vec{r}(t) = \langle f(t), g(t) \rangle$, note that $\sqrt{f'(t)^2 + g'(t)^2} = \|\vec{r}\,'(t)\|$. Therefore we can express the arc length of the graph of a vector–valued function as an integral of the magnitude of its derivative.

Theorem 11.2.7   Arc Length of a Vector–Valued Function

Let $\vec{r}(t)$ be a vector–valued function where $\vec{r}\,'(t)$ is continuous on $[a, b]$. The arc length L of the graph of $\vec{r}(t)$ is

$$L = \int_a^b \|\vec{r}\,'(t)\|\,dt.$$

Note that we are actually integrating a scalar–function here, not a vector–valued function.

The next section takes what we have established thus far and applies it to objects in motion. We will let $\vec{r}(t)$ describe the path of an object in the plane or in space and will discover the information provided by $\vec{r}\,'(t)$ and $\vec{r}\,''(t)$.

Notes:

Exercises 11.2

Terms and Concepts

1. Limits, derivatives and integrals of vector–valued functions are all evaluated _____–wise.

2. The definite integral of a rate of change function gives _____.

3. Why is it generally not useful to graph both $\vec{r}(t)$ and $\vec{r}\,'(t)$ on the same axes?

4. Theorem 11.2.4 contains three product rules. What are the three different types of products used in these rules?

Problems

In Exercises 5 – 8, evaluate the given limit.

5. $\lim\limits_{t \to 5} \left\langle 2t + 1, 3t^2 - 1, \sin t \right\rangle$

6. $\lim\limits_{t \to 3} \left\langle e^t, \dfrac{t^2 - 9}{t + 3} \right\rangle$

7. $\lim\limits_{t \to 0} \left\langle \dfrac{t}{\sin t}, (1 + t)^{\frac{1}{t}} \right\rangle$

8. $\lim\limits_{h \to 0} \dfrac{\vec{r}(t + h) - \vec{r}(t)}{h}$, where $\vec{r}(t) = \left\langle t^2, t, 1 \right\rangle$.

In Exercises 9 – 10, identify the interval(s) on which $\vec{r}(t)$ is continuous.

9. $\vec{r}(t) = \left\langle t^2, 1/t \right\rangle$

10. $\vec{r}(t) = \left\langle \cos t, e^t, \ln t \right\rangle$

In Exercises 11 – 16, find the derivative of the given function.

11. $\vec{r}(t) = \left\langle \cos t, e^t, \ln t \right\rangle$

12. $\vec{r}(t) = \left\langle \dfrac{1}{t}, \dfrac{2t - 1}{3t + 1}, \tan t \right\rangle$

13. $\vec{r}(t) = (t^2) \left\langle \sin t, 2t + 5 \right\rangle$

14. $r(t) = \left\langle t^2 + 1, t - 1 \right\rangle \cdot \left\langle \sin t, 2t + 5 \right\rangle$

15. $\vec{r}(t) = \left\langle t^2 + 1, t - 1, 1 \right\rangle \times \left\langle \sin t, 2t + 5, 1 \right\rangle$

16. $\vec{r}(t) = \left\langle \cosh t, \sinh t \right\rangle$

In Exercises 17 – 20, find $\vec{r}\,'(t)$. Sketch $\vec{r}(t)$ and $\vec{r}\,'(1)$, with the initial point of $\vec{r}\,'(1)$ at $\vec{r}(1)$.

17. $\vec{r}(t) = \left\langle t^2 + t, t^2 - t \right\rangle$

18. $\vec{r}(t) = \left\langle t^2 - 2t + 2, t^3 - 3t^2 + 2t \right\rangle$

19. $\vec{r}(t) = \left\langle t^2 + 1, t^3 - t \right\rangle$

20. $\vec{r}(t) = \left\langle t^2 - 4t + 5, t^3 - 6t^2 + 11t - 6 \right\rangle$

In Exercises 21 – 24, give the equation of the line tangent to the graph of $\vec{r}(t)$ at the given t value.

21. $\vec{r}(t) = \left\langle t^2 + t, t^2 - t \right\rangle$ at $t = 1$.

22. $\vec{r}(t) = \left\langle 3 \cos t, \sin t \right\rangle$ at $t = \pi/4$.

23. $\vec{r}(t) = \left\langle 3 \cos t, 3 \sin t, t \right\rangle$ at $t = \pi$.

24. $\vec{r}(t) = \left\langle e^t, \tan t, t \right\rangle$ at $t = 0$.

In Exercises 25 – 28, find the value(s) of t for which $\vec{r}(t)$ is not smooth.

25. $\vec{r}(t) = \left\langle \cos t, \sin t - t \right\rangle$

26. $\vec{r}(t) = \left\langle t^2 - 2t + 1, t^3 + t^2 - 5t + 3 \right\rangle$

27. $\vec{r}(t) = \left\langle \cos t - \sin t, \sin t - \cos t, \cos(4t) \right\rangle$

28. $\vec{r}(t) = \left\langle t^3 - 3t + 2, -\cos(\pi t), \sin^2(\pi t) \right\rangle$

Exercises 29 – 32 ask you to verify parts of Theorem 11.2.4. In each let $f(t) = t^3$, $\vec{r}(t) = \left\langle t^2, t - 1, 1 \right\rangle$ and $\vec{s}(t) = \left\langle \sin t, e^t, t \right\rangle$. Compute the various derivatives as indicated.

29. Simplify $f(t)\vec{r}(t)$, then find its derivative; show this is the same as $f'(t)\vec{r}(t) + f(t)\vec{r}\,'(t)$.

30. Simplify $\vec{r}(t) \cdot \vec{s}(t)$, then find its derivative; show this is the same as $\vec{r}\,'(t) \cdot \vec{s}(t) + \vec{r}(t) \cdot \vec{s}\,'(t)$.

31. Simplify $\vec{r}(t) \times \vec{s}(t)$, then find its derivative; show this is the same as $\vec{r}\,'(t) \times \vec{s}(t) + \vec{r}(t) \times \vec{s}\,'(t)$.

32. Simplify $\vec{r}\big(f(t)\big)$, then find its derivative; show this is the same as $\vec{r}\,'\big(f(t)\big)f'(t)$.

In Exercises 33 – 36, evaluate the given definite or indefinite integral.

33. $\displaystyle\int \left\langle t^3, \cos t, te^t \right\rangle \, dt$

34. $\displaystyle\int \left\langle \dfrac{1}{1 + t^2}, \sec^2 t \right\rangle \, dt$

35. $\displaystyle\int_0^{\pi} \left\langle -\sin t, \cos t \right\rangle \, dt$

36. $\displaystyle\int_{-2}^{2} \left\langle 2t + 1, 2t - 1 \right\rangle \, dt$

In Exercises 37 – 40, solve the given initial value problems.

37. Find $\vec{r}(t)$, given that $\vec{r}\,'(t) = \langle t, \sin t \rangle$ and $\vec{r}(0) = \langle 2, 2 \rangle$.

38. Find $\vec{r}(t)$, given that $\vec{r}\,'(t) = \langle 1/(t+1), \tan t \rangle$ and $\vec{r}(0) = \langle 1, 2 \rangle$.

39. Find $\vec{r}(t)$, given that $\vec{r}\,''(t) = \langle t^2, t, 1 \rangle$, $\vec{r}\,'(0) = \langle 1, 2, 3 \rangle$ and $\vec{r}(0) = \langle 4, 5, 6 \rangle$.

40. Find $\vec{r}(t)$, given that $\vec{r}\,''(t) = \langle \cos t, \sin t, e^t \rangle$, $\vec{r}\,'(0) = \langle 0, 0, 0 \rangle$ and $\vec{r}(0) = \langle 0, 0, 0 \rangle$.

In Exercises 41 – 44 , find the arc length of $\vec{r}(t)$ on the indi-cated interval.

41. $\vec{r}(t) = \langle 2\cos t, 2\sin t, 3t \rangle$ on $[0, 2\pi]$.

42. $\vec{r}(t) = \langle 5\cos t, 3\sin t, 4\sin t \rangle$ on $[0, 2\pi]$.

43. $\vec{r}(t) = \langle t^3, t^2, t^3 \rangle$ on $[0, 1]$.

44. $\vec{r}(t) = \langle e^{-t}\cos t, e^{-t}\sin t \rangle$ on $[0, 1]$.

45. Prove Theorem 11.2.5; that is, show if $\vec{r}(t)$ has constant length and is differentiable, then $\vec{r}(t) \cdot \vec{r}\,'(t) = 0$. (Hint: use the Product Rule to compute $\frac{d}{dt}\big(\vec{r}(t) \cdot \vec{r}(t)\big)$.)

11.3 The Calculus of Motion

A common use of vector–valued functions is to describe the motion of an object in the plane or in space. A **position function** $\vec{r}(t)$ gives the position of an object at **time** t. This section explores how derivatives and integrals are used to study the motion described by such a function.

Definition 11.3.1   Velocity, Speed and Acceleration

Let $\vec{r}(t)$ be a position function in $\mathbb{R}^2$ or $\mathbb{R}^3$.

1. **Velocity**, denoted $\vec{v}(t)$, is the instantaneous rate of position change; that is, $\vec{v}(t) = \vec{r}\,'(t)$.

2. **Speed** is the magnitude of velocity, $\|\vec{v}(t)\|$.

3. **Acceleration**, denoted $\vec{a}(t)$, is the instantaneous rate of velocity change; that is, $\vec{a}(t) = \vec{v}\,'(t) = \vec{r}\,''(t)$.

Example 11.3.1   Finding velocity and acceleration

An object is moving with position function $\vec{r}(t) = \langle t^2 - t, t^2 + t \rangle$, $-3 \le t \le 3$, where distances are measured in feet and time is measured in seconds.

1. Find $\vec{v}(t)$ and $\vec{a}(t)$.

2. Sketch $\vec{r}(t)$; plot $\vec{v}(-1)$, $\vec{a}(-1)$, $\vec{v}(1)$ and $\vec{a}(1)$, each with their initial point at their corresponding point on the graph of $\vec{r}(t)$.

3. When is the object's speed minimized?

SOLUTION

1. Taking derivatives, we find

$$\vec{v}(t) = \vec{r}\,'(t) = \langle 2t - 1, 2t + 1 \rangle \quad \text{and} \quad \vec{a}(t) = \vec{r}\,''(t) = \langle 2, 2 \rangle .$$

Note that acceleration is constant.

2. $\vec{v}(-1) = \langle -3, -1 \rangle$, $\vec{a}(-1) = \langle 2, 2 \rangle$; $\vec{v}(1) = \langle 1, 3 \rangle$, $\vec{a}(1) = \langle 2, 2 \rangle$. These are plotted with $\vec{r}(t)$ in Figure 11.3.1(a).

We can think of acceleration as "pulling" the velocity vector in a certain direction. At $t = -1$, the velocity vector points down and to the left; at $t = 1$, the velocity vector has been pulled in the $\langle 2, 2 \rangle$ direction and is

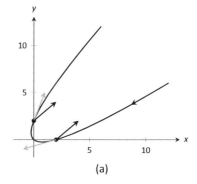

(a)

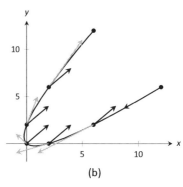

(b)

Figure 11.3.1: Graphing the position, velocity and acceleration of an object in Example 11.3.1.

Notes:

now pointing up and to the right. In Figure 11.3.1(b) we plot more velocity/acceleration vectors, making more clear the effect acceleration has on velocity.

Since $\vec{a}(t)$ is constant in this example, as t grows large $\vec{v}(t)$ becomes almost parallel to $\vec{a}(t)$. For instance, when $t = 10$, $\vec{v}(10) = \langle 19, 21 \rangle$, which is nearly parallel to $\langle 2, 2 \rangle$.

3. The object's speed is given by

$$|| \vec{v}(t) || = \sqrt{(2t-1)^2 + (2t+1)^2} = \sqrt{8t^2 + 2}.$$

To find the minimal speed, we could apply calculus techniques (such as set the derivative equal to 0 and solve for t, etc.) but we can find it by inspection. Inside the square root we have a quadratic which is minimized when $t = 0$. Thus the speed is minimized at $t = 0$, with a speed of $\sqrt{2}$ ft/s.

The graph in Figure 11.3.1(b) also implies speed is minimized here. The filled dots on the graph are located at integer values of t between -3 and 3. Dots that are far apart imply the object traveled a far distance in 1 second, indicating high speed; dots that are close together imply the object did not travel far in 1 second, indicating a low speed. The dots are closest together near $t = 0$, implying the speed is minimized near that value.

Example 11.3.2   Analyzing Motion
Two objects follow an identical path at different rates on $[-1, 1]$. The position function for Object 1 is $\vec{r}_1(t) = \langle t, t^2 \rangle$; the position function for Object 2 is $\vec{r}_2(t) = \langle t^3, t^6 \rangle$, where distances are measured in feet and time is measured in seconds. Compare the velocity, speed and acceleration of the two objects on the path.

SOLUTION   We begin by computing the velocity and acceleration function for each object:

$$\vec{v}_1(t) = \langle 1, 2t \rangle \qquad\qquad \vec{v}_2(t) = \langle 3t^2, 6t^5 \rangle$$
$$\vec{a}_1(t) = \langle 0, 2 \rangle \qquad\qquad \vec{a}_2(t) = \langle 6t, 30t^4 \rangle$$

We immediately see that Object 1 has constant acceleration, whereas Object 2 does not.

At $t = -1$, we have $\vec{v}_1(-1) = \langle 1, -2 \rangle$ and $\vec{v}_2(-1) = \langle 3, -6 \rangle$; the velocity of Object 2 is three times that of Object 1 and so it follows that the speed of Object 2 is three times that of Object 1 ($3\sqrt{5}$ ft/s compared to $\sqrt{5}$ ft/s.)

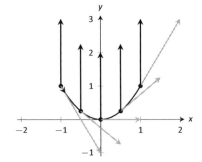

Figure 11.3.2: Plotting velocity and acceleration vectors for Object 1 in Example 11.3.2.

Notes:

At $t = 0$, the velocity of Object 1 is $\vec{v}(1) = \langle 1, 0 \rangle$ and the velocity of Object 2 is $\vec{0}$! This tells us that Object 2 comes to a complete stop at $t = 0$.

In Figure 11.3.2, we see the velocity and acceleration vectors for Object 1 plotted for $t = -1, -1/2, 0, 1/2$ and $t = 1$. Note again how the constant acceleration vector seems to "pull" the velocity vector from pointing down, right to up, right. We could plot the analogous picture for Object 2, but the velocity and acceleration vectors are rather large ($\vec{a}_2(-1) = \langle -6, 30 \rangle$!)

Instead, we simply plot the locations of Object 1 and 2 on intervals of $1/5^{\text{th}}$ of a second, shown in Figure 11.3.3(a) and (b). Note how the x-values of Object 1 increase at a steady rate. This is because the x-component of $\vec{a}(t)$ is 0; there is no acceleration in the x-component. The dots are not evenly spaced; the object is moving faster near $t = -1$ and $t = 1$ than near $t = 0$.

In part (b) of the Figure, we see the points plotted for Object 2. Note the large change in position from $t = -1$ to $t = -0.8$; the object starts moving very quickly. However, it slows considerably at it approaches the origin, and comes to a complete stop at $t = 0$. While it looks like there are 3 points near the origin, there are in reality 5 points there.

Since the objects begin and end at the same location, they have the same displacement. Since they begin and end at the same time, with the same displacement, they have the same average rate of change (i.e, they have the same average velocity). Since they follow the same path, they have the same distance traveled. Even though these three measurements are the same, the objects obviously travel the path in very different ways.

Example 11.3.3 Analyzing the motion of a whirling ball on a string

A young boy whirls a ball, attached to a string, above his head in a counterclockwise circle. The ball follows a circular path and makes 2 revolutions per second. The string has length 2ft.

1. Find the position function $\vec{r}(t)$ that describes this situation.

2. Find the acceleration of the ball and give a physical interpretation of it.

3. A tree stands 10ft in front of the boy. At what t-values should the boy release the string so that the ball hits the tree?

SOLUTION

1. The ball whirls in a circle. Since the string is 2ft long, the radius of the circle is 2. The position function $\vec{r}(t) = \langle 2 \cos t, 2 \sin t \rangle$ describes a circle with radius 2, centered at the origin, but makes a full revolution every 2π seconds, not two revolutions per second. We modify the period of the

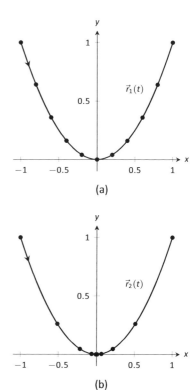

Figure 11.3.3: Comparing the positions of Objects 1 and 2 in Example 11.3.2.

Notes:

trigonometric functions to be 1/2 by multiplying t by 4π. The final position function is thus

$$\vec{r}(t) = \langle 2\cos(4\pi t), 2\sin(4\pi t) \rangle.$$

(Plot this for $0 \le t \le 1/2$ to verify that one revolution is made in 1/2 a second.)

2. To find $\vec{a}(t)$, we take the derivative of $\vec{r}(t)$ twice.

$$\vec{v}(t) = \vec{r}\,'(t) = \langle -8\pi\sin(4\pi t), 8\pi\cos(4\pi t)\rangle$$
$$\vec{a}(t) = \vec{r}\,''(t) = \langle -32\pi^2\cos(4\pi t), -32\pi^2\sin(4\pi t)\rangle$$
$$= -32\pi^2\langle\cos(4\pi t), \sin(4\pi t)\rangle.$$

Note how $\vec{a}(t)$ is parallel to $\vec{r}(t)$, but has a different magnitude and points in the opposite direction. Why is this?

Recall the classic physics equation, "Force = mass × acceleration." A force acting on a mass induces acceleration (i.e., the mass moves); acceleration acting on a mass induces a force (gravity gives our mass a *weight*). Thus force and acceleration are closely related. A moving ball "wants" to travel in a straight line. Why does the ball in our example move in a circle? It is attached to the boy's hand by a string. The string applies a force to the ball, affecting it's motion: the string *accelerates* the ball. This is not acceleration in the sense of "it travels faster;" rather, this acceleration is changing the velocity of the ball. In what direction is this force/acceleration being applied? In the direction of the string, towards the boy's hand.

The magnitude of the acceleration is related to the speed at which the ball is traveling. A ball whirling quickly is rapidly changing direction/velocity. When velocity is changing rapidly, the acceleration must be "large."

3. When the boy releases the string, the string no longer applies a force to the ball, meaning acceleration is $\vec{0}$ and the ball can now move in a straight line in the direction of $\vec{v}(t)$.

Let $t = t_0$ be the time when the boy lets go of the string. The ball will be at $\vec{r}(t_0)$, traveling in the direction of $\vec{v}(t_0)$. We want to find t_0 so that this line contains the point $(0, 10)$ (since the tree is 10ft directly in front of the boy).

There are many ways to find this time value. We choose one that is relatively simple computationally. As shown in Figure 11.3.4, the vector from the release point to the tree is $\langle 0, 10\rangle - \vec{r}(t_0)$. This line segment is tangent to the circle, which means it is also perpendicular to $\vec{r}(t_0)$ itself, so their dot product is 0.

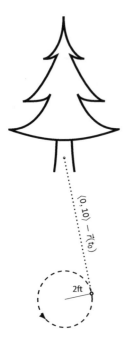

$\langle 0, 10\rangle - \vec{r}(t_0)$

2ft

Figure 11.3.4: Modeling the flight of a ball in Example 11.3.3.

Notes:

$$\vec{r}(t_0) \cdot \big(\langle 0, 10 \rangle - \vec{r}(t_0) \big) = 0$$
$$\langle 2 \cos(4\pi t_0), 2 \sin(4\pi t_0) \rangle \cdot \langle -2 \cos(4\pi t_0), 10 - 2 \sin(4\pi t_0) \rangle = 0$$
$$-4 \cos^2(4\pi t_0) + 20 \sin(4\pi t_0) - 4 \sin^2(4\pi t_0) = 0$$
$$20 \sin(4\pi t_0) - 4 = 0$$
$$\sin(4\pi t_0) = 1/5$$
$$4\pi t_0 = \sin^{-1}(1/5)$$
$$4\pi t_0 \approx 0.2 + 2\pi n,$$

where n is an integer. Solving for t_0 we have:

$$t_0 \approx 0.016 + n/2$$

This is a wonderful formula. Every 1/2 second after $t = 0.016$s the boy can release the string (since the ball makes 2 revolutions per second, he has two chances each second to release the ball).

Example 11.3.4 Analyzing motion in space
An object moves in a spiral with position function $\vec{r}(t) = \langle \cos t, \sin t, t \rangle$, where distances are measured in meters and time is in minutes. Describe the object's speed and acceleration at time t.

SOLUTION With $\vec{r}(t) = \langle \cos t, \sin t, t \rangle$, we have:

$$\vec{v}(t) = \langle -\sin t, \cos t, 1 \rangle \quad \text{and}$$
$$\vec{a}(t) = \langle -\cos t, -\sin t, 0 \rangle .$$

The speed of the object is $\| \vec{v}(t) \| = \sqrt{(-\sin t)^2 + \cos^2 t + 1} = \sqrt{2}$m/min; it moves at a constant speed. Note that the object does not accelerate in the z-direction, but rather moves up at a constant rate of 1m/min.

The objects in Examples 11.3.3 and 11.3.4 traveled at a constant speed. That is, $\| \vec{v}(t) \| = c$ for some constant c. Recall Theorem 11.2.5, which states that if a vector–valued function $\vec{r}(t)$ has constant length, then $\vec{r}(t)$ is perpendicular to its derivative: $\vec{r}(t) \cdot \vec{r}\,'(t) = 0$. In these examples, the velocity function has constant length, therefore we can conclude that the velocity is perpendicular to the acceleration: $\vec{v}(t) \cdot \vec{a}(t) = 0$. A quick check verifies this.

There is an intuitive understanding of this. If acceleration is parallel to velocity, then it is only affecting the object's speed; it does not change the direction of travel. (For example, consider a dropped stone. Acceleration and velocity are

Notes:

parallel – straight down – and the direction of velocity never changes, though speed does increase.) If acceleration is not perpendicular to velocity, then there is some acceleration in the direction of travel, influencing the speed. If speed is constant, then acceleration must be orthogonal to velocity, as it then only affects direction, and not speed.

Key Idea 11.3.1   Objects With Constant Speed

If an object moves with constant speed, then its velocity and acceleration vectors are orthogonal. That is, $\vec{v}(t) \cdot \vec{a}(t) = 0$.

Projectile Motion

An important application of vector–valued position functions is *projectile motion*: the motion of objects under only the influence of gravity. We will measure time in seconds, and distances will either be in meters or feet. We will show that we can completely describe the path of such an object knowing its initial position and initial velocity (i.e., where it *is* and where it *is going*.)

Suppose an object has initial position $\vec{r}(0) = \langle x_0, y_0 \rangle$ and initial velocity $\vec{v}(0) = \langle v_x, v_y \rangle$. It is customary to rewrite $\vec{v}(0)$ in terms of its speed v_0 and direction $\vec{u}$, where $\vec{u}$ is a unit vector. Recall all unit vectors in $\mathbb{R}^2$ can be written as $\langle \cos\theta, \sin\theta \rangle$, where θ is an angle measure counter–clockwise from the x-axis. (We refer to θ as the **angle of elevation**.) Thus $\vec{v}(0) = v_0 \langle \cos\theta, \sin\theta \rangle$.

Since the acceleration of the object is known, namely $\vec{a}(t) = \langle 0, -g \rangle$, where g is the gravitational constant, we can find $\vec{r}(t)$ knowing our two initial conditions. We first find $\vec{v}(t)$:

Note: This text uses $g = 32\text{ft/s}^2$ when using Imperial units, and $g = 9.8\text{m/s}^2$ when using SI units.

$$\vec{v}(t) = \int \vec{a}(t) \, dt$$
$$\vec{v}(t) = \int \langle 0, -g \rangle \, dt$$
$$\vec{v}(t) = \langle 0, -gt \rangle + \vec{C}.$$

Knowing $\vec{v}(0) = v_0 \langle \cos\theta, \sin\theta \rangle$, we have $\vec{C} = v_0 \langle \cos\theta, \sin\theta \rangle$ and so

$$\vec{v}(t) = \langle v_0 \cos\theta, -gt + v_0 \sin\theta \rangle.$$

Notes:

We integrate once more to find $\vec{r}(t)$:

$$\vec{r}(t) = \int \vec{v}(t)\, dt$$

$$\vec{r}(t) = \int \left\langle v_0 \cos\theta, -gt + v_0 \sin\theta \right\rangle\, dt$$

$$\vec{r}(t) = \left\langle \left(v_0 \cos\theta\right)t, -\frac{1}{2}gt^2 + \left(v_0 \sin\theta\right)t \right\rangle + \vec{C}.$$

Knowing $\vec{r}(0) = \langle x_0, y_0 \rangle$, we conclude $\vec{C} = \langle x_0, y_0 \rangle$ and

$$\vec{r}(t) = \left\langle \left(v_0 \cos\theta\right)t + x_0 \ , -\frac{1}{2}gt^2 + \left(v_0 \sin\theta\right)t + y_0 \right\rangle.$$

Key Idea 11.3.2   Projectile Motion

The position function of a projectile propelled from an initial position of $\vec{r}_0 = \langle x_0, y_0 \rangle$, with initial speed v_0, with angle of elevation θ and neglecting all accelerations but gravity is

$$\vec{r}(t) = \left\langle \left(v_0 \cos\theta\right)t + x_0 \ , -\frac{1}{2}gt^2 + \left(v_0 \sin\theta\right)t + y_0 \right\rangle.$$

Letting $\vec{v}_0 = v_0 \langle \cos\theta, \sin\theta \rangle$, $\vec{r}(t)$ can be written as

$$\vec{r}(t) = \left\langle 0, -\frac{1}{2}gt^2 \right\rangle + \vec{v}_0 t + \vec{r}_0.$$

We demonstrate how to use this position function in the next two examples.

Example 11.3.5   Projectile Motion

Sydney shoots her Red Ryder® bb gun across level ground from an elevation of 4ft, where the barrel of the gun makes a 5° angle with the horizontal. Find how far the bb travels before landing, assuming the bb is fired at the advertised rate of 350ft/s and ignoring air resistance.

SOLUTION   A direct application of Key Idea 11.3.2 gives

$$\vec{r}(t) = \left\langle (350 \cos 5°)t, -16t^2 + (350 \sin 5°)t + 4 \right\rangle$$
$$\approx \left\langle 346.67t, -16t^2 + 30.50t + 4 \right\rangle,$$

Notes:

where we set her initial position to be $\langle 0, 4 \rangle$. We need to find *when* the bb lands, then we can find *where*. We accomplish this by setting the *y*-component equal to 0 and solving for *t*:

$$-16t^2 + 30.50t + 4 = 0$$

$$t = \frac{-30.50 \pm \sqrt{30.50^2 - 4(-16)(4)}}{-32}$$

$$t \approx 2.03s.$$

(We discarded a negative solution that resulted from our quadratic equation.)

We have found that the bb lands 2.03s after firing; with $t = 2.03$, we find the *x*-component of our position function is $346.67(2.03) = 703.74$ft. The bb lands about 704 feet away.

Example 11.3.6   Projectile Motion

Alex holds his sister's bb gun at a height of 3ft and wants to shoot a target that is 6ft above the ground, 25ft away. At what angle should he hold the gun to hit his target? (We still assume the muzzle velocity is 350ft/s.)

SOLUTION   The position function for the path of Alex's bb is

$$\vec{r}(t) = \langle (350\cos\theta)t, -16t^2 + (350\sin\theta)t + 3 \rangle.$$

We need to find θ so that $\vec{r}(t) = \langle 25, 6 \rangle$ for some value of *t*. That is, we want to find θ and *t* such that

$$(350\cos\theta)t = 25 \quad \text{and} \quad -16t^2 + (350\sin\theta)t + 3 = 6.$$

This is not trivial (though not "hard"). We start by solving each equation for $\cos\theta$ and $\sin\theta$, respectively.

$$\cos\theta = \frac{25}{350t} \quad \text{and} \quad \sin\theta = \frac{3 + 16t^2}{350t}.$$

Using the Pythagorean Identity $\cos^2\theta + \sin^2\theta = 1$, we have

$$\left(\frac{25}{350t}\right)^2 + \left(\frac{3 + 16t^2}{350t}\right)^2 = 1$$

Multiply both sides by $(350t)^2$:

$$25^2 + (3 + 16t^2)^2 = 350^2 t^2$$

$$256t^4 - 122,404t^2 + 634 = 0.$$

Notes:

This is a quadratic *in* t^2. That is, we can apply the quadratic formula to find t^2, then solve for t itself.

$$t^2 = \frac{122,404 \pm \sqrt{122,404^2 - 4(256)(634)}}{512}$$

$$t^2 = 0.0052, \ 478.135$$

$$t = \pm 0.072, \ \pm 21.866$$

Clearly the negative t values do not fit our context, so we have $t = 0.072$ and $t = 21.866$. Using $\cos\theta = 25/(350t)$, we can solve for θ:

$$\theta = \cos^{-1}\left(\frac{25}{350 \cdot 0.072}\right) \quad \text{and} \quad \cos^{-1}\left(\frac{25}{350 \cdot 21.866}\right)$$

$$\theta = 7.03° \quad \text{and} \quad 89.8°.$$

Alex has two choices of angle. He can hold the rifle at an angle of about $7°$ with the horizontal and hit his target 0.07s after firing, or he can hold his rifle almost straight up, with an angle of $89.8°$, where he'll hit his target about 22s later. The first option is clearly the option he should choose.

Distance Traveled

Consider a driver who sets her cruise–control to 60mph, and travels at this speed for an hour. We can ask:

1. How far did the driver travel?

2. How far from her starting position is the driver?

The first is easy to answer: she traveled 60 miles. The second is impossible to answer with the given information. We do not know if she traveled in a straight line, on an oval racetrack, or along a slowly–winding highway.

This highlights an important fact: to compute distance traveled, we need only to know the speed, given by $\| \vec{v}(t) \|$.

Theorem 11.3.1   Distance Traveled

Let $\vec{v}(t)$ be a velocity function for a moving object. The distance traveled by the object on $[a, b]$ is:

$$\text{distance traveled} = \int_a^b \| \vec{v}(t) \| \, dt.$$

Note that this is just a restatement of Theorem 11.2.7: arc length is the same as distance traveled, just viewed in a different context.

Notes:

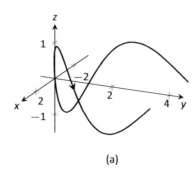

(a)

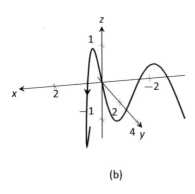

(b)

Figure 11.3.5: The path of the particle, from two perspectives, in Example 11.3.7.

Example 11.3.7   Distance Traveled, Displacement, and Average Speed
A particle moves in space with position function $\vec{r}(t) = \langle t, t^2, \sin(\pi t)\rangle$ on $[-2, 2]$, where t is measured in seconds and distances are in meters. Find:

1. The distance traveled by the particle on $[-2, 2]$.

2. The displacement of the particle on $[-2, 2]$.

3. The particle's average speed.

SOLUTION

1. We use Theorem 11.3.1 to establish the integral:

$$\text{distance traveled} = \int_{-2}^{2} \| \vec{v}(t) \| \, dt$$
$$= \int_{-2}^{2} \sqrt{1 + (2t)^2 + \pi^2 \cos^2(\pi t)} \, dt.$$

   This cannot be solved in terms of elementary functions so we turn to numerical integration, finding the distance to be 12.88m.

2. The displacement is the vector

$$\vec{r}(2) - \vec{r}(-2) = \langle 2, 4, 0\rangle - \langle -2, 4, 0\rangle = \langle 4, 0, 0\rangle.$$

   That is, the particle ends with an x-value increased by 4 and with y- and z-values the same (see Figure 11.3.5).

3. We found above that the particle traveled 12.88m over 4 seconds. We can compute average speed by dividing: 12.88/4 = 3.22m/s.

   We should also consider Definition 5.4.1 of Section 5.4, which says that the average value of a function f on $[a, b]$ is $\frac{1}{b-a}\int_a^b f(x)\,dx$. In our context, the average value of the speed is

$$\text{average speed} = \frac{1}{2 - (-2)} \int_{-2}^{2} \| \vec{v}(t) \| \, dt \approx \frac{1}{4} 12.88 = 3.22\text{m/s}.$$

   Note how the physical context of a particle traveling gives meaning to a more abstract concept learned earlier.

In Definition 5.4.1 of Chapter 5 we defined the average value of a function $f(x)$ on $[a, b]$ to be

$$\frac{1}{b - a} \int_a^b f(x) \, dx.$$

Notes:

Note how in Example 11.3.7 we computed the average speed as

$$\frac{\text{distance traveled}}{\text{travel time}} = \frac{1}{2-(-2)} \int_{-2}^{2} \| \vec{v}(t) \| \, dt;$$

that is, we just found the average value of $\| \vec{v}(t) \|$ on $[-2, 2]$.

Likewise, given position function $\vec{r}(t)$, the average velocity on $[a, b]$ is

$$\frac{\text{displacement}}{\text{travel time}} = \frac{1}{b-a} \int_{a}^{b} \vec{r}'(t) \, dt = \frac{\vec{r}(b) - \vec{r}(a)}{b-a};$$

that is, it is the average value of $\vec{r}'(t)$, or $\vec{v}(t)$, on $[a, b]$.

Key Idea 11.3.3   Average Speed, Average Velocity

Let $\vec{r}(t)$ be a differentiable position function on $[a, b]$.

The **average speed** is:

$$\frac{\text{distance traveled}}{\text{travel time}} = \frac{\int_{a}^{b} \| \vec{v}(t) \| \, dt}{b-a} = \frac{1}{b-a} \int_{a}^{b} \| \vec{v}(t) \| \, dt.$$

The **average velocity** is:

$$\frac{\text{displacement}}{\text{travel time}} = \frac{\int_{a}^{b} \vec{r}'(t) \, dt}{b-a} = \frac{1}{b-a} \int_{a}^{b} \vec{r}'(t) \, dt.$$

The next two sections investigate more properties of the graphs of vector–valued functions and we'll apply these new ideas to what we just learned about motion.

Notes:

Exercises 11.3

Terms and Concepts

1. How is *velocity* different from *speed*?

2. What is the difference between *displacement* and *distance traveled*?

3. What is the difference between *average velocity* and *average speed*?

4. *Distance traveled* is the same as _____ _____, just viewed in a different context.

5. Describe a scenario where an object's average speed is a large number, but the magnitude of the average velocity is not a large number.

6. Explain why it is not possible to have an average velocity with a large magnitude but a small average speed.

Problems

In Exercises 7 – 10 , a position function $\vec{r}(t)$ is given. Find $\vec{v}(t)$ and $\vec{a}(t)$.

7. $\vec{r}(t) = \langle 2t + 1, 5t - 2, 7 \rangle$

8. $\vec{r}(t) = \langle 3t^2 - 2t + 1, -t^2 + t + 14 \rangle$

9. $\vec{r}(t) = \langle \cos t, \sin t \rangle$

10. $\vec{r}(t) = \langle t/10, -\cos t, \sin t \rangle$

In Exercises 11 – 14 , a position function $\vec{r}(t)$ is given. Sketch $\vec{r}(t)$ on the indicated interval. Find $\vec{v}(t)$ and $\vec{a}(t)$, then add $\vec{v}(t_0)$ and $\vec{a}(t_0)$ to your sketch, with their initial points at $\vec{r}(t_0)$, for the given value of t_0.

11. $\vec{r}(t) = \langle t, \sin t \rangle$ on $[0, \pi/2]$; $t_0 = \pi/4$

12. $\vec{r}(t) = \langle t^2, \sin t^2 \rangle$ on $[0, \pi/2]$; $t_0 = \sqrt{\pi/4}$

13. $\vec{r}(t) = \langle t^2 + t, -t^2 + 2t \rangle$ on $[-2, 2]$; $t_0 = 1$

14. $\vec{r}(t) = \left\langle \dfrac{2t + 3}{t^2 + 1}, t^2 \right\rangle$ on $[-1, 1]$; $t_0 = 0$

In Exercises 15 – 24 , a position function $\vec{r}(t)$ of an object is given. Find the speed of the object in terms of t, and find where the speed is minimized/maximized on the indicated interval.

15. $\vec{r}(t) = \langle t^2, t \rangle$ on $[-1, 1]$

16. $\vec{r}(t) = \langle t^2, t^2 - t^3 \rangle$ on $[-1, 1]$

17. $\vec{r}(t) = \langle 5 \cos t, 5 \sin t \rangle$ on $[0, 2\pi]$

18. $\vec{r}(t) = \langle 2 \cos t, 5 \sin t \rangle$ on $[0, 2\pi]$

19. $\vec{r}(t) = \langle \sec t, \tan t \rangle$ on $[0, \pi/4]$

20. $\vec{r}(t) = \langle t + \cos t, 1 - \sin t \rangle$ on $[0, 2\pi]$

21. $\vec{r}(t) = \langle 12t, 5 \cos t, 5 \sin t \rangle$ on $[0, 4\pi]$

22. $\vec{r}(t) = \langle t^2 - t, t^2 + t, t \rangle$ on $[0, 1]$

23. $\vec{r}(t) = \left\langle t, t^2, \sqrt{1 - t^2} \right\rangle$ on $[-1, 1]$

24. **Projectile Motion:** $\vec{r}(t) = \left\langle (v_0 \cos \theta)t, -\dfrac{1}{2}gt^2 + (v_0 \sin \theta)t \right\rangle$ on $\left[0, \dfrac{2v_0 \sin \theta}{g} \right]$

In Exercises 25 – 28 , position functions $\vec{r}_1(t)$ and $\vec{r}_2(s)$ for two objects are given that follow the same path on the respective intervals.

    (a) Show that the positions are the same at the indicated t_0 and s_0 values; i.e., show $\vec{r}_1(t_0) = \vec{r}_2(s_0)$.

    (b) Find the velocity, speed and acceleration of the two objects at t_0 and s_0, respectively.

25. $\vec{r}_1(t) = \langle t, t^2 \rangle$ on $[0, 1]$; $t_0 = 1$
    $\vec{r}_2(s) = \langle s^2, s^4 \rangle$ on $[0, 1]$; $s_0 = 1$

26. $\vec{r}_1(t) = \langle 3 \cos t, 3 \sin t \rangle$ on $[0, 2\pi]$; $t_0 = \pi/2$
    $\vec{r}_2(s) = \langle 3 \cos(4s), 3 \sin(4s) \rangle$ on $[0, \pi/2]$; $s_0 = \pi/8$

27. $\vec{r}_1(t) = \langle 3t, 2t \rangle$ on $[0, 2]$; $t_0 = 2$
    $\vec{r}_2(s) = \langle 6s - 6, 4s - 4 \rangle$ on $[1, 2]$; $s_0 = 2$

28. $\vec{r}_1(t) = \langle t, \sqrt{t} \rangle$ on $[0, 1]$; $t_0 = 1$
    $\vec{r}_2(s) = \langle \sin t, \sqrt{\sin t} \rangle$ on $[0, \pi/2]$; $s_0 = \pi/2$

In Exercises 29 – 32 , find the position function of an object given its acceleration and initial velocity and position.

29. $\vec{a}(t) = \langle 2, 3 \rangle$; $\vec{v}(0) = \langle 1, 2 \rangle$, $\vec{r}(0) = \langle 5, -2 \rangle$

30. $\vec{a}(t) = \langle 2, 3 \rangle$; $\vec{v}(1) = \langle 1, 2 \rangle$, $\vec{r}(1) = \langle 5, -2 \rangle$

31. $\vec{a}(t) = \langle \cos t, -\sin t \rangle$; $\vec{v}(0) = \langle 0, 1 \rangle$, $\vec{r}(0) = \langle 0, 0 \rangle$

32. $\vec{a}(t) = \langle 0, -32 \rangle$; $\vec{v}(0) = \langle 10, 50 \rangle$, $\vec{r}(0) = \langle 0, 0 \rangle$

In Exercises 33 – 36 , find the displacement, distance traveled, average velocity and average speed of the described object on the given interval.

33. An object with position function $\vec{r}(t) = \langle 2 \cos t, 2 \sin t, 3t \rangle$, where distances are measured in feet and time is in seconds, on $[0, 2\pi]$.

34. An object with position function $\vec{r}(t) = \langle 5\cos t, -5\sin t\rangle$, where distances are measured in feet and time is in seconds, on $[0, \pi]$.

35. An object with velocity function $\vec{v}(t) = \langle \cos t, \sin t\rangle$, where distances are measured in feet and time is in seconds, on $[0, 2\pi]$.

36. An object with velocity function $\vec{v}(t) = \langle 1, 2, -1\rangle$, where distances are measured in feet and time is in seconds, on $[0, 10]$.

Exercises 37 – 42 ask you to solve a variety of problems based on the principles of projectile motion.

37. A boy whirls a ball, attached to a 3ft string, above his head in a counter–clockwise circle. The ball makes 2 revolutions per second.

    At what t-values should the boy release the string so that the ball heads directly for a tree standing 10ft in front of him?

38. David faces Goliath with only a stone in a 3ft sling, which he whirls above his head at 4 revolutions per second. They stand 20ft apart.

    (a) At what t-values must David release the stone in his sling in order to hit Goliath?

    (b) What is the speed at which the stone is traveling when released?

    (c) Assume David releases the stone from a height of 6ft and Goliath's forehead is 9ft above the ground. What angle of elevation must David apply to the stone to hit Goliath's head?

39. A hunter aims at a deer which is 40 yards away. Her crossbow is at a height of 5ft, and she aims for a spot on the deer 4ft above the ground. The crossbow fires her arrows at 300ft/s.

    (a) At what angle of elevation should she hold the crossbow to hit her target?

    (b) If the deer is moving perpendicularly to her line of sight at a rate of 20mph, by approximately how much should she lead the deer in order to hit it in the desired location?

40. A baseball player hits a ball at 100mph, with an initial height of 3ft and an angle of elevation of $20°$, at Boston's Fenway Park. The ball flies towards the famed "Green Monster," a wall 37ft high located 310ft from home plate.

    (a) Show that as hit, the ball hits the wall.

    (b) Show that if the angle of elevation is $21°$, the ball clears the Green Monster.

41. A Cessna flies at 1000ft at 150mph and drops a box of supplies to the professor (and his wife) on an island. Ignoring wind resistance, how far horizontally will the supplies travel before they land?

42. A football quarterback throws a pass from a height of 6ft, intending to hit his receiver 20yds away at a height of 5ft.

    (a) If the ball is thrown at a rate of 50mph, what angle of elevation is needed to hit his intended target?

    (b) If the ball is thrown at with an angle of elevation of $8°$, what initial ball speed is needed to hit his target?

11.4   Unit Tangent and Normal Vectors

Unit Tangent Vector

Given a smooth vector–valued function $\vec{r}(t)$, we defined in Definition 11.2.4 that any vector parallel to $\vec{r}\,'(t_0)$ is *tangent* to the graph of $\vec{r}(t)$ at $t = t_0$. It is often useful to consider just the *direction* of $\vec{r}\,'(t)$ and not its magnitude. Therefore we are interested in the unit vector in the direction of $\vec{r}\,'(t)$. This leads to a definition.

Definition 11.4.1   Unit Tangent Vector

Let $\vec{r}(t)$ be a smooth function on an open interval I. The unit tangent vector $\vec{T}(t)$ is

$$\vec{T}(t) = \frac{1}{||\,\vec{r}\,'(t)\,||}\vec{r}\,'(t).$$

Example 11.4.1   Computing the unit tangent vector
Let $\vec{r}(t) = \langle 3\cos t, 3\sin t, 4t\rangle$. Find $\vec{T}(t)$ and compute $\vec{T}(0)$ and $\vec{T}(1)$.

SOLUTION   We apply Definition 11.4.1 to find $\vec{T}(t)$.

$$\vec{T}(t) = \frac{1}{||\,\vec{r}\,'(t)\,||}\vec{r}\,'(t)$$

$$= \frac{1}{\sqrt{\left(-3\sin t\right)^2 + \left(3\cos t\right)^2 + 4^2}}\langle -3\sin t, 3\cos t, 4\rangle$$

$$= \left\langle -\frac{3}{5}\sin t, \frac{3}{5}\cos t, \frac{4}{5}\right\rangle.$$

We can now easily compute $\vec{T}(0)$ and $\vec{T}(1)$:

$$\vec{T}(0) = \left\langle 0, \frac{3}{5}, \frac{4}{5}\right\rangle ; \quad \vec{T}(1) = \left\langle -\frac{3}{5}\sin 1, \frac{3}{5}\cos 1, \frac{4}{5}\right\rangle \approx \langle -0.505, 0.324, 0.8\rangle .$$

These are plotted in Figure 11.4.1 with their initial points at $\vec{r}(0)$ and $\vec{r}(1)$, respectively. (They look rather "short" since they are only length 1.)

The unit tangent vector $\vec{T}(t)$ always has a magnitude of 1, though it is sometimes easy to doubt that is true. We can help solidify this thought in our minds by computing $||\,\vec{T}(1)\,||$:

$$||\,\vec{T}(1)\,|| \approx \sqrt{(-0.505)^2 + 0.324^2 + 0.8^2} = 1.000001.$$

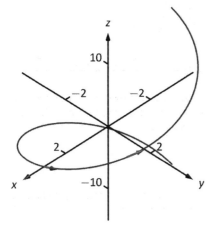

Figure 11.4.1: Plotting unit tangent vectors in Example 11.4.1.

Notes:

We have rounded in our computation of $\vec{T}(1)$, so we don't get 1 exactly. We leave it to the reader to use the exact representation of $\vec{T}(1)$ to verify it has length 1.

In many ways, the previous example was "too nice." It turned out that $\vec{r}\,'(t)$ was always of length 5. In the next example the length of $\vec{r}\,'(t)$ is variable, leaving us with a formula that is not as clean.

Example 11.4.2   **Computing the unit tangent vector**
Let $\vec{r}(t) = \left\langle t^2 - t, t^2 + t \right\rangle$. Find $\vec{T}(t)$ and compute $\vec{T}(0)$ and $\vec{T}(1)$.

   SOLUTION   We find $\vec{r}\,'(t) = \langle 2t - 1, 2t + 1 \rangle$, and

$$\| \vec{r}\,'(t) \| = \sqrt{(2t-1)^2 + (2t+1)^2} = \sqrt{8t^2 + 2}.$$

Therefore

$$\vec{T}(t) = \frac{1}{\sqrt{8t^2+2}} \langle 2t-1, 2t+1 \rangle = \left\langle \frac{2t-1}{\sqrt{8t^2+2}}, \frac{2t+1}{\sqrt{8t^2+2}} \right\rangle.$$

When $t = 0$, we have $\vec{T}(0) = \langle -1/\sqrt{2}, 1/\sqrt{2} \rangle$; when $t = 1$, we have $\vec{T}(1) = \langle 1/\sqrt{10}, 3/\sqrt{10} \rangle$. We leave it to the reader to verify each of these is a unit vector. They are plotted in Figure 11.4.2

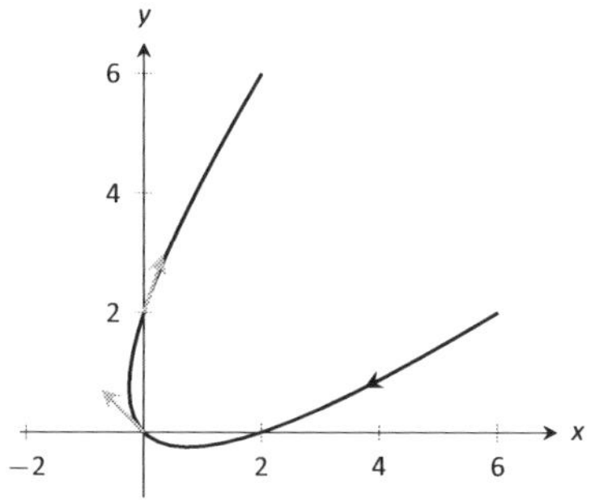

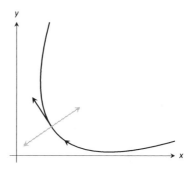

Figure 11.4.2: Plotting unit tangent vectors in Example 11.4.2.

Unit Normal Vector

Just as knowing the direction tangent to a path is important, knowing a direction orthogonal to a path is important. When dealing with real-valued functions, we defined the normal line at a point to the be the line through the point that was perpendicular to the tangent line at that point. We can do a similar thing with vector–valued functions. Given $\vec{r}(t)$ in $\mathbb{R}^2$, we have 2 directions perpendicular to the tangent vector, as shown in Figure 11.4.3. It is good to wonder "Is one of these two directions preferable over the other?"

Given $\vec{r}(t)$ in $\mathbb{R}^3$, there are infinitely many vectors orthogonal to the tangent vector at a given point. Again, we might wonder "Is one of these infinite choices preferable over the others? Is one of these the 'right' choice?"

The answer in both $\mathbb{R}^2$ and $\mathbb{R}^3$ is "Yes, there is one vector that is not only preferable, it is the 'right' one to choose." Recall Theorem 11.2.5, which states that if $\vec{r}(t)$ has constant length, then $\vec{r}(t)$ is orthogonal to $\vec{r}\,'(t)$ for all t. We know $\vec{T}(t)$, the unit tangent vector, has constant length. Therefore $\vec{T}(t)$ is orthogonal to $\vec{T}\,'(t)$.

We'll see that $\vec{T}\,'(t)$ is more than just a convenient choice of vector that is orthogonal to $\vec{r}\,'(t)$; rather, it is the "right" choice. Since all we care about is the direction, we define this newly found vector to be a unit vector.

Figure 11.4.3: Given a direction in the plane, there are always two directions orthogonal to it.

Note: $\vec{T}(t)$ is a unit vector, by definition. This *does not* imply that $\vec{T}\,'(t)$ is also a unit vector.

Notes:

> **Definition 11.4.2   Unit Normal Vector**
>
> Let $\vec{r}(t)$ be a vector–valued function where the unit tangent vector, $\vec{T}(t)$, is smooth on an open interval I. The **unit normal vector** $\vec{N}(t)$ is
>
> $$\vec{N}(t) = \frac{1}{\|\vec{T}'(t)\|}\vec{T}'(t).$$

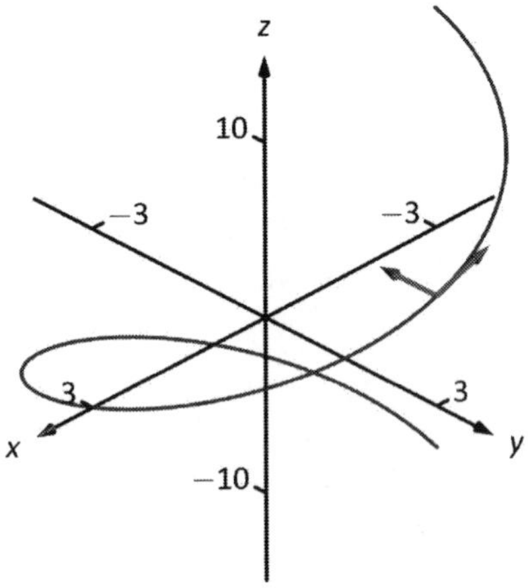

Figure 11.4.4: Plotting unit tangent and normal vectors in Example 11.4.4.

Example 11.4.3   Computing the unit normal vector

Let $\vec{r}(t) = \langle 3\cos t, 3\sin t, 4t\rangle$ as in Example 11.4.1. Sketch both $\vec{T}(\pi/2)$ and $\vec{N}(\pi/2)$ with initial points at $\vec{r}(\pi/2)$.

SOLUTION   In Example 11.4.1, we found $\vec{T}(t) = \left\langle (-3/5)\sin t, (3/5)\cos t, 4/5\right\rangle$. Therefore

$$\vec{T}'(t) = \left\langle -\frac{3}{5}\cos t, -\frac{3}{5}\sin t, 0\right\rangle \quad \text{and} \quad \|\vec{T}'(t)\| = \frac{3}{5}.$$

Thus

$$\vec{N}(t) = \frac{\vec{T}'(t)}{3/5} = \langle -\cos t, -\sin t, 0\rangle.$$

We compute $\vec{T}(\pi/2) = \langle -3/5, 0, 4/5\rangle$ and $\vec{N}(\pi/2) = \langle 0, -1, 0\rangle$. These are sketched in Figure 11.4.4.

The previous example was once again "too nice." In general, the expression for $\vec{T}(t)$ contains fractions of square–roots, hence the expression of $\vec{T}'(t)$ is very messy. We demonstrate this in the next example.

Example 11.4.4   Computing the unit normal vector

Let $\vec{r}(t) = \langle t^2 - t, t^2 + t\rangle$ as in Example 11.4.2. Find $\vec{N}(t)$ and sketch $\vec{r}(t)$ with the unit tangent and normal vectors at $t = -1, 0$ and 1.

SOLUTION   In Example 11.4.2, we found

$$\vec{T}(t) = \left\langle \frac{2t-1}{\sqrt{8t^2+2}}, \frac{2t+1}{\sqrt{8t^2+2}}\right\rangle.$$

Finding $\vec{T}'(t)$ requires two applications of the Quotient Rule:

Notes:

$$T'(t) = \left\langle \frac{\sqrt{8t^2+2}(2) - (2t-1)\left(\frac{1}{2}(8t^2+2)^{-1/2}(16t)\right)}{8t^2+2}, \right.$$

$$\left. \frac{\sqrt{8t^2+2}(2) - (2t+1)\left(\frac{1}{2}(8t^2+2)^{-1/2}(16t)\right)}{8t^2+2} \right\rangle$$

$$= \left\langle \frac{4(2t+1)}{(8t^2+2)^{3/2}}, \frac{4(1-2t)}{(8t^2+2)^{3/2}} \right\rangle$$

This is not a unit vector; to find $\vec{N}(t)$, we need to divide $\vec{T}'(t)$ by it's magnitude.

$$\|\vec{T}'(t)\| = \sqrt{\frac{16(2t+1)^2}{(8t^2+2)^3} + \frac{16(1-2t)^2}{(8t^2+2)^3}}$$

$$= \sqrt{\frac{16(8t^2+2)}{(8t^2+2)^3}}$$

$$= \frac{4}{8t^2+2}.$$

Finally,

$$\vec{N}(t) = \frac{1}{4/(8t^2+2)} \left\langle \frac{4(2t+1)}{(8t^2+2)^{3/2}}, \frac{4(1-2t)}{(8t^2+2)^{3/2}} \right\rangle$$

$$= \left\langle \frac{2t+1}{\sqrt{8t^2+2}}, -\frac{2t-1}{\sqrt{8t^2+2}} \right\rangle.$$

Using this formula for $\vec{N}(t)$, we compute the unit tangent and normal vectors for $t = -1, 0$ and 1 and sketch them in Figure 11.4.5.

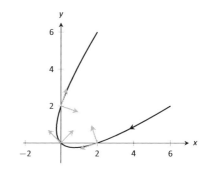

Figure 11.4.5: Plotting unit tangent and normal vectors in Example 11.4.4.

The final result for $\vec{N}(t)$ in Example 11.4.4 is suspiciously similar to $\vec{T}(t)$. There is a clear reason for this. If $\vec{u} = \langle u_1, u_2 \rangle$ is a unit vector in $\mathbb{R}^2$, then the *only* unit vectors orthogonal to $\vec{u}$ are $\langle -u_2, u_1 \rangle$ and $\langle u_2, -u_1 \rangle$. Given $\vec{T}(t)$, we can quickly determine $\vec{N}(t)$ if we know which term to multiply by (-1).

Consider again Figure 11.4.5, where we have plotted some unit tangent and normal vectors. Note how $\vec{N}(t)$ always points "inside" the curve, or to the concave side of the curve. This is not a coincidence; this is true in general. Knowing the direction that $\vec{r}(t)$ "turns" allows us to quickly find $\vec{N}(t)$.

Notes:

> **Theorem 11.4.1   Unit Normal Vectors in $\mathbb{R}^2$**
>
> Let $\vec{r}(t)$ be a vector–valued function in $\mathbb{R}^2$ where $\vec{T}'(t)$ is smooth on an open interval I. Let t_0 be in I and $\vec{T}(t_0) = \langle t_1, t_2 \rangle$ Then $\vec{N}(t_0)$ is either
>
> $$\vec{N}(t_0) = \langle -t_2, t_1 \rangle \quad \text{or} \quad \vec{N}(t_0) = \langle t_2, -t_1 \rangle,$$
>
> whichever is the vector that points to the concave side of the graph of $\vec{r}$.

Application to Acceleration

Let $\vec{r}(t)$ be a position function. It is a fact (stated later in Theorem 11.4.2) that acceleration, $\vec{a}(t)$, lies in the plane defined by $\vec{T}$ and $\vec{N}$. That is, there are scalar functions $a_T(t)$ and $a_N(t)$ such that

$$\vec{a}(t) = a_T(t)\vec{T}(t) + a_N(t)\vec{N}(t).$$

We generally drop the "of t" part of the notation and just write a_T and a_N.

The scalar a_T measures "how much" acceleration is in the direction of travel, that is, it measures the component of acceleration that affects the speed. The scalar a_N measures "how much" acceleration is perpendicular to the direction of travel, that is, it measures the component of acceleration that affects the direction of travel.

We can find a_T using the orthogonal projection of $\vec{a}(t)$ onto $\vec{T}(t)$ (review Definition 10.3.3 in Section 10.3 if needed). Recalling that since $\vec{T}(t)$ is a unit vector, $\vec{T}(t) \cdot \vec{T}(t) = 1$, so we have

Note: Keep in mind that both a_T and a_N are functions of t; that is, the scalar changes depending on t. It is convention to drop the "(t)" notation from $a_T(t)$ and simply write a_T.

$$\text{proj}_{\vec{T}(t)}\, \vec{a}(t) = \frac{\vec{a}(t) \cdot \vec{T}(t)}{\vec{T}(t) \cdot \vec{T}(t)}\vec{T}(t) = \underbrace{\left(\vec{a}(t) \cdot \vec{T}(t) \right)}_{a_T} \vec{T}(t).$$

Thus the amount of $\vec{a}(t)$ in the direction of $\vec{T}(t)$ is $a_T = \vec{a}(t) \cdot \vec{T}(t)$. The same logic gives $a_N = \vec{a}(t) \cdot \vec{N}(t)$.

While this is a fine way of computing a_T, there are simpler ways of finding a_N (as finding $\vec{N}$ itself can be complicated). The following theorem gives alternate formulas for a_T and a_N.

Notes:

Theorem 11.4.2 Acceleration in the Plane Defined by $\vec{T}$ and $\vec{N}$

Let $\vec{r}(t)$ be a position function with acceleration $\vec{a}(t)$ and unit tangent and normal vectors $\vec{T}(t)$ and $\vec{N}(t)$. Then $\vec{a}(t)$ lies in the plane defined by $\vec{T}(t)$ and $\vec{N}(t)$; that is, there exists scalars a_T and a_N such that

$$\vec{a}(t) = a_T\vec{T}(t) + a_N\vec{N}(t).$$

Moreover,

$$a_T = \vec{a}(t) \cdot \vec{T}(t) = \frac{d}{dt}\left(\|\vec{v}(t)\|\right)$$

$$a_N = \vec{a}(t) \cdot \vec{N}(t) = \sqrt{\|\vec{a}(t)\|^2 - a_T^2} = \frac{\|\vec{a}(t) \times \vec{v}(t)\|}{\|\vec{v}(t)\|} = \|\vec{v}(t)\|\,\|\vec{T}'(t)\|$$

Note the second formula for a_T: $\frac{d}{dt}\left(\|\vec{v}(t)\|\right)$. This measures the rate of change of speed, which again is the amount of acceleration in the direction of travel.

Example 11.4.5 Computing a_T and a_N
Let $\vec{r}(t) = \langle 3\cos t, 3\sin t, 4t\rangle$ as in Examples 11.4.1 and 11.4.3. Find a_T and a_N.

SOLUTION The previous examples give $\vec{a}(t) = \langle -3\cos t, -3\sin t, 0\rangle$ and

$$\vec{T}(t) = \left\langle -\frac{3}{5}\sin t, \frac{3}{5}\cos t, \frac{4}{5}\right\rangle \quad \text{and} \quad \vec{N}(t) = \langle -\cos t, -\sin t, 0\rangle.$$

We can find a_T and a_N directly with dot products:

$$a_T = \vec{a}(t) \cdot \vec{T}(t) = \frac{9}{5}\cos t \sin t - \frac{9}{5}\cos t \sin t + 0 = 0.$$

$$a_N = \vec{a}(t) \cdot \vec{N}(t) = 3\cos^2 t + 3\sin^2 t + 0 = 3.$$

Thus $\vec{a}(t) = 0\vec{T}(t) + 3\vec{N}(t) = 3\vec{N}(t)$, which is clearly the case.

What is the practical interpretation of these numbers? $a_T = 0$ means the object is moving at a constant speed, and hence all acceleration comes in the form of direction change.

Example 11.4.6 Computing a_T and a_N
Let $\vec{r}(t) = \langle t^2 - t, t^2 + t\rangle$ as in Examples 11.4.2 and 11.4.4. Find a_T and a_N.

Notes:

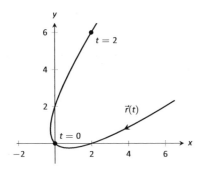

Figure 11.4.6: Graphing $\vec{r}(t)$ in Example 11.4.6.

SOLUTION The previous examples give $\vec{a}(t) = \langle 2, 2 \rangle$ and

$$\vec{T}(t) = \left\langle \frac{2t-1}{\sqrt{8t^2+2}}, \frac{2t+1}{\sqrt{8t^2+2}} \right\rangle \quad \text{and} \quad \vec{N}(t) = \left\langle \frac{2t+1}{\sqrt{8t^2+2}}, -\frac{2t-1}{\sqrt{8t^2+2}} \right\rangle.$$

While we can compute a_N using $\vec{N}(t)$, we instead demonstrate using another formula from Theorem 11.4.2.

$$a_T = \vec{a}(t) \cdot \vec{T}(t) = \frac{4t-2}{\sqrt{8t^2+2}} + \frac{4t+2}{\sqrt{8t^2+2}} = \frac{8t}{\sqrt{8t^2+2}}.$$

$$a_N = \sqrt{\|\vec{a}(t)\|^2 - a_T^2} = \sqrt{8 - \left(\frac{8t}{\sqrt{8t^2+2}}\right)^2} = \frac{4}{\sqrt{8t^2+2}}.$$

When $t = 2$, $a_T = \dfrac{16}{\sqrt{34}} \approx 2.74$ and $a_N = \dfrac{4}{\sqrt{34}} \approx 0.69$. We interpret this to mean that at $t = 2$, the particle is accelerating mostly by increasing speed, not by changing direction. As the path near $t = 2$ is relatively straight, this should make intuitive sense. Figure 11.4.6 gives a graph of the path for reference.

Contrast this with $t = 0$, where $a_T = 0$ and $a_N = 4/\sqrt{2} \approx 2.82$. Here the particle's speed is not changing and all acceleration is in the form of direction change.

Example 11.4.7 Analyzing projectile motion

A ball is thrown from a height of 240ft with an initial speed of 64ft/s and an angle of elevation of $30°$. Find the position function $\vec{r}(t)$ of the ball and analyze a_T and a_N.

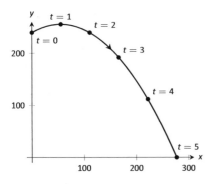

Figure 11.4.7: Plotting the position of a thrown ball, with 1s increments shown.

SOLUTION Using Key Idea 11.3.2 of Section 11.3 we form the position function of the ball:

$$\vec{r}(t) = \left\langle \left(64\cos 30°\right)t, -16t^2 + \left(64\sin 30°\right)t + 240 \right\rangle,$$

which we plot in Figure 11.4.7.

From this we find $\vec{v}(t) = \langle 64\cos 30°, -32t + 64\sin 30° \rangle$ and $\vec{a}(t) = \langle 0, -32 \rangle$. Computing $\vec{T}(t)$ is not difficult, and with some simplification we find

$$\vec{T}(t) = \left\langle \frac{\sqrt{3}}{\sqrt{t^2 - 2t + 4}}, \frac{1-t}{\sqrt{t^2 - 2t + 4}} \right\rangle.$$

With $\vec{a}(t)$ as simple as it is, finding a_T is also simple:

$$a_T = \vec{a}(t) \cdot \vec{T}(t) = \frac{32t - 32}{\sqrt{t^2 - 2t + 4}}.$$

Notes:

We choose to not find $\vec{N}(t)$ and find a_N through the formula $a_N = \sqrt{||\, \vec{a}(t)\,||^2 - a_T^2}$:

$$a_N = \sqrt{32^2 - \left(\frac{32t - 32}{\sqrt{t^2 - 2t + 4}}\right)^2} = \frac{32\sqrt{3}}{\sqrt{t^2 - 2t + 4}}.$$

Figure 11.4.8 gives a table of values of a_T and a_N. When $t = 0$, we see the ball's speed is decreasing; when $t = 1$ the speed of the ball is unchanged. This corresponds to the fact that at $t = 1$ the ball reaches its highest point.

After $t = 1$ we see that a_N is decreasing in value. This is because as the ball falls, it's path becomes straighter and most of the acceleration is in the form of speeding up the ball, and not in changing its direction.

t	a_T	a_N
0	−16	27.7
1	0	32
2	16	27.7
3	24.2	20.9
4	27.7	16
5	29.4	12.7

Figure 11.4.8: A table of values of a_T and a_N in Example 11.4.7.

Our understanding of the unit tangent and normal vectors is aiding our understanding of motion. The work in Example 11.4.7 gave quantitative analysis of what we intuitively knew.

The next section provides two more important steps towards this analysis. We currently describe position only in terms of time. In everyday life, though, we often describe position in terms of distance ("The gas station is about 2 miles ahead, on the left."). The *arc length parameter* allows us to reference position in terms of distance traveled.

We also intuitively know that some paths are straighter than others – and some are curvier than others, but we lack a measurement of "curviness." The arc length parameter provides a way for us to compute *curvature*, a quantitative measurement of how curvy a curve is.

Notes:

Exercises 11.4

Terms and Concepts

1. If $\vec{T}(t)$ is a unit tangent vector, what is $\|\vec{T}(t)\|$?

2. If $\vec{N}(t)$ is a unit normal vector, what is $\vec{N}(t) \cdot \vec{r}\,'(t)$?

3. The acceleration vector $\vec{a}(t)$ lies in the plane defined by what two vectors?

4. a_T measures how much the acceleration is affecting the _____ of an object.

Problems

In Exercises 5 – 8 , given $\vec{r}(t)$, find $\vec{T}(t)$ and evaluate it at the indicated value of t.

5. $\vec{r}(t) = \langle 2t^2, t^2 - t \rangle, \quad t = 1$

6. $\vec{r}(t) = \langle t, \cos t \rangle, \quad t = \pi/4$

7. $\vec{r}(t) = \langle \cos^3 t, \sin^3 t \rangle, \quad t = \pi/4$

8. $\vec{r}(t) = \langle \cos t, \sin t \rangle, \quad t = \pi$

In Exercises 9 – 12 , find the equation of the line tangent to the curve at the indicated t-value using the unit tangent vector. Note: these are the same problems as in Exercises 5 – 8.

9. $\vec{r}(t) = \langle 2t^2, t^2 - t \rangle, \quad t = 1$

10. $\vec{r}(t) = \langle t, \cos t \rangle, \quad t = \pi/4$

11. $\vec{r}(t) = \langle \cos^3 t, \sin^3 t \rangle, \quad t = \pi/4$

12. $\vec{r}(t) = \langle \cos t, \sin t \rangle, \quad t = \pi$

In Exercises 13 – 16 , find $\vec{N}(t)$ using Definition 11.4.2. Confirm the result using Theorem 11.4.1.

13. $\vec{r}(t) = \langle 3\cos t, 3\sin t \rangle$

14. $\vec{r}(t) = \langle t, t^2 \rangle$

15. $\vec{r}(t) = \langle \cos t, 2\sin t \rangle$

16. $\vec{r}(t) = \langle e^t, e^{-t} \rangle$

In Exercises 17 – 20 , a position function $\vec{r}(t)$ is given along with its unit tangent vector $\vec{T}(t)$ evaluated at $t = a$, for some value of a.

(a) Confirm that $\vec{T}(a)$ is as stated.

(b) Using a graph of $\vec{r}(t)$ and Theorem 11.4.1, find $\vec{N}(a)$.

17. $\vec{r}(t) = \langle 3\cos t, 5\sin t \rangle; \quad \vec{T}(\pi/4) = \left\langle -\dfrac{3}{\sqrt{34}}, \dfrac{5}{\sqrt{34}} \right\rangle$.

18. $\vec{r}(t) = \left\langle t, \dfrac{1}{t^2 + 1} \right\rangle; \quad \vec{T}(1) = \left\langle \dfrac{2}{\sqrt{5}}, -\dfrac{1}{\sqrt{5}} \right\rangle$.

19. $\vec{r}(t) = (1 + 2\sin t)\langle \cos t, \sin t \rangle; \quad \vec{T}(0) = \left\langle \dfrac{2}{\sqrt{5}}, \dfrac{1}{\sqrt{5}} \right\rangle$.

20. $\vec{r}(t) = \langle \cos^3 t, \sin^3 t \rangle; \quad \vec{T}(\pi/4) = \left\langle -\dfrac{1}{\sqrt{2}}, \dfrac{1}{\sqrt{2}} \right\rangle$.

In Exercises 21 – 24 , find $\vec{N}(t)$.

21. $\vec{r}(t) = \langle 4t, 2\sin t, 2\cos t \rangle$

22. $\vec{r}(t) = \langle 5\cos t, 3\sin t, 4\sin t \rangle$

23. $\vec{r}(t) = \langle a\cos t, a\sin t, bt \rangle; \quad a > 0$

24. $\vec{r}(t) = \langle \cos(at), \sin(at), t \rangle$

In Exercises 25 – 30 , find a_T and a_N given $\vec{r}(t)$. Sketch $\vec{r}(t)$ on the indicated interval, and comment on the relative sizes of a_T and a_N at the indicated t values.

25. $\vec{r}(t) = \langle t, t^2 \rangle$ on $[-1, 1]$; consider $t = 0$ and $t = 1$.

26. $\vec{r}(t) = \langle t, 1/t \rangle$ on $(0, 4]$; consider $t = 1$ and $t = 2$.

27. $\vec{r}(t) = \langle 2\cos t, 2\sin t \rangle$ on $[0, 2\pi]$; consider $t = 0$ and $t = \pi/2$.

28. $\vec{r}(t) = \langle \cos(t^2), \sin(t^2) \rangle$ on $(0, 2\pi]$; consider $t = \sqrt{\pi/2}$ and $t = \sqrt{\pi}$.

29. $\vec{r}(t) = \langle a\cos t, a\sin t, bt \rangle$ on $[0, 2\pi]$, where $a, b > 0$; consider $t = 0$ and $t = \pi/2$.

30. $\vec{r}(t) = \langle 5\cos t, 4\sin t, 3\sin t \rangle$ on $[0, 2\pi]$; consider $t = 0$ and $t = \pi/2$.

11.5   The Arc Length Parameter and Curvature

In normal conversation we describe position in terms of both *time* and *distance*. For instance, imagine driving to visit a friend. If she calls and asks where you are, you might answer "I am 20 minutes from your house," or you might say "I am 10 miles from your house." Both answers provide your friend with a general idea of where you are.

Currently, our vector–valued functions have defined points with a parameter t, which we often take to represent time. Consider Figure 11.5.1(a), where $\vec{r}(t) = \langle t^2 - t, t^2 + t \rangle$ is graphed and the points corresponding to $t = 0$, 1 and 2 are shown. Note how the arc length between $t = 0$ and $t = 1$ is smaller than the arc length between $t = 1$ and $t = 2$; if the parameter t is time and $\vec{r}$ is position, we can say that the particle traveled faster on $[1, 2]$ than on $[0, 1]$.

Now consider Figure 11.5.1(b), where the same graph is parametrized by a different variable s. Points corresponding to $s = 0$ through $s = 6$ are plotted. The arc length of the graph between each adjacent pair of points is 1. We can view this parameter s as distance; that is, the arc length of the graph from $s = 0$ to $s = 3$ is 3, the arc length from $s = 2$ to $s = 6$ is 4, etc. If one wants to find the point 2.5 units from an initial location (i.e., $s = 0$), one would compute $\vec{r}(2.5)$. This parameter s is very useful, and is called the **arc length parameter**.

How do we find the arc length parameter?

Start with any parametrization of $\vec{r}$. We can compute the arc length of the graph of $\vec{r}$ on the interval $[0, t]$ with

$$\text{arc length} = \int_0^t \| \vec{r}\,'(u) \| \, du.$$

We can turn this into a function: as t varies, we find the arc length s from 0 to t. This function is

$$s(t) = \int_0^t \| \vec{r}\,'(u) \| \, du. \tag{11.1}$$

This establishes a relationship between s and t. Knowing this relationship explicitly, we can rewrite $\vec{r}(t)$ as a function of s: $\vec{r}(s)$. We demonstrate this in an example.

Example 11.5.1   Finding the arc length parameter

Let $\vec{r}(t) = \langle 3t - 1, 4t + 2 \rangle$. Parametrize $\vec{r}$ with the arc length parameter s.

SOLUTION   Using Equation (11.1), we write

$$s(t) = \int_0^t \| \vec{r}\,'(u) \| \, du.$$

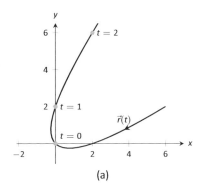

(a)

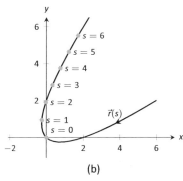

(b)

Figure 11.5.1: Introducing the arc length parameter.

Notes:

We can integrate this, explicitly finding a relationship between s and t:

$$s(t) = \int_0^t \| \vec{r}'(u) \| \, du$$
$$= \int_0^t \sqrt{3^2 + 4^2} \, du$$
$$= \int_0^t 5 \, du$$
$$= 5t.$$

Since $s = 5t$, we can write $t = s/5$ and replace t in $\vec{r}(t)$ with $s/5$:

$$\vec{r}(s) = \langle 3(s/5) - 1, 4(s/5) + 2 \rangle = \left\langle \frac{3}{5}s - 1, \frac{4}{5}s + 2 \right\rangle.$$

Clearly, as shown in Figure 11.5.2, the graph of $\vec{r}$ is a line, where $t = 0$ corresponds to the point $(-1, 2)$. What point on the line is 2 units away from this initial point? We find it with $\vec{r}(2) = \langle 1/5, 18/5 \rangle$.

Is the point $(1/5, 18/5)$ really 2 units away from $(-1, 2)$? We use the Distance Formula to check:

$$d = \sqrt{\left(\frac{1}{5} - (-1) \right)^2 + \left(\frac{18}{5} - 2 \right)^2} = \sqrt{\frac{36}{25} + \frac{64}{25}} = \sqrt{4} = 2.$$

Yes, $\vec{r}(2)$ is indeed 2 units away, in the direction of travel, from the initial point.

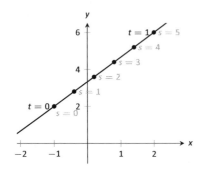

Figure 11.5.2: Graphing $\vec{r}$ in Example 11.5.1 with parameters t and s.

Things worked out very nicely in Example 11.5.1; we were able to establish directly that $s = 5t$. Usually, the arc length parameter is much more difficult to describe in terms of t, a result of integrating a square–root. There are a number of things that we can learn about the arc length parameter from Equation (11.1), though, that are incredibly useful.

First, take the derivative of s with respect to t. The Fundamental Theorem of Calculus (see Theorem 5.4.1) states that

$$\frac{ds}{dt} = s'(t) = \| \vec{r}'(t) \|. \tag{11.2}$$

Letting t represent time and $\vec{r}(t)$ represent position, we see that the rate of change of s with respect to t is speed; that is, the rate of change of "distance traveled" is speed, which should match our intuition.

The Chain Rule states that

$$\frac{d\vec{r}}{dt} = \frac{d\vec{r}}{ds} \cdot \frac{ds}{dt}$$
$$\vec{r}'(t) = \vec{r}'(s) \cdot \| \vec{r}'(t) \|.$$

Notes:

Solving for $\vec{r}''(s)$, we have

$$\vec{r}'(s) = \frac{\vec{r}'(t)}{\|\vec{r}'(t)\|} = \vec{T}(t), \tag{11.3}$$

where $\vec{T}(t)$ is the unit tangent vector. Equation 11.3 is often misinterpreted, as one is tempted to think it states $\vec{r}'(t) = \vec{T}(t)$, but there is a big difference between $\vec{r}'(s)$ and $\vec{r}'(t)$. The key to take from it is that $\vec{r}'(s)$ is a unit vector. In fact, the following theorem states that this characterizes the arc length parameter.

Theorem 11.5.1 Arc Length Parameter

Let $\vec{r}(s)$ be a vector–valued function. The parameter s is the arc length parameter if, and only if, $\|\vec{r}'(s)\| = 1$.

Curvature

Consider points A and B on the curve graphed in Figure 11.5.3(a). One can readily argue that the curve curves more sharply at A than at B. It is useful to use a number to describe how sharply the curve bends; that number is the **curvature** of the curve.

We derive this number in the following way. Consider Figure 11.5.3(b), where unit tangent vectors are graphed around points A and B. Notice how the direction of the unit tangent vector changes quite a bit near A, whereas it does not change as much around B. This leads to an important concept: measuring the rate of change of the unit tangent vector with respect to arc length gives us a measurement of curvature.

Definition 11.5.1 Curvature

Let $\vec{r}(s)$ be a vector–valued function where s is the arc length parameter. The curvature κ of the graph of $\vec{r}(s)$ is

$$\kappa = \left\| \frac{d\vec{T}}{ds} \right\| = \|\vec{T}'(s)\|.$$

If $\vec{r}(s)$ is parametrized by the arc length parameter, then

$$\vec{T}(s) = \frac{\vec{r}'(s)}{\|\vec{r}'(s)\|} \quad \text{and} \quad \vec{N}(s) = \frac{\vec{T}'(s)}{\|\vec{T}'(s)\|}.$$

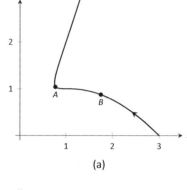

(a)

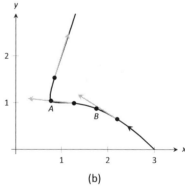

(b)

Figure 11.5.3: Establishing the concept of curvature.

Notes:

Having defined $|| \vec{T}'(s) || = \kappa$, we can rewrite the second equation as

$$\vec{T}'(s) = \kappa\vec{N}(s). \tag{11.4}$$

We already knew that $\vec{T}'(s)$ is in the same direction as $\vec{N}(s)$; that is, we can think of $\vec{T}(s)$ as being "pulled" in the direction of $\vec{N}(s)$. How "hard" is it being pulled? By a factor of κ. When the curvature is large, $\vec{T}(s)$ is being "pulled hard" and the direction of $\vec{T}(s)$ changes rapidly. When κ is small, $T(s)$ is not being pulled hard and hence its direction is not changing rapidly.

We use Definition 11.5.1 to find the curvature of the line in Example 11.5.1.

Example 11.5.2   Finding the curvature of a line
Use Definition 11.5.1 to find the curvature of $\vec{r}(t) = \langle 3t - 1, 4t + 2 \rangle$.

SOLUTION   In Example 11.5.1, we found that the arc length parameter was defined by $s = 5t$, so $\vec{r}(s) = \langle 3s/5 - 1, 4s/5 + 2 \rangle$ parametrized $\vec{r}$ with the arc length parameter. To find κ, we need to find $\vec{T}'(s)$.

$$\vec{T}(s) = \vec{r}'(s) \quad \text{(recall this is a unit vector)}$$
$$= \langle 3/5, 4/5 \rangle .$$

Therefore

$$\vec{T}'(s) = \langle 0, 0 \rangle$$

and

$$\kappa = || \vec{T}'(s) || = 0.$$

It probably comes as no surprise that the curvature of a line is 0. (How "curvy" is a line? It is not curvy at all.)

While the definition of curvature is a beautiful mathematical concept, it is nearly impossible to use most of the time; writing $\vec{r}$ in terms of the arc length parameter is generally very hard. Fortunately, there are other methods of calculating this value that are much easier. There is a tradeoff: the definition is "easy" to understand though hard to compute, whereas these other formulas are easy to compute though it may be hard to understand why they work.

Notes:

Theorem 11.5.2   Formulas for Curvature

Let C be a smooth curve in the plane or in space.

1. If C is defined by $y = f(x)$, then

$$\kappa = \frac{|f''(x)|}{\left(1 + \left(f'(x)\right)^2\right)^{3/2}}.$$

2. If C is defined as a vector–valued function in the plane, $\vec{r}(t) = \langle x(t), y(t) \rangle$, then

$$\kappa = \frac{|x'y'' - x''y'|}{\left((x')^2 + (y')^2\right)^{3/2}}.$$

3. If C is defined in space by a vector–valued function $\vec{r}(t)$, then

$$\kappa = \frac{\|\vec{T}'(t)\|}{\|\vec{r}'(t)\|} = \frac{\|\vec{r}'(t) \times \vec{r}''(t)\|}{\|\vec{r}'(t)\|^3} = \frac{\vec{a}(t) \cdot \vec{N}(t)}{\|\vec{v}(t)\|^2}.$$

We practice using these formulas.

Example 11.5.3   Finding the curvature of a circle

Find the curvature of a circle with radius r, defined by $\vec{c}(t) = \langle r \cos t, r \sin t \rangle$.

SOLUTION   Before we start, we should expect the curvature of a circle to be constant, and not dependent on t. (Why?)

We compute κ using the second part of Theorem 11.5.2.

$$\kappa = \frac{|(-r \sin t)(-r \sin t) - (-r \cos t)(r \cos t)|}{\left((-r \sin t)^2 + (r \cos t)^2\right)^{3/2}}$$

$$= \frac{r^2(\sin^2 t + \cos^2 t)}{\left(r^2(\sin^2 t + \cos^2 t)\right)^{3/2}}$$

$$= \frac{r^2}{r^3} = \frac{1}{r}.$$

We have found that a circle with radius r has curvature $\kappa = 1/r$.

Example 11.5.3 gives a great result. Before this example, if we were told

---

Notes:

"The curve has a curvature of 5 at point A," we would have no idea what this really meant. Is 5 "big" – does is correspond to a really sharp turn, or a not-so-sharp turn? Now we can think of 5 in terms of a circle with radius 1/5. Knowing the units (inches vs. miles, for instance) allows us to determine how sharply the curve is curving.

Let a point P on a smooth curve C be given, and let κ be the curvature of the curve at P. A circle that:

- passes through P,

- lies on the concave side of C,

- has a common tangent line as C at P and

- has radius $r = 1/\kappa$ (hence has curvature κ)

is the **osculating circle**, or **circle of curvature**, to C at P, and r is the **radius of curvature**. Figure 11.5.4 shows the graph of the curve seen earlier in Figure 11.5.3 and its osculating circles at A and B. A sharp turn corresponds to a circle with a small radius; a gradual turn corresponds to a circle with a large radius. Being able to think of curvature in terms of the radius of a circle is very useful. (The word "osculating" comes from a Latin word related to kissing; an osculating circle "kisses" the graph at a particular point. Many beautiful ideas in mathematics have come from studying the osculating circles to a curve.)

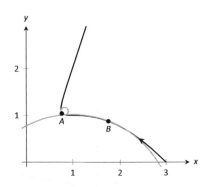

Figure 11.5.4: Illustrating the osculating circles for the curve seen in Figure 11.5.3.

Example 11.5.4   Finding curvature
Find the curvature of the parabola defined by $y = x^2$ at the vertex and at $x = 1$.

SOLUTION   We use the first formula found in Theorem 11.5.2.

$$\kappa(x) = \frac{|2|}{\left(1 + (2x)^2\right)^{3/2}}$$
$$= \frac{2}{\left(1 + 4x^2\right)^{3/2}}.$$

At the vertex ($x = 0$), the curvature is $\kappa = 2$. At $x = 1$, the curvature is $\kappa = 2/(5)^{3/2} \approx 0.179$. So at $x = 0$, the curvature of $y = x^2$ is that of a circle of radius 1/2; at $x = 1$, the curvature is that of a circle with radius $\approx 1/0.179 \approx 5.59$. This is illustrated in Figure 11.5.5. At $x = 3$, the curvature is 0.009; the graph is nearly straight as the curvature is very close to 0.

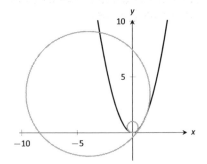

Figure 11.5.5: Examining the curvature of $y = x^2$.

Example 11.5.5   Finding curvature
Find where the curvature of $\vec{r}(t) = \langle t, t^2, 2t^3 \rangle$ is maximized.

Notes:

SOLUTION We use the third formula in Theorem 11.5.2 as $\vec{r}(t)$ is defined in space. We leave it to the reader to verify that

$$\vec{r}\,'(t) = \langle 1, 2t, 6t^2 \rangle, \quad \vec{r}\,''(t) = \langle 0, 2, 12t \rangle, \quad \text{and} \quad \vec{r}\,'(t) \times \vec{r}\,''(t) = \langle 12t^2, -12t, 2 \rangle.$$

Thus

$$\kappa(t) = \frac{\| \vec{r}\,'(t) \times \vec{r}\,''(t) \|}{\| \vec{r}\,'(t) \|^3}$$

$$= \frac{\| \langle 12t^2, -12t, 2 \rangle \|}{\| \langle 1, 2t, 6t^2 \rangle \|^3}$$

$$= \frac{\sqrt{144t^4 + 144t^2 + 4}}{\left(\sqrt{1 + 4t^2 + 36t^4} \right)^3}$$

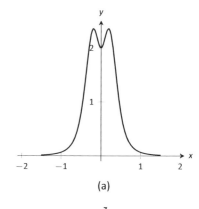

(a)

While this is not a particularly "nice" formula, it does explicitly tell us what the curvature is at a given t value. To maximize $\kappa(t)$, we should solve $\kappa'(t) = 0$ for t. This is doable, but *very* time consuming. Instead, consider the graph of $\kappa(t)$ as given in Figure 11.5.6(a). We see that κ is maximized at two t values; using a numerical solver, we find these values are $t \approx \pm 0.189$. In part (b) of the figure we graph $\vec{r}(t)$ and indicate the points where curvature is maximized.

Curvature and Motion

Let $\vec{r}(t)$ be a position function of an object, with velocity $\vec{v}(t) = \vec{r}\,'(t)$ and acceleration $\vec{a}(t) = \vec{r}\,''(t)$. In Section 11.4 we established that acceleration is in the plane formed by $\vec{T}(t)$ and $\vec{N}(t)$, and that we can find scalars a_T and a_N such that

$$\vec{a}(t) = a_T \vec{T}(t) + a_N \vec{N}(t).$$

Theorem 11.4.2 gives formulas for a_T and a_N:

$$a_T = \frac{d}{dt}\left(\| \vec{v}(t) \| \right) \quad \text{and} \quad a_N = \frac{\| \vec{v}(t) \times \vec{a}(t) \|}{\| \vec{v}(t) \|}.$$

We understood that the amount of acceleration in the direction of $\vec{T}$ relates only to how the speed of the object is changing, and that the amount of acceleration in the direction of $\vec{N}$ relates to how the direction of travel of the object is changing. (That is, if the object travels at constant speed, $a_T = 0$; if the object travels in a constant direction, $a_N = 0$.)

In Equation (11.2) at the beginning of this section, we found $s'(t) = \| \vec{v}(t) \|$. We can combine this fact with the above formula for a_T to write

$$a_T = \frac{d}{dt}\left(\| \vec{v}(t) \| \right) = \frac{d}{dt}\left(s'(t) \right) = s''(t).$$

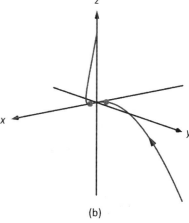

(b)

Figure 11.5.6: Understanding the curvature of a curve in space.

Notes:

Since $s'(t)$ is speed, $s''(t)$ is the rate at which speed is changing with respect to time. We see once more that the component of acceleration in the direction of travel relates only to speed, not to a change in direction.

Now compare the formula for a_N above to the formula for curvature in Theorem 11.5.2:

$$a_N = \frac{||\, \vec{v}(t) \times \vec{a}(t)\, ||}{||\, \vec{v}(t)\, ||} \quad \text{and} \quad \kappa = \frac{||\, \vec{r}'(t) \times \vec{r}''(t)\, ||}{||\, \vec{r}'(t)\, ||^3} = \frac{||\, \vec{v}(t) \times \vec{a}(t)\, ||}{||\, \vec{v}(t)\, ||^3}.$$

Thus

$$a_N = \kappa ||\, \vec{v}(t)\, ||^2 \tag{11.5}$$
$$= \kappa \left(s'(t) \right)^2$$

This last equation shows that the component of acceleration that changes the object's direction is dependent on two things: the curvature of the path and the speed of the object.

Imagine driving a car in a clockwise circle. You will naturally feel a force pushing you towards the door (more accurately, the door is pushing you as the car is turning and you want to travel in a straight line). If you keep the radius of the circle constant but speed up (i.e., increasing $s'(t)$), the door pushes harder against you (a_N has increased). If you keep your speed constant but tighten the turn (i.e., increase κ), once again the door will push harder against you.

Putting our new formulas for a_T and a_N together, we have

$$\vec{a}(t) = s''(t)\vec{T}(t) + \kappa ||\, \vec{v}(t)\, ||^2 \vec{N}(t).$$

This is not a particularly practical way of finding a_T and a_N, but it reveals some great concepts about how acceleration interacts with speed and the shape of a curve.

Example 11.5.6   Curvature and road design

The minimum radius of the curve in a highway cloverleaf is determined by the operating speed, as given in the table in Figure 11.5.7. For each curve and speed, compute a_N.

SOLUTION   Using Equation (11.5), we can compute the acceleration normal to the curve in each case. We start by converting each speed from "miles per hour" to "feet per second" by multiplying by 5280/3600.

Operating Speed (mph)	Minimum Radius (ft)
35	310
40	430
45	540

Figure 11.5.7: Operating speed and minimum radius in highway cloverleaf design.

Notes:

$$35\text{mph}, 310\text{ft} \Rightarrow 51.33\text{ft/s}, \quad \kappa = 1/310$$

$$a_N = \kappa \, \| \, \vec{v}(t) \, \|^2$$

$$= \frac{1}{310} (51.33)^2$$

$$= 8.50\text{ft/s}^2.$$

$$40\text{mph}, 430\text{ft} \Rightarrow 58.67\text{ft/s}, \quad \kappa = 1/430$$

$$a_N = \frac{1}{430} (58.67)^2$$

$$= 8.00\text{ft/s}^2.$$

$$45\text{mph}, 540\text{ft} \Rightarrow 66\text{ft/s}, \quad \kappa = 1/540$$

$$a_N = \frac{1}{540} (66)^2$$

$$= 8.07\text{ft/s}^2.$$

Note that each acceleration is similar; this is by design. Considering the classic "Force = mass × acceleration" formula, this acceleration must be kept small in order for the tires of a vehicle to keep a "grip" on the road. If one travels on a turn of radius 310ft at a rate of 50mph, the acceleration is double, at 17.35ft/s^2. If the acceleration is too high, the frictional force created by the tires may not be enough to keep the car from sliding. Civil engineers routinely compute a "safe" design speed, then subtract 5-10mph to create the posted speed limit for additional safety.

We end this chapter with a reflection on what we've covered. We started with vector–valued functions, which may have seemed at the time to be just another way of writing parametric equations. However, we have seen that the vector perspective has given us great insight into the behavior of functions and the study of motion. Vector–valued position functions convey displacement, distance traveled, speed, velocity, acceleration and curvature information, each of which has great importance in science and engineering.

Notes:

Exercises 11.5

Terms and Concepts

1. It is common to describe position in terms of both _____ and/or _____.

2. A measure of the "curviness" of a curve is _____.

3. Give two shapes with constant curvature.

4. Describe in your own words what an "osculating circle" is.

5. Complete the identity: $\vec{T}'(s) = $ _____ $\vec{N}(s)$.

6. Given a position function $\vec{r}(t)$, how are a_T and a_N affected by the curvature?

Problems

In Exercises 7 – 10 , a position function $\vec{r}(t)$ is given, where $t = 0$ corresponds to the initial position. Find the arc length parameter s, and rewrite $\vec{r}(t)$ in terms of s; that is, find $\vec{r}(s)$.

7. $\vec{r}(t) = \langle 2t, t, -2t \rangle$

8. $\vec{r}(t) = \langle 7\cos t, 7\sin t \rangle$

9. $\vec{r}(t) = \langle 3\cos t, 3\sin t, 2t \rangle$

10. $\vec{r}(t) = \langle 5\cos t, 13\sin t, 12\cos t \rangle$

In Exercises 11 – 22 , a curve C is described along with 2 points on C.

(a) Using a sketch, determine at which of these points the curvature is greater.

(b) Find the curvature κ of C, and evaluate κ at each of the 2 given points.

11. C is defined by $y = x^3 - x$; points given at $x = 0$ and $x = 1/2$.

12. C is defined by $y = \dfrac{1}{x^2 + 1}$; points given at $x = 0$ and $x = 2$.

13. C is defined by $y = \cos x$; points given at $x = 0$ and $x = \pi/2$.

14. C is defined by $y = \sqrt{1 - x^2}$ on $(-1, 1)$; points given at $x = 0$ and $x = 1/2$.

15. C is defined by $\vec{r}(t) = \langle \cos t, \sin(2t) \rangle$; points given at $t = 0$ and $t = \pi/4$.

16. C is defined by $\vec{r}(t) = \langle \cos^2 t, \sin t \cos t \rangle$; points given at $t = 0$ and $t = \pi/3$.

17. C is defined by $\vec{r}(t) = \langle t^2 - 1, t^3 - t \rangle$; points given at $t = 0$ and $t = 5$.

18. C is defined by $\vec{r}(t) = \langle \tan t, \sec t \rangle$; points given at $t = 0$ and $t = \pi/6$.

19. C is defined by $\vec{r}(t) = \langle 4t + 2, 3t - 1, 2t + 5 \rangle$; points given at $t = 0$ and $t = 1$.

20. C is defined by $\vec{r}(t) = \langle t^3 - t, t^3 - 4, t^2 - 1 \rangle$; points given at $t = 0$ and $t = 1$.

21. C is defined by $\vec{r}(t) = \langle 3\cos t, 3\sin t, 2t \rangle$; points given at $t = 0$ and $t = \pi/2$.

22. C is defined by $\vec{r}(t) = \langle 5\cos t, 13\sin t, 12\cos t \rangle$; points given at $t = 0$ and $t = \pi/2$.

In Exercises 23 – 26 , find the value of x or t where curvature is maximized.

23. $y = \dfrac{1}{6}x^3$

24. $y = \sin x$

25. $\vec{r}(t) = \langle t^2 + 2t, 3t - t^2 \rangle$

26. $\vec{r}(t) = \langle t, 4/t, 3/t \rangle$

In Exercises 27 – 30 , find the radius of curvature at the indicated value.

27. $y = \tan x$, at $x = \pi/4$

28. $y = x^2 + x - 3$, at $x = \pi/4$

29. $\vec{r}(t) = \langle \cos t, \sin(3t) \rangle$, at $t = 0$

30. $\vec{r}(t) = \langle 5\cos(3t), t \rangle$, at $t = 0$

In Exercises 31 – 34 , find the equation of the osculating circle to the curve at the indicated t-value.

31. $\vec{r}(t) = \langle t, t^2 \rangle$, at $t = 0$

32. $\vec{r}(t) = \langle 3\cos t, \sin t \rangle$, at $t = 0$

33. $\vec{r}(t) = \langle 3\cos t, \sin t \rangle$, at $t = \pi/2$

34. $\vec{r}(t) = \langle t^2 - t, t^2 + t \rangle$, at $t = 0$

A: Solutions To Selected Problems

Chapter 8

Section 8.1

1. Answers will vary.

3. Answers will vary.

5. $2, \frac{8}{3}, \frac{8}{3}, \frac{32}{15}, \frac{64}{45}$

7. $-\frac{1}{3}, -2, -\frac{81}{5}, -\frac{512}{3}, -\frac{15625}{7}$

9. $a_n = 3n + 1$

11. $a_n = 10 \cdot 2^{n-1}$

13. $1/7$

15. 0

17. diverges

19. converges to 0

21. diverges

23. converges to e

25. converges to 0

27. converges to 2

29. bounded

31. bounded

33. neither bounded above or below

35. monotonically increasing

37. never monotonic

39. Let $\{a_n\}$ be given such that $\lim_{n \to \infty} |a_n| = 0$. By the definition of the limit of a sequence, given any $\varepsilon > 0$, there is a m such that for all $n > m$, $\big| |a_n| - 0 \big| < \varepsilon$. Since $\big| |a_n| - 0 \big| = |a_n - 0|$, this directly implies that for all $n > m$, $|a_n - 0| < \varepsilon$, meaning that $\lim_{n \to \infty} a_n = 0$.

41. A sketch of one proof method:

    Let any $\varepsilon > 0$ be given. Since $\{a_n\}$ and $\{b_n\}$ converge, there exists an $N > 0$ such that for all $n \geq N$, both a_n and b_n are within $\varepsilon/2$ of L; we can conclude that they are at most ε apart from each other. Since $a_n \leq c_n \leq b_n$, one can show that c_n is within ε of L, showing that $\{c_n\}$ also converges to L.

Section 8.2

1. Answers will vary.

3. One sequence is the sequence of terms $\{a_n\}$. The other is the sequence of nth partial sums, $\{S_n\} = \{\sum_{i=1}^{n} a_i\}$.

5. F

7. (a) $-1, -\frac{1}{2}, -\frac{5}{6}, -\frac{7}{12}, -\frac{47}{60}$

   (b) Plot omitted

9. (a) $-1, 0, -1, 0, -1$

   (b) Plot omitted

11. (a) $1, \frac{3}{2}, \frac{5}{3}, \frac{41}{24}, \frac{103}{60}$

    (b) Plot omitted

13. (a) $-0.9, -0.09, -0.819, -0.1629, -0.75339$

    (b) Plot omitted

15. $\lim_{n \to \infty} a_n = 3$; by Theorem 8.2.4 the series diverges.

17. $\lim_{n \to \infty} a_n = \infty$; by Theorem 8.2.4 the series diverges.

19. $\lim_{n \to \infty} a_n = 1/2$; by Theorem 8.2.4 the series diverges.

21. Converges; p-series with $p = 5$.

23. Diverges; geometric series with $r = 6/5$.

25. Diverges; fails n^{th} term test

27. F

29. Diverges; by Theorem 8.2.3 this is half the Harmonic Series, which diverges by growing without bound. "Half of growing without bound" is still growing without bound.

31. (a) $S_n = \frac{1 - (1/4)^n}{3/4}$

    (b) Converges to $4/3$.

33. (a) $S_n = \left(\frac{n(n+1)}{2} \right)^2$

    (b) Diverges

35. (a) $S_n = 5\frac{1 - 1/2^n}{1/2}$

    (b) Converges to 10.

37. (a) $S_n = \frac{1 - (-1/3)^n}{4/3}$

    (b) Converges to $3/4$.

39. (a) With partial fractions, $a_n = \frac{3}{2}\left(\frac{1}{n} - \frac{1}{n+2} \right)$. Thus
    $$S_n = \frac{3}{2}\left(\frac{3}{2} - \frac{1}{n+1} - \frac{1}{n+2} \right).$$

    (b) Converges to $9/4$

41. (a) $S_n = \ln\left(1/(n+1) \right)$

    (b) Diverges (to $-\infty$).

43. (a) $a_n = \frac{1}{n(n+3)}$; using partial fractions, the resulting telescoping sum reduces to
    $$S_n = \frac{1}{3}\left(1 + \frac{1}{2} + \frac{1}{3} - \frac{1}{n+1} - \frac{1}{n+2} - \frac{1}{n+3} \right)$$

    (b) Converges to $11/18$.

45. (a) With partial fractions, $a_n = \frac{1}{2}\left(\frac{1}{n-1} - \frac{1}{n+1} \right)$. Thus
    $$S_n = \frac{1}{2}\left(3/2 - \frac{1}{n} - \frac{1}{n+1} \right).$$

    (b) Converges to $3/4$.

47. (a) The nth partial sum of the odd series is $1 + \frac{1}{3} + \frac{1}{5} + \cdots + \frac{1}{2n-1}$. The nth partial sum of the even series is $\frac{1}{2} + \frac{1}{4} + \frac{1}{6} + \cdots + \frac{1}{2n}$. Each term of the even series is less than the corresponding term of the odd series, giving us our result.

    (b) The nth partial sum of the odd series is $1 + \frac{1}{3} + \frac{1}{5} + \cdots + \frac{1}{2n-1}$. The nth partial sum of 1 plus the even series is $1 + \frac{1}{2} + \frac{1}{4} + \cdots + \frac{1}{2(n-1)}$. Each term of the even series is now greater than or equal to the corresponding term of the odd series, with equality only on the first term. This gives us the result.

    (c) If the odd series converges, the work done in (a) shows the even series converges also. (The sequence of the nth partial sum of the even series is bounded and monotonically increasing.) Likewise, (b) shows that if the even series converges, the odd series will, too. Thus if either series converges, the other does.

    Similarly, (a) and (b) can be used to show that if either series diverges, the other does, too.

(d) If both the even and odd series converge, then their sum would be a convergent series. This would imply that the Harmonic Series, their sum, is convergent. It is not. Hence each series diverges.

Section 8.3

1. continuous, positive and decreasing

3. The Integral Test (we do not have a continuous definition of $n!$ yet) and the Limit Comparison Test (same as above, hence we cannot take its derivative).

5. Converges

7. Diverges

9. Converges

11. Converges

13. Converges; compare to $\sum_{n=1}^{\infty} \frac{1}{n^2}$, as $1/(n^2 + 3n - 5) \le 1/n^2$ for all $n > 1$.

15. Diverges; compare to $\sum_{n=1}^{\infty} \frac{1}{n}$, as $1/n \le \ln n/n$ for all $n \ge 3$.

17. Diverges; compare to $\sum_{n=1}^{\infty} \frac{1}{n}$. Since $n = \sqrt{n^2} > \sqrt{n^2 - 1}$, $1/n \le 1/\sqrt{n^2 - 1}$ for all $n \ge 2$.

19. Diverges; compare to $\sum_{n=1}^{\infty} \frac{1}{n}$:

$$\frac{1}{n} = \frac{n^2}{n^3} < \frac{n^2 + n + 1}{n^3} < \frac{n^2 + n + 1}{n^3 - 5},$$

for all $n \ge 1$.

21. Diverges; compare to $\sum_{n=1}^{\infty} \frac{1}{n}$. Note that

$$\frac{n}{n^2 - 1} = \frac{n^2}{n^2 - 1} \cdot \frac{1}{n} > \frac{1}{n},$$

as $\frac{n^2}{n^2 - 1} > 1$, for all $n \ge 2$.

23. Converges; compare to $\sum_{n=1}^{\infty} \frac{1}{n^2}$.

25. Diverges; compare to $\sum_{n=1}^{\infty} \frac{\ln n}{n}$.

27. Diverges; compare to $\sum_{n=1}^{\infty} \frac{1}{n}$.

29. Diverges; compare to $\sum_{n=1}^{\infty} \frac{1}{n}$. Just as $\lim_{n \to 0} \frac{\sin n}{n} = 1$,

$$\lim_{n \to \infty} \frac{\sin(1/n)}{1/n} = 1.$$

31. Converges; compare to $\sum_{n=1}^{\infty} \frac{1}{n^{3/2}}$.

33. Converges; Integral Test

35. Diverges; the nth Term Test and Direct Comparison Test can be used.

37. Converges; the Direct Comparison Test can be used with sequence $1/3^n$.

39. Diverges; the nth Term Test can be used, along with the Integral Test.

41. (a) Converges; use Direct Comparison Test as $\frac{a_n}{n} < n$.

    (b) Converges; since original series converges, we know $\lim_{n \to \infty} a_n = 0$. Thus for large n, $a_n a_{n+1} < a_n$.

    (c) Converges; similar logic to part (b) so $(a_n)^2 < a_n$.

    (d) May converge; certainly $na_n > a_n$ but that does not mean it does not converge.

    (e) Does not converge, using logic from (b) and n^{th} Term Test.

Section 8.4

1. algebraic, or polynomial.

3. Integral Test, Limit Comparison Test, and Root Test

5. Converges

7. Converges

9. The Ratio Test is inconclusive; the p-Series Test states it diverges.

11. Converges

13. Converges; note the summation can be rewritten as $\sum_{n=1}^{\infty} \frac{2^n n!}{3^n n!}$, to which the Ratio Test or Geometric Series Test can be applied.

15. Converges

17. Converges

19. Diverges

21. Diverges. The Root Test is inconclusive, but the nth-Term Test shows divergence. (The terms of the sequence approach e^2, not 0, as $n \to \infty$.)

23. Converges

25. Diverges; Limit Comparison Test with $1/n$.

27. Converges; Ratio Test or Limit Comparison Test with $1/3^n$.

29. Diverges; n^{th}-Term Test or Limit Comparison Test with 1.

31. Diverges; Direct Comparison Test with $1/n$

33. Converges; Root Test

Section 8.5

1. The signs of the terms do not alternate; in the given series, some terms are negative and the others positive, but they do not necessarily alternate.

3. Many examples exist; one common example is $a_n = (-1)^n/n$.

5. (a) converges
   (b) converges (p-Series)
   (c) absolute

7. (a) diverges (limit of terms is not 0)
   (b) diverges
   (c) n/a; diverges

9. (a) converges
   (b) diverges (Limit Comparison Test with $1/n$)
   (c) conditional

11. (a) diverges (limit of terms is not 0)
    (b) diverges
    (c) n/a; diverges

13. (a) diverges (terms oscillate between ± 1)
    (b) diverges
    (c) n/a; diverges

15. (a) converges

(b) converges (Geometric Series with $r = 2/3$)

(c) absolute

17. (a) converges

(b) converges (Ratio Test)

(c) absolute

19. (a) converges

(b) diverges (p-Series Test with $p = 1/2$)

(c) conditional

21. $S_5 = -1.1906$; $S_6 = -0.6767$;

$$-1.1906 \le \sum_{n=1}^{\infty} \frac{(-1)^n}{\ln(n+1)} \le -0.6767$$

23. $S_6 = 0.3681$; $S_7 = 0.3679$;

$$0.3681 \le \sum_{n=0}^{\infty} \frac{(-1)^n}{n!} \le 0.3679$$

25. $n = 5$

27. Using the theorem, we find $n = 499$ guarantees the sum is within 0.001 of $\pi/4$. (Convergence is actually faster, as the sum is within ε of $\pi/24$ when $n \ge 249$.)

Section 8.6

1. 1

3. 5

5. $1 + 2x + 4x^2 + 8x^3 + 16x^4$

7. $1 + x + \frac{x^2}{2} + \frac{x^3}{6} + \frac{x^4}{24}$

9. (a) $R = \infty$

(b) $(-\infty, \infty)$

11. (a) $R = 1$

(b) $(2, 4]$

13. (a) $R = 2$

(b) $(-2, 2)$

15. (a) $R = 1/5$

(b) $(4/5, 6/5)$

17. (a) $R = 1$

(b) $(-1, 1)$

19. (a) $R = \infty$

(b) $(-\infty, \infty)$

21. (a) $R = 1$

(b) $[-1, 1]$

23. (a) $R = 0$

(b) $x = 0$

25. (a) $f'(x) = \sum_{n=1}^{\infty} n^2 x^{n-1}$; $(-1, 1)$

(b) $\int f(x)\,dx = C + \sum_{n=0}^{\infty} \frac{n}{n+1} x^{n+1}$; $(-1, 1)$

27. (a) $f'(x) = \sum_{n=1}^{\infty} \frac{n}{2^n} x^{n-1}$; $(-2, 2)$

(b) $\int f(x)\,dx = C + \sum_{n=0}^{\infty} \frac{1}{(n+1)2^n} x^{n+1}$; $[-2, 2)$

29. (a) $f'(x) = \sum_{n=1}^{\infty} \frac{(-1)^n x^{2n-1}}{(2n-1)!} = \sum_{n=0}^{\infty} \frac{(-1)^{n+1} x^{2n+1}}{(2n+1)!}$; $(-\infty, \infty)$

(b) $\int f(x)\,dx = C + \sum_{n=0}^{\infty} \frac{(-1)^n x^{2n+1}}{(2n+1)!}$; $(-\infty, \infty)$

31. $1 + 3x + \frac{9}{2}x^2 + \frac{9}{2}x^3 + \frac{27}{8}x^4$

33. $1 + x + x^2 + x^3 + x^4$

35. $0 + x + 0x^2 - \frac{1}{6}x^3 + 0x^4$

Section 8.7

1. The Maclaurin polynomial is a special case of Taylor polynomials. Taylor polynomials are centered at a specific x-value; when that x-value is 0, it is a Maclauring polynomial.

3. $p_2(x) = 6 + 3x - 4x^2$.

5. $p_3(x) = 1 - x + \frac{1}{2}x^2 - \frac{1}{6}x^3$

7. $p_5(x) = x + x^2 + \frac{1}{2}x^3 + \frac{1}{6}x^4 + \frac{1}{24}x^5$

9. $p_4(x) = \frac{2x^4}{3} + \frac{4x^3}{3} + 2x^2 + 2x + 1$

11. $p_4(x) = x^4 - x^3 + x^2 - x + 1$

13. $p_4(x) = 1 + \frac{1}{2}(-1+x) - \frac{1}{8}(-1+x)^2 + \frac{1}{16}(-1+x)^3 - \frac{5}{128}(-1+x)^4$

15. $p_6(x) = \frac{1}{\sqrt{2}} - \frac{-\frac{\pi}{4}+x}{\sqrt{2}} - \frac{(-\frac{\pi}{4}+x)^2}{2\sqrt{2}} + \frac{(-\frac{\pi}{4}+x)^3}{6\sqrt{2}} + \frac{(-\frac{\pi}{4}+x)^4}{24\sqrt{2}} - \frac{(-\frac{\pi}{4}+x)^5}{120\sqrt{2}} - \frac{(-\frac{\pi}{4}+x)^6}{720\sqrt{2}}$

17. $p_5(x) = \frac{1}{2} - \frac{x-2}{4} + \frac{1}{8}(x-2)^2 - \frac{1}{16}(x-2)^3 + \frac{1}{32}(x-2)^4 - \frac{1}{64}(x-2)^5$

19. $p_3(x) = \frac{1}{2} + \frac{1+x}{2} + \frac{1}{4}(1+x)^2$

21. $p_3(x) = x - \frac{x^3}{6}$; $p_3(0.1) = 0.09983$. Error is bounded by $\pm\frac{1}{4!} \cdot 0.1^4 \approx \pm 0.000004167$.

23. $p_2(x) = 3 + \frac{1}{6}(-9+x) - \frac{1}{216}(-9+x)^2$; $p_2(10) = 3.16204$. The third derivative of $f(x) = \sqrt{x}$ is bounded on $(8, 11)$ by 0.003. Error is bounded by $\pm\frac{0.003}{3!} \cdot 1^3 = \pm 0.0005$.

25. The nth derivative of $f(x) = e^x$ is bounded by 3 on intervals containing 0 and 1. Thus $|R_n(1)| \le \frac{3}{(n+1)!} 1^{(n+1)}$. When $n = 7$, this is less than 0.0001.

27. The nth derivative of $f(x) = \cos x$ is bounded by 1 on intervals containing 0 and $\pi/3$. Thus $|R_n(\pi/3)| \le \frac{1}{(n+1)!}(\pi/3)^{(n+1)}$. When $n = 7$, this is less than 0.0001. Since the Maclaurin polynomial of $\cos x$ only uses even powers, we can actually just use $n = 6$.

29. The nth term is $\frac{1}{n!}x^n$.

31. The nth term is: when n even, 0; when n is odd, $\frac{(-1)^{(n-1)/2}}{n!}x^n$.

33. The nth term is $(-1)^n x^n$.

35. $1 + x + \frac{1}{2}x^2 + \frac{1}{6}x^3 + \frac{1}{24}x^4$

37. $1 + 2x - 2x^2 + 4x^3 - 10x^4$

Section 8.8

1. A Taylor polynomial is a **polynomial**, containing a finite number of terms. A Taylor series is a **series**, the summation of an infinite number of terms.

3. All derivatives of e^x are e^x which evaluate to 1 at $x = 0$. The Taylor series starts $1 + x + \frac{1}{2}x^2 + \frac{1}{3!}x^3 + \frac{1}{4!}x^4 + \cdots$; the Taylor series is $\sum_{n=0}^{\infty} \frac{x^n}{n!}$

5. The nth derivative of $1/(1-x)$ is $f^{(n)}(x) = (n)!/(1-x)^{n+1}$, which evaluates to $n!$ at $x = 0$.
The Taylor series starts $1 + x + x^2 + x^3 + \cdots$;
the Taylor series is $\displaystyle\sum_{n=0}^{\infty} x^n$

7. The Taylor series starts
$0 - (x - \pi/2) + 0x^2 + \frac{1}{6}(x - \pi/2)^3 + 0x^4 - \frac{1}{120}(x - \pi/2)^5$;
the Taylor series is $\displaystyle\sum_{n=0}^{\infty} (-1)^{n+1} \frac{(x - \pi/2)^{2n+1}}{(2n+1)!}$

9. $f^{(n)}(x) = (-1)^n e^{-x}$; at $x = 0$, $f^{(n)}(0) = -1$ when n is odd and $f^{(n)}(0) = 1$ when n is even.
The Taylor series starts $1 - x + \frac{1}{2}x^2 - \frac{1}{3!}x^3 + \cdots$;
the Taylor series is $\displaystyle\sum_{n=0}^{\infty} (-1)^n \frac{x^n}{n!}$.

11. $f^{(n)}(x) = (-1)^{n+1} \frac{n!}{(x+1)^{n+1}}$; at $x = 1$, $f^{(n)}(1) = (-1)^{n+1} \frac{n!}{2^{n+1}}$
The Taylor series starts
$\frac{1}{2} + \frac{1}{4}(x - 1) - \frac{1}{8}(x - 1)^2 + \frac{1}{16}(x - 1)^3 \cdots$;
the Taylor series is $\displaystyle\sum_{n=0}^{\infty} (-1)^{n+1} \frac{(x - 1)^n}{2^{n+1}}$.

13. Given a value x, the magnitude of the error term $R_n(x)$ is bounded by
$$\left| R_n(x) \right| \leq \frac{\max \left| f^{(n+1)}(z) \right|}{(n+1)!} \left| x^{(n+1)} \right|,$$
where z is between 0 and x.
If $x > 0$, then $z < x$ and $f^{(n+1)}(z) = e^z < e^x$. If $x < 0$, then $x < z < 0$ and $f^{(n+1)}(z) = e^z < 1$. So given a fixed x value, let $M = \max\{e^x, 1\}$; $f^{(n)}(z) < M$. This allows us to state
$$\left| R_n(x) \right| \leq \frac{M}{(n+1)!} \left| x^{(n+1)} \right|.$$
For any x, $\displaystyle\lim_{n \to \infty} \frac{M}{(n+1)!} \left| x^{(n+1)} \right| = 0$. Thus by the Squeeze Theorem, we conclude that $\displaystyle\lim_{n \to \infty} R_n(x) = 0$ for all x, and hence
$$e^x = \sum_{n=0}^{\infty} \frac{x^n}{n!} \quad \text{for all } x.$$

15. Given a value x, the magnitude of the error term $R_n(x)$ is bounded by
$$\left| R_n(x) \right| \leq \frac{\max \left| f^{(n+1)}(z) \right|}{(n+1)!} \left| (x - 1)^{(n+1)} \right|,$$
where z is between 1 and x.
Note that $\left| f^{(n+1)}(x) \right| = \frac{n!}{x^{n+1}}$.
Per the statement of the problem, we only consider the case $1 < x < 2$.
If $1 < x < 2$, then $1 < z < x$ and $f^{(n+1)}(z) = \frac{n!}{z^{n+1}} < n!$. Thus
$$\left| R_n(x) \right| \leq \frac{n!}{(n+1)!} \left| (x - 1)^{(n+1)} \right| = \frac{(x - 1)^{n+1}}{n + 1} < \frac{1}{n + 1}.$$
Thus
$$\lim_{n \to \infty} \left| R_n(x) \right| < \lim_{n \to \infty} \frac{1}{n + 1} = 0,$$
hence
$$\ln x = \sum_{n=1}^{\infty} (-1)^{n+1} \frac{(x - 1)^n}{n} \quad \text{on } (1, 2).$$

17. Given $\cos x = \displaystyle\sum_{n=0}^{\infty} (-1)^n \frac{x^{2n}}{(2n)!}$,
$$\cos(-x) = \sum_{n=0}^{\infty} (-1)^n \frac{(-x)^{2n}}{(2n)!} = \sum_{n=0}^{\infty} (-1)^n \frac{x^{2n}}{(2n)!} = \cos x, \text{ as all}$$
powers in the series are even.

19. Given $\sin x = \displaystyle\sum_{n=0}^{\infty} (-1)^n \frac{x^{2n+1}}{(2n+1)!}$,
$$\frac{d}{dx}(\sin x) = \frac{d}{dx} \left(\sum_{n=0}^{\infty} (-1)^n \frac{x^{2n+1}}{(2n+1)!} \right) =$$
$$\sum_{n=0}^{\infty} (-1)^n \frac{(2n+1)x^{2n}}{(2n+1)!} = \sum_{n=0}^{\infty} (-1)^n \frac{x^{2n}}{(2n)!} = \cos x. \text{ (The}$$
summation still starts at $n = 0$ as there was no constant term in the expansion of $\sin x$).

21. $1 + \frac{x}{2} - \frac{x^2}{8} + \frac{x^3}{16} - \frac{5x^4}{128}$

23. $1 + \frac{x}{3} - \frac{x^2}{9} + \frac{5x^3}{81} - \frac{10x^4}{243}$

25. $\displaystyle\sum_{n=0}^{\infty} (-1)^n \frac{(x^2)^{2n}}{(2n)!} = \sum_{n=0}^{\infty} (-1)^n \frac{x^{4n}}{(2n)!}$.

27. $\displaystyle\sum_{n=0}^{\infty} (-1)^n \frac{(2x + 3)^{2n+1}}{(2n+1)!}$.

29. $x + x^2 + \frac{x^3}{3} - \frac{x^5}{30}$

31. $\displaystyle\int_0^{\sqrt{\pi}} \sin(x^2)\, dx \approx \int_0^{\sqrt{\pi}} \left(x^2 - \frac{x^6}{6} + \frac{x^{10}}{120} - \frac{x^{14}}{5040} \right) dx = 0.8877$

Chapter 9

Section 9.1

1. When defining the conics as the intersections of a plane and a double napped cone, degenerate conics are created when the plane intersects the tips of the cones (usually taken as the origin). Nondegenerate conics are formed when this plane does not contain the origin.

3. Hyperbola

5. With a horizontal transverse axis, the x^2 term has a positive coefficient; with a vertical transverse axis, the y^2 term has a positive coefficient.

7. $y = \frac{1}{2}(x - 3)^2 + \frac{3}{2}$

9. $x = -\frac{1}{4}(y - 5)^2 + 2$

11. $y = -\frac{1}{4}(x - 1)^2 + 2$

13. $y = 4x^2$

15. focus: $(0, 1)$; directrix: $y = -1$. The point P is 2 units from each.

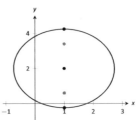

17.

19. $\frac{(x+1)^2}{9} + \frac{(y-2)^2}{4} = 1$; foci at $(-1 \pm \sqrt{5}, 2)$; $e = \sqrt{5}/3$

21. $\frac{x^2}{9} + \frac{y^2}{5} = 1$

23. $\frac{(x-2)^2}{45} + \frac{y^2}{49} = 1$

25. $\frac{(x-1)^2}{2} + (y - 2)^2 = 1$

27. $\frac{x^2}{4} + \frac{(y-3)^2}{6} = 1$

29. $x^2 - \frac{y^2}{3} = 1$

31. $\frac{(y-3)^2}{4} - \frac{(x-1)^2}{9} = 1$

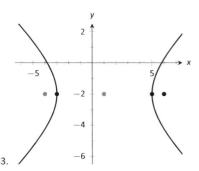

33.

35. $\frac{x^2}{4} - \frac{y^2}{5} = 1$

37. $\frac{(x-3)^2}{16} - \frac{(y-3)^2}{9} = 1$

39. $\frac{x^2}{4} - \frac{y^2}{3} = 1$

41. $(y-2)^2 - \frac{x^2}{10} = 1$

43. (a) $c = \sqrt{12 - 4} = 2\sqrt{2}$.

   (b) The sum of distances for each point is $2\sqrt{12} \approx 6.9282$.

45. The sound originated from a point approximately 31m to the left of B and 1340m above it.

Section 9.2

1. T

3. rectangular

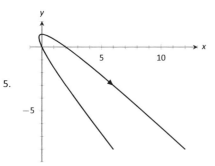

5.

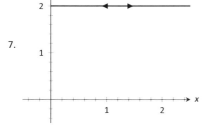

7.

9.

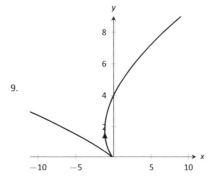

11.

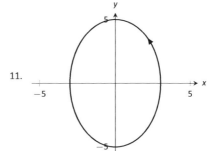

13.

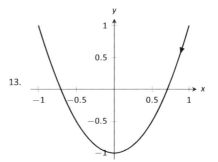

15.

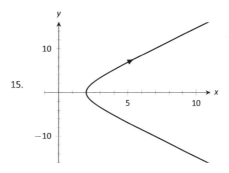

17.

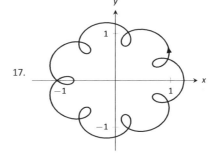

19. (a) Traces the parabola $y = x^2$, moves from left to right.

   (b) Traces the parabola $y = x^2$, but only from $-1 \le x \le 1$; traces this portion back and forth infinitely.

(c) Traces the parabola $y = x^2$, but only for $0 < x$. Moves left to right.

(d) Traces the parabola $y = x^2$, moves from right to left.

21. $y = -1.5x + 8.5$

23. $\frac{(x-1)^2}{16} + \frac{(y+2)^2}{9} = 1$

25. $y = 2x + 3$

27. $y = e^{2x} - 1$

29. $x^2 - y^2 = 1$

31. $y = \frac{b}{a}(x - x_0) + y_0$; line through (x_0, y_0) with slope b/a.

33. $\frac{(x-h)^2}{a^2} + \frac{(y-k)^2}{b^2} = 1$; ellipse centered at (h, k) with horizontal axis of length $2a$ and vertical axis of length $2b$.

35. $x = (t + 11)/6$, $y = (t^2 - 97)/12$. At $t = 1$, $x = 2$, $y = -8$.
$y' = 6x - 11$; when $x = 2$, $y' = 1$.

37. $x = \cos^{-1} t$, $y = \sqrt{1 - t^2}$. At $t = 1$, $x = 0$, $y = 0$.
$y' = \cos x$; when $x = 0$, $y' = 1$.

39. $t = \pm 1$

41. $t = \pi/2, 3\pi/2$

43. $t = -1$

45. $t = \ldots \pi/2,\ 3\pi/2,\ 5\pi/2,\ \ldots$

47. $x = 4t$, $y = -16t^2 + 64t$

49. $x = 10t$, $y = -16t^2 + 320t$

51. $x = 3\cos(2\pi t) + 1$, $y = 3\sin(2\pi t) + 1$; other answers possible

53. $x = 5\cos t$, $y = \sqrt{24}\sin t$; other answers possible

55. $x = 2\tan t$, $y = \pm 6\sec t$; other answers possible

Section 9.3

1. F

3. F

5. (a) $\frac{dy}{dx} = 2t$

   (b) Tangent line: $y = 2(x - 1) + 1$; normal line:
   $y = -1/2(x - 1) + 1$

7. (a) $\frac{dy}{dx} = \frac{2t+1}{2t-1}$

   (b) Tangent line: $y = 3x + 2$; normal line: $y = -1/3x + 2$

9. (a) $\frac{dy}{dx} = \csc t$

   (b) $t = \pi/4$: Tangent line: $y = \sqrt{2}(x - \sqrt{2}) + 1$; normal line:
   $y = -1/\sqrt{2}(x - \sqrt{2}) + 1$

11. (a) $\frac{dy}{dx} = \frac{\cos t \sin(2t) + \sin t \cos(2t)}{-\sin t \sin(2t) + 2\cos t \cos(2t)}$

    (b) Tangent line: $y = x - \sqrt{2}$; normal line: $y = -x - \sqrt{2}$

13. $t = 0$

15. $t = -1/2$

17. The graph does not have a horizontal tangent line.

19. The solution is non-trivial; use identities $\sin(2t) = 2\sin t \cos t$ and $\cos(2t) = \cos^2 t - \sin^2 t$ to rewrite
$g'(t) = 2\sin t(2\cos^2 t - \sin^2 t)$. On $[0, 2\pi]$, $\sin t = 0$ when
$t = 0, \pi, 2\pi$, and $2\cos^2 t - \sin^2 t = 0$ when
$t = \tan^{-1}(\sqrt{2})$, $\pi \pm \tan^{-1}(\sqrt{2})$, $2\pi - \tan^{-1}(\sqrt{2})$.

21. $t_0 = 0$; $\lim_{t \to 0} \frac{dy}{dx} = 0$.

23. $t_0 = 1$; $\lim_{t \to 1} \frac{dy}{dx} = \infty$.

25. $\frac{d^2y}{dx^2} = 2$; always concave up

27. $\frac{d^2y}{dx^2} = -\frac{4}{(2t-1)^3}$; concave up on $(-\infty, 1/2)$; concave down on $(1/2, \infty)$.

29. $\frac{d^2y}{dx^2} = -\cot^3 t$; concave up on $(-\infty, 0)$; concave down on $(0, \infty)$.

31. $\frac{d^2y}{dx^2} = \frac{4(13 + 3\cos(4t))}{(\cos t + 3\cos(3t))^3}$, obtained with a computer algebra system; concave up on $\left(-\tan^{-1}(\sqrt{2}/2), \tan^{-1}(\sqrt{2}/2)\right)$, concave down on $\left(-\pi/2, -\tan^{-1}(\sqrt{2}/2)\right) \cup \left(\tan^{-1}(\sqrt{2}/2), \pi/2\right)$

33. $L = 6\pi$

35. $L = 2\sqrt{34}$

37. $L \approx 2.4416$ (actual value: $L = 2.42211$)

39. $L \approx 4.19216$ (actual value: $L = 4.18308$)

41. The answer is 16π for both (of course), but the integrals are different.

43. $SA \approx 8.50101$ (actual value $SA = 8.02851$)

Section 9.4

1. Answers will vary.

3. T

5.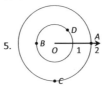

7. $A = P(2.5, \pi/4)$ and $P(-2.5, 5\pi/4)$;
$B = P(-1, 5\pi/6)$ and $P(1, 11\pi/6)$;
$C = P(3, 4\pi/3)$ and $P(-3, \pi/3)$;
$D = P(1.5, 2\pi/3)$ and $P(-1.5, 5\pi/3)$;

9. $A = (\sqrt{2}, \sqrt{2})$
$B = (\sqrt{2}, -\sqrt{2})$
$C = P(\sqrt{5}, -0.46)$
$D = P(\sqrt{5}, 2.68)$

11.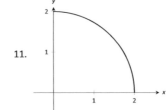

13.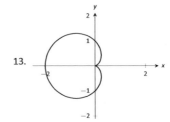

15.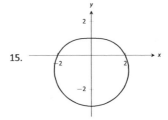

17.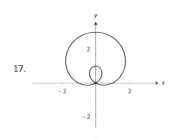

19.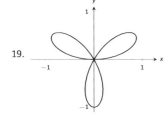

21.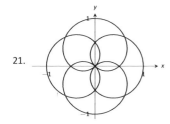

23.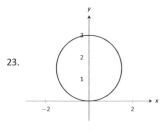

25.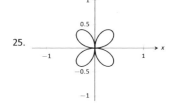

27.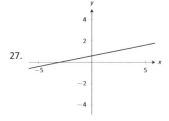

29.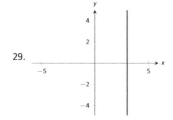

31. $(x-3)^2 + y^2 = 3$

33. $(x-1/2)^2 + (y-1/2)^2 = 1/2$

35. $x = 3$

37. $x^4 + x^2 y^2 - y^2 = 0$

39. $x^2 + y^2 = 4$

41. $\theta = \pi/4$

43. $r = 5\sec\theta$

45. $r = \cos\theta / \sin^2\theta$

47. $r = \sqrt{7}$

49. $P(\sqrt{3}/2, \pi/6), P(0, \pi/2), P(-\sqrt{3}/2, 5\pi/6)$

51. $P(0,0) = P(0, \pi/2), P(\sqrt{2}, \pi/4)$

53. $P(\sqrt{2}/2, \pi/12), P(-\sqrt{2}/2, 5\pi/12), P(\sqrt{2}/2, 3\pi/4)$

55. For all points, $r = 1; \theta =$
$\pi/12, 5\pi/12, 7\pi/12, 11\pi/12, 13\pi/12, 17\pi/12, 19\pi/12, 23\pi/12$.

57. Answers will vary. If m and n do not have any common factors, then an interval of $2n\pi$ is needed to sketch the entire graph.

Section 9.5

1. Using $x = r\cos\theta$ and $y = r\sin\theta$, we can write $x = f(\theta)\cos\theta$, $y = f(\theta)\sin\theta$.

3. (a) $\frac{dy}{dx} = -\cot\theta$

    (b) tangent line: $y = -(x - \sqrt{2}/2) + \sqrt{2}/2$; normal line: $y = x$

5. (a) $\frac{dy}{dx} = \frac{\cos\theta(1+2\sin\theta)}{\cos^2\theta - \sin\theta(1+\sin\theta)}$

    (b) tangent line: $x = 3\sqrt{3}/4$; normal line: $y = 3/4$

7. (a) $\frac{dy}{dx} = \frac{\theta\cos\theta + \sin\theta}{\cos\theta - \theta\sin\theta}$

    (b) tangent line: $y = -2/\pi x + \pi/2$; normal line:
$y = \pi/2x + \pi/2$

9. (a) $\frac{dy}{dx} = \frac{4\sin(\theta)\cos(4\theta) + \sin(4\theta)\cos(\theta)}{4\cos(\theta)\cos(4\theta) - \sin(\theta)\sin(4\theta)}$

    (b) tangent line: $y = 5\sqrt{3}(x + \sqrt{3}/4) - 3/4$; normal line:
$y = -1/5\sqrt{3}(x + \sqrt{3}/4) - 3/4$

11. horizontal: $\theta = \pi/2, 3\pi/2$;
vertical: $\theta = 0, \pi, 2\pi$

13. horizontal: $\theta = \tan^{-1}(1/\sqrt{5}), \pi/2, \pi - \tan^{-1}(1/\sqrt{5}), \pi + \tan^{-1}(1/\sqrt{5}), 3\pi/2, 2\pi - \tan^{-1}(1/\sqrt{5})$;
vertical: $\theta = 0, \tan^{-1}(\sqrt{5}), \pi - \tan^{-1}(\sqrt{5}), \pi, \pi + \tan^{-1}(\sqrt{5}), 2\pi - \tan^{-1}(\sqrt{5})$

15. In polar: $\theta = 0 \cong \theta = \pi$
In rectangular: $y = 0$

17. area = 4π

19. area = $\pi/12$

21. area = $3\pi/2$

23. area = $2\pi + 3\sqrt{3}/2$

25. area = 1

27. area = $\frac{1}{32}(4\pi - 3\sqrt{3})$

29. 4π

31. area = $\sqrt{2}\pi$

33. $L \approx 2.2592$; (actual value $L = 2.22748$)

35. $SA = 16\pi$

37. $SA = 32\pi/5$

39. $SA = 36\pi$

Chapter 10

Section 10.1

1. right hand

3. curve (a parabola); surface (a cylinder)

5. a hyperboloid of two sheets

7. $\| \overline{AB} \| = \sqrt{6}; \| \overline{BC} \| = \sqrt{17}; \| \overline{AC} \| = \sqrt{11}$. Yes, it is a right triangle as $\| \overline{AB} \|^2 + \| \overline{AC} \|^2 = \| \overline{BC} \|^2$.

9. Center at $(4, -1, 0)$; radius = 3

11. Interior of a sphere with radius 1 centered at the origin.

13. The first octant of space; all points (x, y, z) where each of x, y and z are non-negative. (Analogous to the first quadrant in the plane.)

15.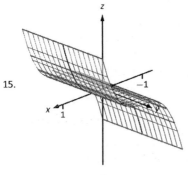

17.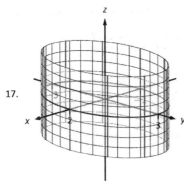

19. $x^2 + z^2 = \frac{1}{(1+y^2)^2}$

21. $z = (\sqrt{x^2 + y^2})^2 = x^2 + y^2$

23. (a) $x = y^2 + \frac{z^2}{9}$

25. (b) $x^2 + \frac{y^2}{9} + \frac{z^2}{4} = 1$

27.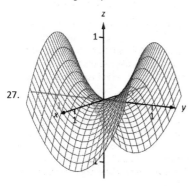

29.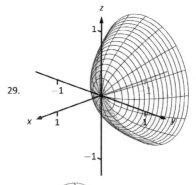

31.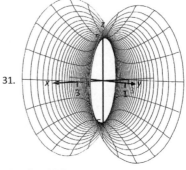

Section 10.2

1. Answers will vary.

3. A vector with magnitude 1.

5. Their respective unit vectors are parallel; unit vectors $\vec{u}_1$ and $\vec{u}_2$ are parallel if $\vec{u}_1 = \pm \vec{u}_2$.

7. $\vec{PQ} = \langle 1, 6 \rangle = 1\vec{i} + 6\vec{j}$

9. $\vec{PQ} = \langle 6, -1, 6 \rangle = 6\vec{i} - \vec{j} + 6\vec{k}$

11. (a) $\vec{u} + \vec{v} = \langle 2, -1 \rangle; \vec{u} - \vec{v} = \langle 0, -3 \rangle; 2\vec{u} - 3\vec{v} = \langle -1, -7 \rangle$.
    (c) $\vec{x} = \langle 1/2, 2 \rangle$.

13.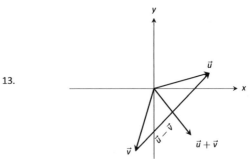

15.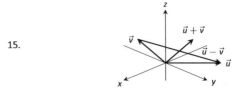

17. $\| \vec{u} \| = \sqrt{5}, \| \vec{v} \| = \sqrt{13}, \| \vec{u} + \vec{v} \| = \sqrt{26}, \| \vec{u} - \vec{v} \| = \sqrt{10}$

19. $\| \vec{u} \| = \sqrt{5}, \| \vec{v} \| = 3\sqrt{5}, \| \vec{u} + \vec{v} \| = 2\sqrt{5}, \| \vec{u} - \vec{v} \| = 4\sqrt{5}$

21. When $\vec{u}$ and $\vec{v}$ have the same direction. (Note: parallel is not enough.)

23. $\vec{u} = \langle 0.6, 0.8 \rangle$

25. $\vec{u} = \langle 1/\sqrt{3}, -1/\sqrt{3}, 1/\sqrt{3} \rangle$

27. $\vec{u} = \langle \cos 120°, \sin 120° \rangle = \langle -1/2, \sqrt{3}/2 \rangle$.

29. The force on each chain is $100/\sqrt{3} \approx 57.735$lb.

31. The force on the chain with angle θ is approx. 45.124lb; the force on the chain with angle φ is approx. 59.629lb.

33. $\theta = 45°$; the weight is lifted 0.29 ft (about 3.5in).

35. $\theta = 45°$; the weight is lifted 2.93 ft.

Section 10.3

1. Scalar

3. By considering the sign of the dot product of the two vectors. If the dot product is positive, the angle is acute; if the dot product is negative, the angle is obtuse.

5. -22

7. 3

9. not defined

11. Answers will vary.

13. $\theta = 0.3218 \approx 18.43°$

15. $\theta = \pi/4 = 45°$

17. Answers will vary; two possible answers are $\langle -7, 4 \rangle$ and $\langle 14, -8 \rangle$.

19. Answers will vary; two possible answers are $\langle 1, 0, -1 \rangle$ and $\langle 4, 5, -9 \rangle$.

21. $\text{proj}_{\vec{v}}\, \vec{u} = \langle -1/2, 3/2 \rangle$.

23. $\text{proj}_{\vec{v}}\, \vec{u} = \langle -1/2, -1/2 \rangle$.

25. $\text{proj}_{\vec{v}}\, \vec{u} = \langle 1, 2, 3 \rangle$.

27. $\vec{u} = \langle -1/2, 3/2 \rangle + \langle 3/2, 1/2 \rangle$.

29. $\vec{u} = \langle -1/2, -1/2 \rangle + \langle -5/2, 5/2 \rangle$.

31. $\vec{u} = \langle 1, 2, 3 \rangle + \langle 0, 3, -2 \rangle$.

33. 1.96lb

35. 141.42ft–lb

37. 500ft–lb

39. 500ft–lb

Section 10.4

1. vector

3. "Perpendicular" is one answer.

5. Torque

7. $\vec{u} \times \vec{v} = \langle 12, -15, 3 \rangle$

9. $\vec{u} \times \vec{v} = \langle -5, -31, 27 \rangle$

11. $\vec{u} \times \vec{v} = \langle 0, -2, 0 \rangle$

13. $\vec{u} \times \vec{v} = \langle 0, 0, ad - bc \rangle$

15. $\vec{i} \times \vec{k} = -\vec{j}$

17. Answers will vary.

19. 5

21. 0

23. $\sqrt{14}$

25. 3

27. $5\sqrt{2}/2$

29. 1

31. 7

33. 2

35. $\pm \frac{1}{\sqrt{6}} \langle 1, 1, -2 \rangle$

37. $\langle 0, \pm 1, 0 \rangle$

39. 87.5ft–lb

41. $200/3 \approx 66.67$ft–lb

43. With $\vec{u} = \langle u_1, u_2, u_3 \rangle$ and $\vec{v} = \langle v_1, v_2, v_3 \rangle$, we have

$$\vec{u} \cdot (\vec{u} \times \vec{v}) = \langle u_1, u_2, u_3 \rangle \cdot (\langle u_2 v_3 - u_3 v_2, -(u_1 v_3 - u_3 v_1), u_1 v_2 - u_2 v_1 \rangle)$$
$$= u_1(u_2 v_3 - u_3 v_2) - u2(u_1 v_3 - u_3 v_1) + u3(u_1 v_2 - u_2 v_1)$$
$$= 0.$$

Section 10.5

1. A point on the line and the direction of the line.

3. parallel, skew

5. vector: $\ell(t) = \langle 2, -4, 1 \rangle + t \langle 9, 2, 5 \rangle$
   parametric: $x = 2 + 9t, y = -4 + 2t, z = 1 + 5t$
   symmetric: $(x - 2)/9 = (y + 4)/2 = (z - 1)/5$

7. Answers can vary: vector: $\ell(t) = \langle 2, 1, 5 \rangle + t \langle 5, -3, -1 \rangle$
   parametric: $x = 2 + 5t, y = 1 - 3t, z = 5 - t$
   symmetric: $(x - 2)/5 = -(y - 1)/3 = -(z - 5)$

9. Answers can vary; here the direction is given by $\vec{d}_1 \times \vec{d}_2$: vector:
   $\ell(t) = \langle 0, 1, 2 \rangle + t \langle -10, 43, 9 \rangle$
   parametric: $x = -10t, y = 1 + 43t, z = 2 + 9t$
   symmetric: $-x/10 = (y - 1)/43 = (z - 2)/9$

11. Answers can vary; here the direction is given by $\vec{d}_1 \times \vec{d}_2$: vector:
    $\ell(t) = \langle 7, 2, -1 \rangle + t \langle 1, -1, 2 \rangle$
    parametric: $x = 7 + t, y = 2 - t, z = -1 + 2t$
    symmetric: $x - 7 = 2 - y = (z + 1)/2$

13. vector: $\ell(t) = \langle 1, 1 \rangle + t \langle 2, 3 \rangle$
    parametric: $x = 1 + 2t, y = 1 + 3t$
    symmetric: $(x - 1)/2 = (y - 1)/3$

15. parallel

17. intersecting; $\vec{\ell}_1(3) = \vec{\ell}_2(4) = \langle 9, -5, 13 \rangle$

19. skew

21. same

23. $\sqrt{41}/3$

25. $5\sqrt{2}/2$

27. $3/\sqrt{2}$

29. Since both P and Q are on the line, $\vec{PQ}$ is parallel to $\vec{d}$. Thus $\vec{PQ} \times \vec{d} = \vec{0}$, giving a distance of 0.

31. (a) The distance formula cannot be used because since $\vec{d}_1$ and $\vec{d}_2$ are parallel, $\vec{c}$ is $\vec{0}$ and we cannot divide by $|| \vec{0} ||$.

    (b) Since $\vec{d}_1$ and $\vec{d}_2$ are parallel, $\vec{P_1 P_2}$ lies in the plane formed by the two lines. Thus $\vec{P_1 P_2} \times \vec{d}_2$ is orthogonal to this plane, and $\vec{c} = (\vec{P_1 P_2} \times \vec{d}_2) \times \vec{d}_2$ is parallel to the plane, but still orthogonal to both $\vec{d}_1$ and $\vec{d}_2$. We desire the length of the projection of $\vec{P_1 P_2}$ onto $\vec{c}$, which is what the formula provides.

    (c) Since the lines are parallel, one can measure the distance between the lines at any location on either line (just as to find the distance between straight railroad tracks, one can use a measuring tape anywhere along the track, not just at one specific place.) Let $P = P_1$ and $Q = P_2$ as given by the equations of the lines, and apply the formula for distance between a point and a line.

Section 10.6

1. A point in the plane and a normal vector (i.e., a direction orthogonal to the plane).

3. Answers will vary.

5. Answers will vary.

7. Standard form: $3(x-2) - (y-3) + 7(z-4) = 0$
   general form: $3x - y + 7z = 31$

9. Answers may vary;
   Standard form: $8(x-1) + 4(y-2) - 4(z-3) = 0$
   general form: $8x + 4y - 4z = 4$

11. Answers may vary;
    Standard form: $-7(x-2) + 2(y-1) + (z-2) = 0$
    general form: $-7x + 2y + z = -10$

13. Answers may vary;
    Standard form: $2(x-1) - (y-1) = 0$
    general form: $2x - y = 1$

15. Answers may vary;
    Standard form: $2(x-2) - (y+6) - 4(z-1) = 0$
    general form: $2x - y - 4z = 6$

17. Answers may vary;
    Standard form: $(x-5) + (y-7) + (z-3) = 0$
    general form: $x + y + z = 15$

19. Answers may vary;
    Standard form: $3(x+4) + 8(y-7) - 10(z-2) = 0$
    general form: $3x + 8y - 10z = 24$

21. Answers may vary:
    $$\ell = \begin{cases} x = 14t \\ y = -1 - 10t \\ z = 2 - 8t \end{cases}$$

23. $(-3, -7, -5)$

25. No point of intersection; the plane and line are parallel.

27. $\sqrt{5/7}$

29. $1/\sqrt{3}$

31. If P is any point in the plane, and Q is also in the plane, then $\overrightarrow{PQ}$ lies parallel to the plane and is orthogonal to $\vec{n}$, the normal vector. Thus $\vec{n} \cdot \overrightarrow{PQ} = 0$, giving the distance as 0.

Chapter 11

Section 11.1

1. parametric equations

3. displacement

5.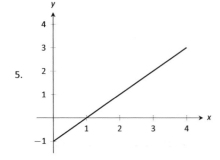

7.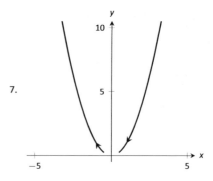

9.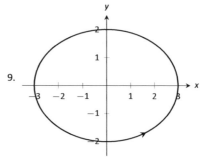

11.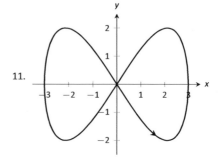

13.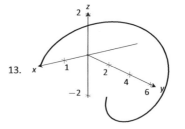

15.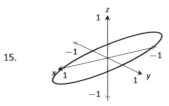

17. $\| \vec{r}(t) \| = \sqrt{t^2 + t^4} = |t|\sqrt{t^2 + 1}.$

19. $\| \vec{r}(t) \| = \sqrt{4\cos^2 t + 4\sin^2 t + t^2} = \sqrt{t^2 + 4}.$

21. Answers may vary, though most direct solution is $\vec{r}(t) = \langle 2\cos t + 1, 2\sin t + 2 \rangle.$

23. Answers may vary, though most direct solution is $\vec{r}(t) = \langle 1.5\cos t, 5\sin t \rangle.$

25. Answers may vary, though most direct solutions are
$\vec{r}(t) = \langle t, 5(t-2) + 3 \rangle$ and
$\vec{r}(t) = \langle t + 2, 5t + 3 \rangle$.

27. Specific forms may vary, though most direct solutions are
$\vec{r}(t) = \langle 1, 2, 3 \rangle + t \langle 3, 3, 3 \rangle$ and
$\vec{r}(t) = \langle 3t + 1, 3t + 2, 3t + 3 \rangle$.

29. Answers may vary, though most direct solution is
$\vec{r}(t) = \langle 2 \cos t, 2 \sin t, 2t \rangle$.

31. $\langle 1, 0 \rangle$

33. $\langle 0, 0, 1 \rangle$

Section 11.2

1. component

3. It is difficult to identify the points on the graphs of $\vec{r}(t)$ and $\vec{r}'(t)$ that correspond to each other.

5. $\langle 11, 74, \sin 5 \rangle$

7. $\langle 1, e \rangle$

9. $(-\infty, 0) \bigcup (0, \infty)$

11. $\vec{r}'(t) = \langle -\sin t, e^t, 1/t \rangle$

13. $\vec{r}'(t) = (2t) \langle \sin t, 2t + 5 \rangle + (t^2) \langle \cos t, 2 \rangle = \langle 2t \sin t + t^2 \cos t, 6t^2 + 10t \rangle$

15. $\vec{r}'(t) = \langle 2t, 1, 0 \rangle \times \langle \sin t, 2t + 5, 1 \rangle + \langle t^2 + 1, t - 1, 1 \rangle \times \langle \cos t, 2, 0 \rangle = \langle -1, \cos t - 2t, 6t^2 + 10t + 2 + \cos t - \sin t - t \cos t \rangle$

17.

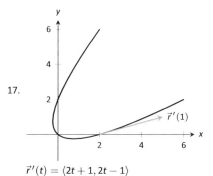

$\vec{r}'(t) = \langle 2t + 1, 2t - 1 \rangle$

19.

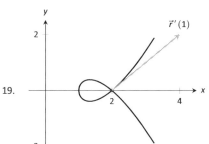

$\vec{r}'(t) = \langle 2t, 3t^2 - 1 \rangle$

21. $\ell(t) = \langle 2, 0 \rangle + t \langle 3, 1 \rangle$

23. $\ell(t) = \langle -3, 0, \pi \rangle + t \langle 0, -3, 1 \rangle$

25. $t = 2n\pi$, where n is an integer; so
$t = \ldots -4\pi, -2\pi, 0, 2\pi, 4\pi, \ldots$

27. $\vec{r}(t)$ is not smooth at $t = 3\pi/4 + n\pi$, where n is an integer

29. Both derivatives return $\langle 5t^4, 4t^3 - 3t^2, 3t^2 \rangle$.

31. Both derivatives return
$\langle 2t - e^t - 1, \cos t - 3t^2, (t^2 + 2t)e^t - (t - 1) \cos t - \sin t \rangle$.

33. $\langle \frac{1}{4}t^4, \sin t, te^t - e^t \rangle + \vec{C}$

35. $\langle -2, 0 \rangle$

37. $\vec{r}(t) = \langle \frac{1}{2}t^2 + 2, -\cos t + 3 \rangle$

39. $\vec{r}(t) = \langle t^4/12 + t + 4, \ t^3/6 + 2t + 5, \ t^2/2 + 3t + 6 \rangle$

41. $2\sqrt{13}\pi$

43. $\frac{1}{54}\left((22)^{3/2} - 8\right)$

45. As $\vec{r}(t)$ has constant length, $\vec{r}(t) \cdot \vec{r}(t) = c^2$ for some constant c. Thus

$$\vec{r}(t) \cdot \vec{r}(t) = c^2$$
$$\frac{d}{dt}\left(\vec{r}(t) \cdot \vec{r}(t)\right) = \frac{d}{dt}\left(c^2\right)$$
$$\vec{r}'(t) \cdot \vec{r}(t) + \vec{r}(t) \cdot \vec{r}'(t) = 0$$
$$2\vec{r}(t) \cdot \vec{r}'(t) = 0$$
$$\vec{r}(t) \cdot \vec{r}'(t) = 0.$$

Section 11.3

1. Velocity is a vector, indicating an objects direction of travel and its rate of distance change (i.e., its speed). Speed is a scalar.

3. The average velocity is found by dividing the displacement by the time traveled – it is a vector. The average speed is found by dividing the distance traveled by the time traveled – it is a scalar.

5. One example is traveling at a constant speed s in a circle, ending at the starting position. Since the displacement is $\vec{0}$, the average velocity is $\vec{0}$, hence $\|\vec{0}\| = 0$. But traveling at constant speed s means the average speed is also $s > 0$.

7. $\vec{v}(t) = \langle 2, 5, 0 \rangle, \vec{a}(t) = \langle 0, 0, 0 \rangle$

9. $\vec{v}(t) = \langle -\sin t, \cos t \rangle, \vec{a}(t) = \langle -\cos t, -\sin t \rangle$

11. $\vec{v}(t) = \langle 1, \cos t \rangle, \vec{a}(t) = \langle 0, -\sin t \rangle$

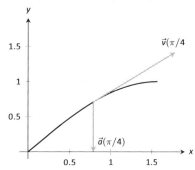

13. $\vec{v}(t) = \langle 2t + 1, -2t + 2 \rangle, \vec{a}(t) = \langle 2, -2 \rangle$

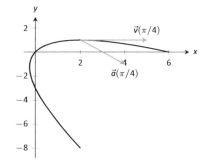

15. $\|\vec{v}(t)\| = \sqrt{4t^2 + 1}$.
Min at $t = 0$; Max at $t = \pm 1$.

17. $\|\vec{v}(t)\| = 5$.
Speed is constant, so there is no difference between min/max

19. $\|\vec{v}(t)\| = |\sec t|\sqrt{\tan^2 t + \sec^2 t}$.
   min: $t = 0$; max: $t = \pi/4$

21. $\|\vec{v}(t)\| = 13$.
   speed is constant, so there is no difference between min/max

23. $\|\vec{v}(t)\| = \sqrt{4t^2 + 1 + t^2/(1 - t^2)}$.
   min: $t = 0$; max: there is no max; speed approaches ∞ as $t \to \pm 1$

25. (a) $\vec{r}_1(1) = \langle 1, 1 \rangle$; $\vec{r}_2(1) = \langle 1, 1 \rangle$
   (b) $\vec{v}_1(1) = \langle 1, 2 \rangle$; $\|\vec{v}_1(1)\| = \sqrt{5}$; $\vec{a}_1(1) = \langle 0, 2 \rangle$
   $\vec{v}_2(1) = \langle 2, 4 \rangle$; $\|\vec{v}_2(1)\| = 2\sqrt{5}$; $\vec{a}_2(1) = \langle 2, 12 \rangle$

27. (a) $\vec{r}_1(2) = \langle 6, 4 \rangle$; $\vec{r}_2(2) = \langle 6, 4 \rangle$
   (b) $\vec{v}_1(2) = \langle 3, 2 \rangle$; $\|\vec{v}_1(2)\| = \sqrt{13}$; $\vec{a}_1(2) = \langle 0, 0 \rangle$
   $\vec{v}_2(2) = \langle 6, 4 \rangle$; $\|\vec{v}_2(2)\| = 2\sqrt{13}$; $\vec{a}_2(2) = \langle 0, 0 \rangle$

29. $\vec{v}(t) = \langle 2t + 1, 3t + 2 \rangle$, $\vec{r}(t) = \langle t^2 + t + 5, 3t^2/2 + 2t - 2 \rangle$

31. $\vec{v}(t) = \langle \sin t, \cos t \rangle$, $\vec{r}(t) = \langle 1 - \cos t, \sin t \rangle$

33. Displacement: $\langle 0, 0, 6\pi \rangle$; distance traveled: $2\sqrt{13}\pi \approx 22.65$ft; average velocity: $\langle 0, 0, 3 \rangle$; average speed: $\sqrt{13} \approx 3.61$ft/s

35. Displacement: $\langle 0, 0 \rangle$; distance traveled: $2\pi \approx 6.28$ft; average velocity: $\langle 0, 0 \rangle$; average speed: 1ft/s

37. At t-values of $\sin^{-1}(9/30)/(4\pi) + n/2 \approx 0.024 + n/2$ seconds, where n is an integer.

39. (a) Holding the crossbow at an angle of 0.013 radians, $\approx 0.745°$ will hit the target 0.4s later. (Another solution exists, with an angle of $89°$, landing 18.75s later, but this is impractical.)

   (b) In the .4 seconds the arrow travels, a deer, traveling at 20mph or 29.33ft/s, can travel 11.7ft. So she needs to lead the deer by 11.7ft.

41. The position function is $\vec{r}(t) = \langle 220t, -16t^2 + 1000 \rangle$. The y-component is 0 when $t = 7.9$; $\vec{r}(7.9) = \langle 1739.25, 0 \rangle$, meaning the box will travel about 1740ft horizontally before it lands.

Section 11.4

1. 1

3. $\vec{T}(t)$ and $\vec{N}(t)$.

5. $\vec{T}(t) = \left\langle \frac{4t}{\sqrt{20t^2 - 4t + 1}}, \frac{2t - 1}{\sqrt{20t^2 - 4t + 1}} \right\rangle$; $\vec{T}(1) = \langle 4/\sqrt{17}, 1/\sqrt{17} \rangle$

7. $\vec{T}(t) = \frac{\cos t \sin t}{\sqrt{\cos^2 t \sin^2 t}} \langle -\cos t, \sin t \rangle$. (Be careful; this cannot be simplified as just $\langle -\cos t, \sin t \rangle$ as $\sqrt{\cos^2 t \sin^2 t} \neq \cos t \sin t$, but rather $|\cos t \sin t|$.) $\vec{T}(\pi/4) = \langle -\sqrt{2}/2, \sqrt{2}/2 \rangle$

9. $\ell(t) = \langle 2, 0 \rangle + t \langle 4/\sqrt{17}, 1/\sqrt{17} \rangle$; in parametric form,
   $\ell(t) = \begin{cases} x &=& 2 + 4t/\sqrt{17} \\ y &=& t/\sqrt{17} \end{cases}$

11. $\ell(t) = \langle \sqrt{2}/4, \sqrt{2}/4 \rangle + t \langle -\sqrt{2}/2, \sqrt{2}/2 \rangle$; in parametric form,
   $\ell(t) = \begin{cases} x &=& \sqrt{2}/4 - \sqrt{2}t/2 \\ y &=& \sqrt{2}/4 + \sqrt{2}t/2 \end{cases}$

13. $\vec{T}(t) = \langle -\sin t, \cos t \rangle$; $\vec{N}(t) = \langle -\cos t, -\sin t \rangle$

15. $\vec{T}(t) = \left\langle -\frac{\sin t}{\sqrt{4\cos^2 t + \sin^2 t}}, \frac{2\cos t}{\sqrt{4\cos^2 t + \sin^2 t}} \right\rangle$;
   $\vec{N}(t) = \left\langle -\frac{2\cos t}{\sqrt{4\cos^2 t + \sin^2 t}}, -\frac{\sin t}{\sqrt{4\cos^2 t + \sin^2 t}} \right\rangle$

17. (a) Be sure to show work
   (b) $\vec{N}(\pi/4) = \langle -5/\sqrt{34}, -3/\sqrt{34} \rangle$

19. (a) Be sure to show work

(b) $\vec{N}(0) = \left\langle -\frac{1}{\sqrt{5}}, \frac{2}{\sqrt{5}} \right\rangle$

21. $\vec{T}(t) = \frac{1}{\sqrt{5}} \langle 2, \cos t, -\sin t \rangle$; $\vec{N}(t) = \langle 0, -\sin t, -\cos t \rangle$

23. $\vec{T}(t) = \frac{1}{\sqrt{a^2 + b^2}} \langle -a\sin t, a\cos t, b \rangle$; $\vec{N}(t) = \langle -\cos t, -\sin t, 0 \rangle$

25. $a_T = \frac{4t}{\sqrt{1 + 4t^2}}$ and $a_N = \sqrt{4 - \frac{16t^2}{1 + 4t^2}}$
   At $t = 0$, $a_T = 0$ and $a_N = 2$;
   At $t = 1$, $a_T = 4/\sqrt{5}$ and $a_N = 2/\sqrt{5}$.
   At $t = 0$, all acceleration comes in the form of changing the direction of velocity and not the speed; at $t = 1$, more acceleration comes in changing the speed than in changing direction.

27. $a_T = 0$ and $a_N = 2$
   At $t = 0$, $a_T = 0$ and $a_N = 2$;
   At $t = \pi/2$, $a_T = 0$ and $a_N = 2$.
   The object moves at constant speed, so all acceleration comes from changing direction, hence $a_T = 0$. $\vec{a}(t)$ is always parallel to $\vec{N}(t)$, but twice as long, hence $a_N = 2$.

29. $a_T = 0$ and $a_N = a$
   At $t = 0$, $a_T = 0$ and $a_N = a$;
   At $t = \pi/2$, $a_T = 0$ and $a_N = a$.
   The object moves at constant speed, meaning that a_T is always 0. The object "rises" along the z-axis at a constant rate, so all acceleration comes in the form of changing direction circling the z-axis. The greater the radius of this circle the greater the acceleration, hence $a_N = a$.

Section 11.5

1. time and/or distance

3. Answers may include lines, circles, helixes

5. κ

7. $s = 3t$, so $\vec{r}(s) = \langle 2s/3, s/3, -2s/3 \rangle$

9. $s = \sqrt{13}t$, so $\vec{r}(s) = \langle 3\cos(s/\sqrt{13}), 3\sin(s/\sqrt{13}), 2s/\sqrt{13} \rangle$

11. $\kappa = \frac{|6x|}{\left(1 + (3x^2 - 1)^2\right)^{3/2}}$;
   $\kappa(0) = 0$, $\kappa(1/2) = \frac{192}{17\sqrt{17}} \approx 2.74$.

13. $\kappa = \frac{|\cos x|}{(1 + \sin^2 x)^{3/2}}$;
   $\kappa(0) = 1$, $\kappa(\pi/2) = 0$

15. $\kappa = \frac{|2\cos t \cos(2t) + 4\sin t \sin(2t)|}{\left(4\cos^2(2t) + \sin^2 t\right)^{3/2}}$;
   $\kappa(0) = 1/4$, $\kappa(\pi/4) = 8$

17. $\kappa = \frac{|6t^2 + 2|}{\left(4t^2 + (3t^2 - 1)^2\right)^{3/2}}$;
   $\kappa(0) = 2$, $\kappa(5) = \frac{19}{1394\sqrt{1394}} \approx 0.0004$

19. $\kappa = 0$;
   $\kappa(0) = 0$, $\kappa(1) = 0$

21. $\kappa = \frac{3}{13}$;
   $\kappa(0) = 3/13$, $\kappa(\pi/2) = 3/13$

23. maximized at $x = \pm \frac{\sqrt{2}}{\sqrt[4]{5}}$

25. maximized at $t = 1/4$

27. radius of curvature is $5\sqrt{5}/4$.

29. radius of curvature is 9.

31. $x^2 + (y - 1/2)^2 = 1/4$, or $\vec{c}(t) = \langle 1/2\cos t, 1/2\sin t + 1/2 \rangle$

33. $x^2 + (y + 8)^2 = 81$, or $\vec{c}(t) = \langle 9\cos t, 9\sin t - 8 \rangle$

Index

Differentiation Rules

1. $\dfrac{d}{dx}(cx) = c$

2. $\dfrac{d}{dx}(u \pm v) = u' \pm v'$

3. $\dfrac{d}{dx}(u \cdot v) = uv' + u'v$

4. $\dfrac{d}{dx}\left(\dfrac{u}{v}\right) = \dfrac{vu' - uv'}{v^2}$

5. $\dfrac{d}{dx}(u(v)) = u'(v)v'$

6. $\dfrac{d}{dx}(c) = 0$

7. $\dfrac{d}{dx}(x) = 1$

8. $\dfrac{d}{dx}(x^n) = nx^{n-1}$

9. $\dfrac{d}{dx}(e^x) = e^x$

10. $\dfrac{d}{dx}(a^x) = \ln a \cdot a^x$

11. $\dfrac{d}{dx}(\ln x) = \dfrac{1}{x}$

12. $\dfrac{d}{dx}(\log_a x) = \dfrac{1}{\ln a} \cdot \dfrac{1}{x}$

13. $\dfrac{d}{dx}(\sin x) = \cos x$

14. $\dfrac{d}{dx}(\cos x) = -\sin x$

15. $\dfrac{d}{dx}(\csc x) = -\csc x \cot x$

16. $\dfrac{d}{dx}(\sec x) = \sec x \tan x$

17. $\dfrac{d}{dx}(\tan x) = \sec^2 x$

18. $\dfrac{d}{dx}(\cot x) = -\csc^2 x$

19. $\dfrac{d}{dx}(\sin^{-1} x) = \dfrac{1}{\sqrt{1 - x^2}}$

20. $\dfrac{d}{dx}(\cos^{-1} x) = \dfrac{-1}{\sqrt{1 - x^2}}$

21. $\dfrac{d}{dx}(\csc^{-1} x) = \dfrac{-1}{|x|\sqrt{x^2 - 1}}$

22. $\dfrac{d}{dx}(\sec^{-1} x) = \dfrac{1}{|x|\sqrt{x^2 - 1}}$

23. $\dfrac{d}{dx}(\tan^{-1} x) = \dfrac{1}{1 + x^2}$

24. $\dfrac{d}{dx}(\cot^{-1} x) = \dfrac{-1}{1 + x^2}$

25. $\dfrac{d}{dx}(\cosh x) = \sinh x$

26. $\dfrac{d}{dx}(\sinh x) = \cosh x$

27. $\dfrac{d}{dx}(\tanh x) = \operatorname{sech}^2 x$

28. $\dfrac{d}{dx}(\operatorname{sech} x) = -\operatorname{sech} x \tanh x$

29. $\dfrac{d}{dx}(\operatorname{csch} x) = -\operatorname{csch} x \coth x$

30. $\dfrac{d}{dx}(\coth x) = -\operatorname{csch}^2 x$

31. $\dfrac{d}{dx}(\cosh^{-1} x) = \dfrac{1}{\sqrt{x^2 - 1}}$

32. $\dfrac{d}{dx}(\sinh^{-1} x) = \dfrac{1}{\sqrt{x^2 + 1}}$

33. $\dfrac{d}{dx}(\operatorname{sech}^{-1} x) = \dfrac{-1}{x\sqrt{1 - x^2}}$

34. $\dfrac{d}{dx}(\operatorname{csch}^{-1} x) = \dfrac{-1}{|x|\sqrt{1 + x^2}}$

35. $\dfrac{d}{dx}(\tanh^{-1} x) = \dfrac{1}{1 - x^2}$

36. $\dfrac{d}{dx}(\coth^{-1} x) = \dfrac{1}{1 - x^2}$

Integration Rules

1. $\displaystyle\int c \cdot f(x)\, dx = c \int f(x)\, dx$

2. $\displaystyle\int f(x) \pm g(x)\, dx =$
$\displaystyle\int f(x)\, dx \pm \int g(x)\, dx$

3. $\displaystyle\int 0\, dx = C$

4. $\displaystyle\int 1\, dx = x + C$

5. $\displaystyle\int x^n\, dx = \dfrac{1}{n+1}x^{n+1} + C,\ n \neq -1$

6. $\displaystyle\int e^x\, dx = e^x + C$

7. $\displaystyle\int \ln x\, dx = x \ln x - x + C$

8. $\displaystyle\int a^x\, dx = \dfrac{1}{\ln a} \cdot a^x + C$

9. $\displaystyle\int \dfrac{1}{x}\, dx = \ln|x| + C$

10. $\displaystyle\int \cos x\, dx = \sin x + C$

11. $\displaystyle\int \sin x\, dx = -\cos x + C$

12. $\displaystyle\int \tan x\, dx = -\ln|\cos x| + C$

13. $\displaystyle\int \sec x\, dx = \ln|\sec x + \tan x| + C$

14. $\displaystyle\int \csc x\, dx = -\ln|\csc x + \cot x| + C$

15. $\displaystyle\int \cot x\, dx = \ln|\sin x| + C$

16. $\displaystyle\int \sec^2 x\, dx = \tan x + C$

17. $\displaystyle\int \csc^2 x\, dx = -\cot x + C$

18. $\displaystyle\int \sec x \tan x\, dx = \sec x + C$

19. $\displaystyle\int \csc x \cot x\, dx = -\csc x + C$

20. $\displaystyle\int \cos^2 x\, dx = \dfrac{1}{2}x + \dfrac{1}{4}\sin(2x) + C$

21. $\displaystyle\int \sin^2 x\, dx = \dfrac{1}{2}x - \dfrac{1}{4}\sin(2x) + C$

22. $\displaystyle\int \dfrac{1}{x^2 + a^2}\, dx = \dfrac{1}{a}\tan^{-1}\left(\dfrac{x}{a}\right) + C$

23. $\displaystyle\int \dfrac{1}{\sqrt{a^2 - x^2}}\, dx = \sin^{-1}\left(\dfrac{x}{a}\right) + C$

24. $\displaystyle\int \dfrac{1}{x\sqrt{x^2 - a^2}}\, dx = \dfrac{1}{a}\sec^{-1}\left(\dfrac{|x|}{a}\right) + C$

25. $\displaystyle\int \cosh x\, dx = \sinh x + C$

26. $\displaystyle\int \sinh x\, dx = \cosh x + C$

27. $\displaystyle\int \tanh x\, dx = \ln(\cosh x) + C$

28. $\displaystyle\int \coth x\, dx = \ln|\sinh x| + C$

29. $\displaystyle\int \dfrac{1}{\sqrt{x^2 - a^2}}\, dx = \ln\left|x + \sqrt{x^2 - a^2}\right| + C$

30. $\displaystyle\int \dfrac{1}{\sqrt{x^2 + a^2}}\, dx = \ln\left|x + \sqrt{x^2 + a^2}\right| + C$

31. $\displaystyle\int \dfrac{1}{a^2 - x^2}\, dx = \dfrac{1}{2a}\ln\left|\dfrac{a + x}{a - x}\right| + C$

32. $\displaystyle\int \dfrac{1}{x\sqrt{a^2 - x^2}}\, dx = \dfrac{1}{a}\ln\left(\dfrac{x}{a + \sqrt{a^2 - x^2}}\right) + C$

33. $\displaystyle\int \dfrac{1}{x\sqrt{x^2 + a^2}}\, dx = \dfrac{1}{a}\ln\left|\dfrac{x}{a + \sqrt{x^2 + a^2}}\right| + C$

The Unit Circle

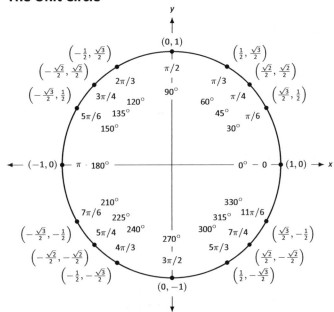

Definitions of the Trigonometric Functions

Unit Circle Definition

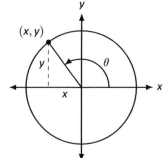

$$\sin\theta = y \qquad \cos\theta = x$$

$$\csc\theta = \frac{1}{y} \qquad \sec\theta = \frac{1}{x}$$

$$\tan\theta = \frac{y}{x} \qquad \cot\theta = \frac{x}{y}$$

Right Triangle Definition

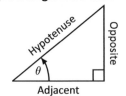

$$\sin\theta = \frac{O}{H} \qquad \csc\theta = \frac{H}{O}$$

$$\cos\theta = \frac{A}{H} \qquad \sec\theta = \frac{H}{A}$$

$$\tan\theta = \frac{O}{A} \qquad \cot\theta = \frac{A}{O}$$

Common Trigonometric Identities

Pythagorean Identities

$$\sin^2 x + \cos^2 x = 1$$
$$\tan^2 x + 1 = \sec^2 x$$
$$1 + \cot^2 x = \csc^2 x$$

Cofunction Identities

$$\sin\left(\frac{\pi}{2} - x\right) = \cos x \qquad \csc\left(\frac{\pi}{2} - x\right) = \sec x$$

$$\cos\left(\frac{\pi}{2} - x\right) = \sin x \qquad \sec\left(\frac{\pi}{2} - x\right) = \csc x$$

$$\tan\left(\frac{\pi}{2} - x\right) = \cot x \qquad \cot\left(\frac{\pi}{2} - x\right) = \tan x$$

Double Angle Formulas

$$\sin 2x = 2\sin x \cos x$$
$$\cos 2x = \cos^2 x - \sin^2 x$$
$$= 2\cos^2 x - 1$$
$$= 1 - 2\sin^2 x$$
$$\tan 2x = \frac{2\tan x}{1 - \tan^2 x}$$

Sum to Product Formulas

$$\sin x + \sin y = 2\sin\left(\frac{x+y}{2}\right)\cos\left(\frac{x-y}{2}\right)$$

$$\sin x - \sin y = 2\sin\left(\frac{x-y}{2}\right)\cos\left(\frac{x+y}{2}\right)$$

$$\cos x + \cos y = 2\cos\left(\frac{x+y}{2}\right)\cos\left(\frac{x-y}{2}\right)$$

$$\cos x - \cos y = -2\sin\left(\frac{x+y}{2}\right)\sin\left(\frac{x-y}{2}\right)$$

Power–Reducing Formulas

$$\sin^2 x = \frac{1 - \cos 2x}{2}$$

$$\cos^2 x = \frac{1 + \cos 2x}{2}$$

$$\tan^2 x = \frac{1 - \cos 2x}{1 + \cos 2x}$$

Even/Odd Identities

$$\sin(-x) = -\sin x$$
$$\cos(-x) = \cos x$$
$$\tan(-x) = -\tan x$$
$$\csc(-x) = -\csc x$$
$$\sec(-x) = \sec x$$
$$\cot(-x) = -\cot x$$

Product to Sum Formulas

$$\sin x \sin y = \frac{1}{2}\left(\cos(x-y) - \cos(x+y)\right)$$

$$\cos x \cos y = \frac{1}{2}\left(\cos(x-y) + \cos(x+y)\right)$$

$$\sin x \cos y = \frac{1}{2}\left(\sin(x+y) + \sin(x-y)\right)$$

Angle Sum/Difference Formulas

$$\sin(x \pm y) = \sin x \cos y \pm \cos x \sin y$$

$$\cos(x \pm y) = \cos x \cos y \mp \sin x \sin y$$

$$\tan(x \pm y) = \frac{\tan x \pm \tan y}{1 \mp \tan x \tan y}$$

Areas and Volumes

Triangles

$h = a \sin \theta$

Area $= \frac{1}{2}bh$

Law of Cosines:
$c^2 = a^2 + b^2 - 2ab \cos \theta$

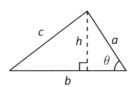

Parallelograms

Area $= bh$

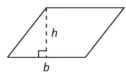

Trapezoids

Area $= \frac{1}{2}(a + b)h$

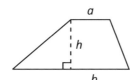

Circles

Area $= \pi r^2$

Circumference $= 2\pi r$

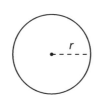

Sectors of Circles

θ in radians

Area $= \frac{1}{2}\theta r^2$

$s = r\theta$

Right Circular Cone

Volume $= \frac{1}{3}\pi r^2 h$

Surface Area $=$
$\pi r\sqrt{r^2 + h^2} + \pi r^2$

Right Circular Cylinder

Volume $= \pi r^2 h$

Surface Area $=$
$2\pi rh + 2\pi r^2$

Sphere

Volume $= \frac{4}{3}\pi r^3$

Surface Area $= 4\pi r^2$

General Cone

Area of Base $= A$

Volume $= \frac{1}{3}Ah$

General Right Cylinder

Area of Base $= A$

Volume $= Ah$

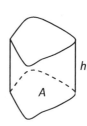

Algebra

Factors and Zeros of Polynomials

Let $p(x) = a_n x^n + a_{n-1} x^{n-1} + \cdots + a_1 x + a_0$ be a polynomial. If $p(a) = 0$, then a is a *zero* of the polynomial and a solution of the equation $p(x) = 0$. Furthermore, $(x - a)$ is a *factor* of the polynomial.

Fundamental Theorem of Algebra

An nth degree polynomial has n (not necessarily distinct) zeros. Although all of these zeros may be imaginary, a real polynomial of odd degree must have at least one real zero.

Quadratic Formula

If $p(x) = ax^2 + bx + c$, and $0 \leq b^2 - 4ac$, then the real zeros of p are $x = (-b \pm \sqrt{b^2 - 4ac})/2a$

Special Factors

$$x^2 - a^2 = (x - a)(x + a) \qquad x^3 - a^3 = (x - a)(x^2 + ax + a^2)$$
$$x^3 + a^3 = (x + a)(x^2 - ax + a^2) \qquad x^4 - a^4 = (x^2 - a^2)(x^2 + a^2)$$
$$(x + y)^n = x^n + nx^{n-1}y + \frac{n(n-1)}{2!}x^{n-2}y^2 + \cdots + nxy^{n-1} + y^n$$
$$(x - y)^n = x^n - nx^{n-1}y + \frac{n(n-1)}{2!}x^{n-2}y^2 - \cdots \pm nxy^{n-1} \mp y^n$$

Binomial Theorem

$$(x + y)^2 = x^2 + 2xy + y^2 \qquad (x - y)^2 = x^2 - 2xy + y^2$$
$$(x + y)^3 = x^3 + 3x^2y + 3xy^2 + y^3 \qquad (x - y)^3 = x^3 - 3x^2y + 3xy^2 - y^3$$
$$(x + y)^4 = x^4 + 4x^3y + 6x^2y^2 + 4xy^3 + y^4 \qquad (x - y)^4 = x^4 - 4x^3y + 6x^2y^2 - 4xy^3 + y^4$$

Rational Zero Theorem

If $p(x) = a_n x^n + a_{n-1} x^{n-1} + \cdots + a_1 x + a_0$ has integer coefficients, then every *rational zero* of p is of the form $x = r/s$, where r is a factor of a_0 and s is a factor of a_n.

Factoring by Grouping

$$acx^3 + adx^2 + bcx + bd = ax^2(cs + d) + b(cx + d) = (ax^2 + b)(cx + d)$$

Arithmetic Operations

$$ab + ac = a(b + c) \qquad \frac{a}{b} + \frac{c}{d} = \frac{ad + bc}{bd} \qquad \frac{a + b}{c} = \frac{a}{c} + \frac{b}{c}$$

$$\frac{\left(\frac{a}{b}\right)}{\left(\frac{c}{d}\right)} = \left(\frac{a}{b}\right)\left(\frac{d}{c}\right) = \frac{ad}{bc} \qquad \frac{\left(\frac{a}{b}\right)}{c} = \frac{a}{bc} \qquad \frac{a}{\left(\frac{b}{c}\right)} = \frac{ac}{b}$$

$$a\left(\frac{b}{c}\right) = \frac{ab}{c} \qquad \frac{a - b}{c - d} = \frac{b - a}{d - c} \qquad \frac{ab + ac}{a} = b + c$$

Exponents and Radicals

$$a^0 = 1, \ a \neq 0 \qquad (ab)^x = a^x b^x \qquad a^x a^y = a^{x+y} \qquad \sqrt{a} = a^{1/2} \qquad \frac{a^x}{a^y} = a^{x-y} \qquad \sqrt[n]{a} = a^{1/n}$$

$$\left(\frac{a}{b}\right)^x = \frac{a^x}{b^x} \qquad \sqrt[n]{a^m} = a^{m/n} \qquad a^{-x} = \frac{1}{a^x} \qquad \sqrt[n]{ab} = \sqrt[n]{a}\sqrt[n]{b} \qquad (a^x)^y = a^{xy} \qquad \sqrt[n]{\frac{a}{b}} = \frac{\sqrt[n]{a}}{\sqrt[n]{b}}$$

Additional Formulas

Summation Formulas:

$$\sum_{i=1}^{n} c = cn \qquad\qquad \sum_{i=1}^{n} i = \frac{n(n+1)}{2}$$

$$\sum_{i=1}^{n} i^2 = \frac{n(n+1)(2n+1)}{6} \qquad\qquad \sum_{i=1}^{n} i^3 = \left(\frac{n(n+1)}{2}\right)^2$$

Trapezoidal Rule:

$$\int_{a}^{b} f(x)\,dx \approx \frac{\Delta x}{2}\left[f(x_1) + 2f(x_2) + 2f(x_3) + \ldots + 2f(x_n) + f(x_{n+1})\right]$$

with Error $\leq \dfrac{(b-a)^3}{12n^2}\left[\max\left|f''(x)\right|\right]$

Simpson's Rule:

$$\int_{a}^{b} f(x)\,dx \approx \frac{\Delta x}{3}\left[f(x_1) + 4f(x_2) + 2f(x_3) + 4f(x_4) + \ldots + 2f(x_{n-1}) + 4f(x_n) + f(x_{n+1})\right]$$

with Error $\leq \dfrac{(b-a)^5}{180n^4}\left[\max\left|f^{(4)}(x)\right|\right]$

Arc Length:

$$L = \int_{a}^{b} \sqrt{1 + f'(x)^2}\ dx$$

Surface of Revolution:

$$S = 2\pi \int_{a}^{b} f(x)\sqrt{1 + f'(x)^2}\ dx$$

(where $f(x) \geq 0$)

$$S = 2\pi \int_{a}^{b} x\sqrt{1 + f'(x)^2}\ dx$$

(where $a, b \geq 0$)

Work Done by a Variable Force:

$$W = \int_{a}^{b} F(x)\ dx$$

Force Exerted by a Fluid:

$$F = \int_{a}^{b} w\,d(y)\,\ell(y)\ dy$$

Taylor Series Expansion for $f(x)$:

$$p_n(x) = f(c) + f'(c)(x-c) + \frac{f''(c)}{2!}(x-c)^2 + \frac{f'''(c)}{3!}(x-c)^3 + \ldots + \frac{f^{(n)}(c)}{n!}(x-c)^n$$

Maclaurin Series Expansion for $f(x)$, where $c = 0$:

$$p_n(x) = f(0) + f'(0)x + \frac{f''(0)}{2!}x^2 + \frac{f'''(0)}{3!}x^3 + \ldots + \frac{f^{(n)}(0)}{n!}x^n$$

Summary of Tests for Series:

Test	Series	Condition(s) of Convergence	Condition(s) of Divergence	Comment
nth-Term	$\displaystyle\sum_{n=1}^{\infty} a_n$		$\displaystyle\lim_{n\to\infty} a_n \neq 0$	This test cannot be used to show convergence.
Geometric Series	$\displaystyle\sum_{n=0}^{\infty} r^n$	$\|r\| < 1$	$\|r\| \geq 1$	$\text{Sum} = \dfrac{1}{1-r}$
Telescoping Series	$\displaystyle\sum_{n=1}^{\infty} (b_n - b_{n+a})$	$\displaystyle\lim_{n\to\infty} b_n = L$		$\text{Sum} = \left(\displaystyle\sum_{n=1}^{a} b_n\right) - L$
p-Series	$\displaystyle\sum_{n=1}^{\infty} \dfrac{1}{(an+b)^p}$	$p > 1$	$p \leq 1$	
Integral Test	$\displaystyle\sum_{n=0}^{\infty} a_n$	$\displaystyle\int_{1}^{\infty} a(n)\,dn$ is convergent	$\displaystyle\int_{1}^{\infty} a(n)\,dn$ is divergent	$a_n = a(n)$ must be continuous
Direct Comparison	$\displaystyle\sum_{n=0}^{\infty} a_n$	$\displaystyle\sum_{n=0}^{\infty} b_n$ converges and $0 \leq a_n \leq b_n$	$\displaystyle\sum_{n=0}^{\infty} b_n$ diverges and $0 \leq b_n \leq a_n$	
Limit Comparison	$\displaystyle\sum_{n=0}^{\infty} a_n$	$\displaystyle\sum_{n=0}^{\infty} b_n$ converges and $\displaystyle\lim_{n\to\infty} a_n/b_n \geq 0$	$\displaystyle\sum_{n=0}^{\infty} b_n$ diverges and $\displaystyle\lim_{n\to\infty} a_n/b_n > 0$	Also diverges if $\displaystyle\lim_{n\to\infty} a_n/b_n = \infty$
Ratio Test	$\displaystyle\sum_{n=0}^{\infty} a_n$	$\displaystyle\lim_{n\to\infty} \dfrac{a_{n+1}}{a_n} < 1$	$\displaystyle\lim_{n\to\infty} \dfrac{a_{n+1}}{a_n} > 1$	$\{a_n\}$ must be positive. Also diverges if $\displaystyle\lim_{n\to\infty} a_{n+1}/a_n = \infty$
Root Test	$\displaystyle\sum_{n=0}^{\infty} a_n$	$\displaystyle\lim_{n\to\infty} (a_n)^{1/n} < 1$	$\displaystyle\lim_{n\to\infty} (a_n)^{1/n} > 1$	$\{a_n\}$ must be positive. Also diverges if $\displaystyle\lim_{n\to\infty} (a_n)^{1/n} = \infty$

Made in the USA
San Bernardino, CA
04 April 2019